Human Factors Methods for Design

Christopher Nemeth's *Human Factors Methods for Design* is an easy-to-use, in-depth guide to human factors research in the development process. The text provides both professionals and students with the orientation, process and methods to perform human factors or ergonomic research.

The book offers a selection of features unique to human factors books and resources, such as:

- An overview of human factors
- Orientation to applied research in product and service development
- Observation methods
- Essential experiment design and analysis
- Briefs, reports and presentations

It also leads the reader through the steps that are crucial to:

- Defining an opportunity, or understanding a problem
- Developing goals (desirable results that the solution is intended to achieve), and criteria (parameters for how to meet goals)
- Collecting and organizing information
- Using collected information to create solutions
- Evaluating potential solutions

Human Factors Methods for Design is a valuable introduction for students, an ongoing reference for practicing professionals, and a stimulating primer for readers who are interested in the use of technology to serve human needs.

Christopher P. Nemeth, Ph.D., CHFP, manages his own design, human factors and expert witness consulting practice. Previously, he was a Research Associate on the staff of the Herman Miller Research Corporation and was Manager, Research and Development for Milcare, Herman Miller's health care products subsidiary. His academic career includes adjunct faculty positions with Northwestern University and Illinois Institute of Technology. He is currently a Research Associate in the Cognitive Technologies Laboratory at the University of Chicago.

Essential reading

Introduction to Usability
Patrick Jordan, Contemporary Trends Institute, UK
Taylor & Francis   Hbk 0-7484-0794-4   Pbk 0-7484-0762-6

Human Factors in Product Design
Patrick Jordan, Contemporary Trends Institute, UK and William Green, University of Canberra, Australia
Taylor & Francis   Hbk 0-7484-0829-0

Pleasure with Products
Beyond usability
Edited by William Green, University of Canberra, Australia and Patrick Jordan, Contemporary Trends Institute, UK
Taylor & Francis   Hbk 0-415-23704-1

Designing Pleasurable Products
Patrick Jordan, Contemporary Trends Institute, UK
Taylor & Francis   Hbk 0-7484-0844-4   Pbk 0-415-29887-3

Design and the Social Sciences
Making connections
Edited by Jorge Frascara, University of Alberta, Canada
Taylor & Francis   Hbk 0-415-27376-5

Human Factors in Consumer Products
Edited by Neville Stanton, Brunel University, UK
Taylor & Francis   Pbk 0-7484-0603-4

Information and ordering details

For price availability and ordering visit our website www.ergonomicsarena.com
Alternatively our books are available from all good bookshops.

Human Factors Methods for Design
Making Systems Human-Centered

Christopher P. Nemeth

CRC PRESS

Boca Raton   London   New York   Washington, D.C.

For my colleagues

Library of Congress Cataloging-in-Publication Data

Catalog record for this book is available from the Library of Congress

British Library Cataloguing in Publication Data

Catalog record for this book is available from the British Library

Visit the CRC Press Web site at www.crcpress.com

© 2004 by CRC Press LLC

No claim to original U.S. Government works
International Standard Book Number 0-415-29799-0 (paperback)
International Standard Book Number 0-415-29798-2 (hardback)
Printed in the United States of America 1 2 3 4 5 6 7 8 9 0
Printed on acid-free paper

Contents

List of illustrations

Figures

Tables

Foreword

> ... and our goal [for the nation] is a simple one—by the time I leave office I want every single American to be able to set the clock on his VCR.
>
> (George H. Bush, speaking of his vision for the nation in a speech given on March 2, 1990.)

New technology has always brought out utopian dreams about future tasks. Developers recognize the potential for benefits and charge ahead in pursuit of the next technological advance. But contrary to the dreams, the message from users who must do something with these devices—a message carried in their voices, their performance, their errors, and their workarounds—is one of *complexity*. On finding this pattern multiple times, researchers began to call for 'human-centered design' (Hoffman, R.R., Ford, K.M., Feltovich, P.J., Woods, D.D., Klein, G. and Feltovich, A. 2002).

Developers confident in their vision of the possibilities of new technology didn't blink an eye, 'I think about the benefits for users, and if I can realize the envisioned system, the user will assuredly get those anticipated benefits; therefore, I do human-centered design.' In other words, developers don't create new complexities for users on purpose, and they don't lose track of the user's perspective simply because their eye is stuck on an economic ball. When new technology creates surprising complexities, developers are honestly perplexed and they turn back to what they know best—'just a little more technology will be enough next time' (Winograd, T. and Woods, D.D. 1997).

I noticed and experienced this pattern of surprising effects of technology change and perplexed developers who then create the next round of complexities for users, repeatedly. In frustration I exclaimed, 'The road to technology-centered systems is paved with user-centered intentions' (Woods, D.D. and Tinapple, D. 1999).

Good intentions (and good technology) without the engine of methods to integrate the user's perspective or to anticipate the effects of technology change on user activities are insufficient. Chris Nemeth provides a comprehensive treatment of the large repertoire that has been developed to marry user-centered technique to technological power. By combining technology with these methods, all of our visions of useful, usable and desirable systems can come true.

David Woods
Institute for Ergonomics
The Ohio State University

Preface

I will confess up front that I have a bias: people matter. This bears mention here, because a number of current popular trends lead away from, rather than toward, the better use of technology to serve human needs and wants: technophilia, pseudoscience, revenge effects and perception management.

Technophilia—Those who are enchanted with technology can and do overlook what is needed to relate it to human behavior and physiology. Heath and Luff (2000:3) find that a common feature of new technology projects is 'disregard for the ways in which people organize their work, coupled with a disdain for the ordinary resources on which they rely.' The failure of multi-million dollar projects show research and development shortcomings when it comes to problem research. The host of videocassette recorders with digital clocks that now blink '12:00' in homes and offices around the world are mute testimony to product development that could have—but did not—make those products useful for their users.

Pseudoscience—Sham science threatens to erode what have up to now been fairly well defined boundaries between fact and fancy. There are a number of factors that lead society in this direction. Urban, information-based society is inherently complex and human-made. Much of reality in this context is reported by others and is not experienced first-hand. The farther members of society are from first-hand contact with the world, the harder it is to verify what one sees and hears. Society's ephemeral, quickly changing character tends to open the scope of reality from what 'is' to what 'might be.' This parallels a popular trend toward subjective interpretation in lieu of objective learning. 'Throughout our culture,' the critic Michiko Kakutani writes, 'the old notions of "truth" and "knowledge" are in danger of being replaced by the new ones of "opinion," "perception," and "credibility"' (Leo 1998).

Revenge effects—Some hold the opinion that the world is more complex than humans can manage and that the use of technology creates as many problems as it solves. This point of view occurs in writings such as those of Edward Tenner: 'When we alter one environment slightly, large and unexpected variations can occur' (Rothstein 1996). Certainly, imprudent use of technology is unwise. The assumption that technology cannot be used effectively, though, is self-defeating.

Perception management—Growing public awareness of human factors has marketing value. Rather than make authentic improvements, some firms have taken an easier route by adding 'ergonomics' to their roster of product benefits. In such hands, 'ergonomics' becomes a handy marketing overlay, rather than an inherent part of how we create our world.

The collective effect of these trends is the potential erosion of 'public trust in such scientific tools as statistical analysis, controlled laboratory and clinical experiments, the rational analysis of political oratory and the study of history, anthropology and every other field dependent on disciplined, rational thought' (Browne 1995).

There are more pragmatic implications for research and development in particular. A Harvard Business School study found that 'Despite the billions spent on research, development and marketing, the failure rate of high tech products is staggering' (*Harvard Business Review* 1994:10). The authors recommend 'empathic design' as a more efficient and effective course that firms should follow in order to develop more successful products. Empathic design relies on field observation to develop a deeper understanding of the user environment, then intuitive exploration to 'extrapolate about the way that environment may evolve in the future and imagine a need that their technology will satisfy.' This combination of insight and foresight results in products that are well suited to human needs and wants.

We deserve to create a better life for ourselves and others through the effective use of technology, on human terms. Research and development must rely on the thorough and accurate development and use of authentic knowledge to create a world that works. For more on how to do that, read on.

What is this book?

I have taught a survey course in human factors at the Illinois Institute of Technology in Chicago for over eight years. During that time, I could not find a book that served the needs of those who worked on the development of products for human use. This is that book.

There is no shortage of available human factors information—the *what* of development. But there is no single guide on *how* to use that information to create things for people. *Human Factors Methods for Design* is a field guide to how to solve human factors challenges.

As the Gertrude Stein quotation before Chapter 1 suggests, asking the right question is the key to the right answer. This book is designed to enable the reader to probe into areas that will lead toward well-considered concepts.

Human Factors Methods for Design provides selected information on human traits that have particular import for research and development and provides methods for research and development work. 'Methods' are a powerful means to collect, use and evaluate information. They can be used in a wide variety of applications.

In this text, 'product' refers to the result of a development effort: artifacts (such as a tool or machine), services and systems. These products do not exist in a vacuum. They typically exist within a real world environment that includes humans and other products. For that reason, the meaning of the word 'product' in this book includes artifacts, software, services and systems.

This book provides a way to design products that extend, amplify and enhance human capabilities. Use it to define and solve problems, to develop new solutions and to evaluate their effectiveness.

Why 'human factors'?

Life is not easy. That is why manufacturers make machines, why architects create spaces and buildings and why entire industries create systems such as airline transportation. Even with all of that help, people are still chagrined by what Matthews (1997) describes as the 'cussedness of everyday events.' At the least, we deserve things that work. At best, we deserve products we can rely on to make life better, safer, healthier and more satisfying.

Research and development teams produce products to extend human abilities. Machines, such as equipment and tools, can excavate ore and transport cargo. Systems, such as data communication and electrical power, can collect and display large amounts of complex information. Services provide knowledge far beyond what a single individual could amass. Environments (such as buildings and mobile modular habitats) create comfortable temporary or permanent shelter for work, recreation, education and survival. Products are developed and exist to serve human needs. That is why people are willing to pay money for them.

But how do we know what people need? *Human Factors Methods for Design* provides the means to create products that succeed at three different levels: functional, legal and ethical.

Functional—Human abilities and limitations form the basis for determining what works well when it comes to products that include people. Products that are conceived and produced without regard to human abilities and limitations will not function well. At a minimum, that dysfunction can erode sales. Aggravation over poorly conceived or poorly executed features can cause customers to choose a competing product. At a further extreme, dysfunction can cause the sort of disorientation and frustration that many consider typical of daily urban life. Ultimately, dysfunction can cause catastrophic failure. Unsafe products cause the loss of life and property. Daily newspapers contain regular accounts of such shortcomings.

Legal—Inattention to hazards that can cause the loss of property, health or life is actionable in a court of law. Lawsuits that are brought in order to remedy loss in such instances can claim and be awarded significant damages. In the United States, liability litigation continues to grow in frequency and scale.

Ethical—Development professionals strive to create products that enhance, rather than impede, human endeavor.

Who should read this book?

You should read this book if you have ever left a team meeting to kick-off a new project and scratched your head asking 'Where do I start?' or 'How do I do this?' *Human Factors Methods for Design* is intended for working professionals, scholars, undergraduate and graduate students in fields that involve people, including:

- The design disciplines, including:
  - Architecture, including space planning
  - Product, or 'industrial,' design
  - Communications, or 'graphic,' design including multimedia
  - Environmental design
- Human factors/ergonomics
- Design and development of machines
- Design and development of software and computers, including interface development
- Systems engineering
- Engineering management.

It is also written for any general reader who has an interest in the human-made ('operational') environment and product development methodologies.

What will I get from reading this book?

The reader will learn how the interaction of human factors research and user-centered design produces a viable product. Using *Human Factors Methods for Design* will provide:

- An understanding of the basics of human behavior and physiology in the context of development. It will provide an orientation for readers who are new to human factors and ergonomics.
- Ways to recognize, approach, analyze and solve problems, including methods for observation as well as simple experimentation.
- A means to determine why one solution alternative is better than another. A chapter on cost-effectiveness is included to enable the reader to prioritize solution alternatives according to their apparent value.

- A process for effectively communicating development team intentions, activities and results.

The book complements other topic areas, particularly:

- Design methodology
- Product planning
- Industrial design
- Multi-disciplinary teamwork in product development
- Design management.

On a broader scale, the book can lead the reader to better judgment. A methodical approach encourages a researcher to dig deeper for what actually matters in a problem area. That can result in research and development work that is well informed and balanced. Used regularly, methodical practice can cultivate insight, awareness and, hopefully, wisdom.

What makes this book different?

Recent human factors titles cover three major areas of interest:

- *Operations*—Workplace health, safety and productivity, which are the domain of much recent interest in 'ergonomics'
- *Human traits*—Sensory, physical capabilities and limits
- *Development*—The creation of solutions to problems, or opportunities to implement new technology.

Many excellent survey texts have been written to cover all aspects of the human factors field, including:

Kantowitz, B. and Sorkin, R. (1983) *Human Factors: Understanding People–System Relationships*. John Wiley & Sons: New York.
Salvendy, G. (ed.) (1997) *Handbook of Human Factors and Ergonomics*, 2nd edn. John Wiley & Sons: New York.
Sanders, M.S. and McCormick, E.J. (1993) *Human Factors in Engineering and Design*, 7th edn. McGraw-Hill: New York.

Additional texts in the field include:

Konz, S. and Johnson, S. (2000). *Work Design: Industrial Ergonomics*, 5th edn. (Scottsdale, AZ: Holcomb Hathaway).
Grandjean, E. (1980). *Fitting the Task to the Man*, 3rd edn. (London: Taylor and Francis).
Sheridan, T.B. and Ferrell, W.R. (1974). *Man-Machine Systems: Information Control and Decision Models of Human Performance*. (Cambridge, MA: The MIT Press).

By contrast, *Human Factors Methods for Design* focuses on the processes that are used to create products for human use (including systems, services and environments). Rather than serve as a definitive compendium, it reads more like a road map. Why? Because most development professionals need to 'get on with it.'

The content and design of this book are geared to make the information easy to find and to use. *Human Factors Methods for Design* is divided into three parts.

- Part I—Human factors practice: The context for performing research and development including a description of the operation environment, human behavior, physiology, internal and external influences on problem solving and the process of conducting research.
- Part II—Human factors methods: The procedures, techniques, or ways of performing human-oriented research and development. Each chapter in Part II is organized into separate sections that describe a particular method. Each section provides a context, a description of how to perform the method and an example of how to apply it. Descriptions are specific so that the reader can understand how to perform the steps that the method requires.
- Part III—Application: The steps to build relationships among key audiences who are important to translating research work into reality, including effecting change, communicating results and examples of how to apply the knowledge in Parts I and II to practice.

Previous books in the field were based on the most important needs of their time: 1940s military sensor, platform and weapons systems development, 1960s aerospace systems analysis and 1980s software interface development. Examples in this work are based on what readers need to understand present day research and development applications. Examples also reflect more common types of applications, as many more readers are likely to design web sites than aircraft.

Which parts of this book should I read?

Some readers may prefer to go directly to specific portions of this work instead of reading it in its entirety. It is organized in order to make that easy to do, particularly for those readers who are focused on a particular problem, or are new to human factors, design, development, methods, or companies.

Problem-focused—The reader who needs to solve a problem may know of a method, or may need to choose one.

- *To find a method already known by name*—Review the Table of Contents. Then, turn to the method's section in Part II. Each section explains the method's benefits and limitations, necessary preparation, materials and equipment, the procedure, results its use produces and gives a brief example.
- *To figure out what method to use*—Read Chapter 5 for descriptions of the kinds of methods that are contained in Part II and guidance on how to choose them. Read the first page of each chapter in Part II for a list of the methods that the chapter contains, as well as the questions it answers. Read the step-by-step descriptions for how to perform each method. Read Chapter 14 for practical examples of the ways that methods can be used to analyze problems and to develop solutions to them.

New to human factors—Read Chapter 1 for an overview of the field of human factors and a description of how it evolved. Read Chapter 2 to understand human limits and abilities, the basis for human factors participation in product development and evaluation.

New to design—Read Chapter 1 to learn about how we create the operational environment. Read Chapter 4 to become familiar with the design process and how it relates to human factors. Read Chapter 5 to learn how design and human factors methods are used in parallel through research and development. Read Section 8.1 to learn about simulation, which combines the visualization methods of design and the evaluation methods of human factors. Read Chapter 14 for examples of design projects and how methods are used.

New to research and development—Readers who are familiar with operations may not be familiar with problem analysis. Read Chapter 3 to understand how humans think about problems and Chapter 4 to learn about external influences on research and development.

New to methods—Read Chapter 5 in order to understand what research methods are and how to select them for use in research and development. Skim Part II (Chapters 6 through 11) for an overview of the range of methods. Read Chapter 14 to learn how to apply the methods that are described in Part II.

New to business—Read Chapter 4 to learn about what influences development, Chapter 12 to learn about effecting change and Chapter 13 to find out about the communications that are important to convey information.

If you're enrolled in a course and your instructor assigns sections to read, by all means follow those directions!

How do human factors methods relate to design methods?

The methods of human factors and design are used in parallel throughout the development process. Human factors methods are used to define problems by collecting product requirements that are contained in regulations, standards and guidelines. This forms the basis of a design brief. Designers respond to the brief and redefine the problem. Human factors specialists provide information on human limits and capabilities which designers use to develop concepts. The process continues back and forth, as each informs the other.

Human factors methods are used to build a base of new knowledge (analysis) and to determine usefulness (evaluation). Design methods are used to spawn a broad range of solution speculations (idea generation) and weave them into a cohesive whole that can be portrayed in physical form (visualization). Without human factors methods, the design effort can lack rigor and produce sub-optimal results. Without design methods, new product concepts can be narrow, vague, unimaginative and meek extensions of that which already exists.

A number of authors have written design methods books and I recommend them to you:

Roosenberg, N. and Eekels, J. (1995) *Product Design: Fundamentals and Methods*. John Wiley & Sons: Chichester, UK.
Jones, J.C. (1992) *Design Methods*. Van Nostrand Reinhold: New York.
Cross, N. (1994) *Engineering Design Methods*. John Wiley & Sons: Chichester.
Hall, A.D. (1989) *Metasystems Methodology: A New Synthesis and Unification* (Ifsr International Series on Systems Science and Engineering,Vol. 3). Pergamon Press: New York.

In addition to these books, Charles Owen, Distinguished Emeritus Professor of Design at Illinois Institute of Technology in Chicago, IL has also written a series of papers that define the practice of structured planning, a computer-supported approach to idea generation and synthesis.

Most of us work as an investigative team of one, lacking the necessary resources for sophisticated analysis. We do not intend to publish scientific papers. We do, though, need to decide what to do and how to do it. Each of us needs what amounts to a portable kit of tools in order to deal with problems. As a field guide to problem solving, *Human Factors Methods for Design* serves as a reference for the process of how to figure things out.

Acknowledgments

Work by Alphonse Chapanis, David Meister and Donald Norman and David Woods has been a key influence on my career development.

Many of my colleagues enthusiastically gave their energy, time, guidance and assistance to make this work possible. All of the good things in this book are due to their contributions (while any errors are mine alone) and I offer them my sincere thanks.

Two colleagues reviewed the entire manuscript throughout its development. Robert P. Bateman, Ph.D. provided invaluable insights and recommendations on its structure and content. Lynette Melnar, Ph.D. shared insightful guidance on its logic and expression.

Three colleagues reviewed selected portions of the manuscript and offered a wealth of insight. Jon Frye, Ph.D. reviewed the chapters that discuss descriptive statistics, controlled studies, and cost-effectiveness. Tammy Ramirez, Ph.D. reviewed the chapter on human factors in research and development. Alvah White reviewed the chapter on surveys: questionnaires, interviews and self-reports.

My first human factors instructor, Harold Wakeley, Ph.D., has shared insightful guidance from the outset. Sarah Kramer and Tony Moore of Taylor and Francis have been a joy to work with.

Colleagues who have also provided resources, guidance, and thoughtful support include: Neil Barnett, Tana Bishop, Ph.D., Ellen Carnahan, Nigel Cross, William Cushman, Ph.D., Frank Dahlkemper, Ray Daniels, Jan DeLandtsheer, Ed Halpern, Ph.D., Victor J. Heckler, Ph.D., Jeanne Herrick, John Heskett, Brent Hoffmann, Jamie Horwitz, Ph.D., Serope Kalpakjian, Kathy Kaye, Keith McKee, Ph.D., David Meister, Ph.D., Hugh Miller, Ph.D., Donald Norman, Ph.D., John Oesch, Laura Oswald, Ph.D., Charles Owen, Richard Pew, Ph.D., Sharon Poggenpohl, Chuck Pycha, COL Valerie Rice, Ph.D., Scott Shappell, Ph.D. and Terry Winograd, Ph.D.

My partner Jeanine has my deepest thanks for her understanding and support.

These individuals and organizations have granted permission to use the following material that is protected by copyright:

Figure 1.3   Life cycle—Illinois Institute of Technology
Figure 1.4   Development process—Robert Bateman
Figure 1.5   Ergonomic design process for (consumer) products—John Wiley and Sons
Figure 2.3   Generic error modeling system—Cambridge University Press
Figure 2.4   Psychological varieties of unsafe acts—Cambridge University Press
Figure 3.1   The skills–rules–knowledge formulation of cognitive control—Institute of Electrical and Electronic Engineers (IEEE)
Figure 3.2   Problem solution model using skills–rules–knowledge behaviors—John Wiley and Sons
Figure 4.1   Visualization media—Illinois Institute of Technology
Figure 6.5   Cause and effect diagram— John Wiley and Sons

Figure 7.1   Flow diagram—The Smithsonian Institution
Figure 7.2   Flow process chart—John Wiley and Sons
Figure 7.4   Decision action analysis diagram—Campaign for Real Ale (CAMRA)
Figure 7.8   Broad-based methodology for allocation of functions—Taylor and Francis
Figure 7.9   Task description format—John Wiley and Sons
Figure 7.10   Task analysis format—Elsevier Science
Figure 7.11   Human reliability analysis event tree—Elsevier Science
Figure 7.12   THERP event tree—Elsevier Science
Figure 7.13   Relationship of functions to work modules—PTR Prentice-Hall
Figure 7.14   Work module design—PTR Prentice–Hall
Figure 8.2   Fault tree diagram—John Wiley and Sons
Figure 8.4   Operational sequence diagram—McGraw-Hill
Figure 8.5   Subjective workload assessment technique— Institute of Electrical and Electronic Engineers (IEEE)
Figure 8.6   NASA task load index—Sandy Hart, NASA Ames Research Center
Figure 11.1   Bus ticket processing unit—Chicago Transit Authority
Figure 11.7   Statistical error decision matrix—Human Factors and Ergonomics Society
Figure 15.1   Bus workstation plan view—Chicago Transit Authority
Figure 15.2   ADA-compliant bus workstation plan view—Chicago Transit Authority

Abbreviations

AA	Activity analysis
ADA	Americans with Disabilities Act
AI	Artificial intelligence
ANOVA	Analysis of variance
ANSI	American National Standards Institute
AoSS	Analysis of similar systems
AT	Assembler-technician
AT&T	American Telephone and Telegraph
BART	Bay Area rapid transit (San Francisco, CA)
BCPE	Board of Certification in Professional Ergonomics
°C	Degrees centigrade
CDM	Critical decision method
CD ROM	Compact disc, Read-only memory
CIS	Critical incident study
COTS	Commercial off-the-shelf
CPM	Critical path method
CPSC	Consumer Products Safety Commission
CPU	(computer) Central processing unit
CRT	Cathode ray tube
CTA	Cognitive task analysis
DA	Decision analysis
DAA	Decision–action analysis
DARPA	Defense Advanced Research Projects Agency (U.S.)
dB	Decibel
DT&E	Development test & evaluation
EA	Error analysis
EID	Ecological interface design
EP	Electronic programmer
EPA	Environmental Protection Agency (U.S.)
°F	Degrees Fahrenheit
FA	Function allocation
FAST	Functional analysis systems technique
FEA	Finite element analysis
FFA	Function flow analysis
FMEA	Failure modes and effects analysis
FRD	Functional requirements definition
FTA	Fault tree analysis
GOMS	Goals, operators, methods and selection rules
GUI	Graphical user interface

HA	Hazard analysis
HCI	Human–computer interface
HEP	Human error probability
HFACS	Human factors accident classification system
HF/E	Human factors and ergonomics
HFES	Human Factors and Ergonomics Society (U.S.)
HPE	Human performance evaluation
HRA	Human reliability analysis
HTML	Hypertext mark-up language
Hz	Hertz (cycles per second)
IBM	International Business Machines
IES	Illumination Engineering Society
IV	Independent variables
IVHS	Intelligent vehicle highway system
JCAHO	Joint Commission on Accreditation of Healthcare Organizations
K-T Analysis	Kepner-Tregoe problem solving and decision making
LA	Link analysis
MANOVA	Multiple analysis of variance
MORT	Management oversight and risk tree analysis
MTTF	Mean time to failure
MTTR	Mean time to repair
NASA	National Aeronautics and Space Administration
NGO	Non-governmental organizations
NIST	National Institute of Science and Technology (U.S.)
NPP	Nuclear power plant
OA	Operational analysis
OE	Operational environment
OEM	Original equipment manufacturer
OSA	Operational sequence analysis
OSHA	Occupational Safety and Health Administration
OT&E	Operational test and evaluation
OTEE	Operator-task-equipment-environment
OTS	Off-the-shelf
PDS	Product development specification
PERT	Program evaluation and review technique
PLF	Parachute landing fall
PSF	Performance shaping factors
R&D	Research and development
RCA	Root cause analysis
RFP	Request for proposal
RPD	Recognition primed decisions
SA	Situational awareness
SAINT	Systems analysis of integrated network of tasks
SAS	Statistical analysis system
SME	Subject matter expert
S-R-K	Skills–rule–knowledge
S-O-R	Stimulus-organism-response
SPSS	Statistical package for the social sciences

STSS	Short term sensory store
SUV	Sports utility vehicle
SWAT	Subjective workload assessment technique
TA	Task analysis
TAG	Task-action-grammar
TD	Task description
THERP	Technique for human error rate prediction
TLA	Time line analysis
TLX	NASA task load index
URL	Universal resource locator
U.S.	United States
USFDA	U.S. Food and Drug Administration
VIE	Valence-instrumentality-expectancy
VPA	Verbal protocol analysis
VR	Virtual reality
WA	Workload assessment
WD	Work design

This is a work in progress. I invite you to share your thoughts on furthering human factors research for design.

Christopher Nemeth
nemeth@id.iit.edu
Evanston, IL, October 2003

… What is the question?

(Gertrude Stein—
Her final words, in response to the query 'What is the answer?')

Part I

Human factors practice

1 The human-made environment

Every day, people around the world who are at work and play enjoy the use of products, buildings and services. Much of the human experience each day relies on items that were made to extend human abilities and to overcome human limits. The sum of all the products, services, environments and systems that humans have produced comprises the operational environment (OE). The inner environment of the individual is separate and apart from the rest of physical existence that comprises the outside environment. 'This artificial world', psychologist and computer scientist Herbert Simon (1998:113) contends, 'is centered precisely on this interface between the inner and outer environments; it is concerned with attaining goals by adapting the former to the latter.'

Historically, craft-based societies relied on experienced individuals to fabricate a small number of items that were gradually refined through trial and error. Each time another item (e.g. clothing, tool, house) was produced, it reflected the collective knowledge of the artisan(s) who made it. Mechanization broke the traditional link between design knowledge and production that is inherent in craft. Machines could be used to manufacture interchangeable parts more cheaply and in greater numbers than could be produced by individual artisans.

The advent of mass production posed questions about economies of scale, variety, efficiency and cost. Attention was focused on the nature of physical work. Decisions had to be made on the appropriate rate of work for both machines and the people who used them. What was needed was an objective measurement and order of work through an understanding of work's natural laws, or 'ergonomics.' Since that time, the evolution of ergonomics and human factors has continued to serve in a complimentary role to design and engineering. In that sense, human factors and ergonomics inform the process of creating the operational environment.

An increasing percentage of daily life in developed countries is spent in the operational environment. As a result, the nature of the OE shapes and influences perceptions in many ways that we are (and are not) aware of. For example, the development of automation and computing systems has created a growing number of products that incorporate software programs to perform routine tasks. Many are now able to govern themselves in limited ways under the guidance of artificial intelligence (AI) software.

Products, spaces and systems that work well routinely go unnoticed. Yet, if an accident occurs which causes loss of property or life, if productivity sags, or if a breakdown causes usually reliable services to fail, the most frequently asked question is 'why?' This is followed by 'what could have been done to avoid this?' Often, the answer lies with the way the product was designed, the environment in which it was used or the way it was used.

Understanding systems and how they work is a key to the creation of products that work well. Most products (even simple ones) are comprised of multiple components and are accompanied by related services. As a result, it is most helpful to use a system model to understand them. This chapter describes systems, their traits, and the process, and the roles of professionals, involved in their creation.

> *Humans are in and of their systems; they are not apart from them.*
> *(Julie Christensen)*

1.1 The systems model

A system is any collection of elements that is organized to achieve a purposeful result. An operating room is a health care system, designed to restore its users to better health. A carnival is an entertainment system that is designed for its customers' enjoyment. A factory is a production system that is created to produce artifacts. An aircraft is a transportation system that is created to deliver cargo and people to desired sites.

Where a system begins and ends is not always clear. Even when systems can be clearly delineated, their interdependence (see Section 1.5.2) and interactions can have a significant effect on other systems.

Goals and objectives are used to develop a system. However, it is human performance that defines and animates the system.

A system's purpose is a desirable goal state that system operation is to achieve. The performance of system elements yields results that are intended to fulfill the system's purpose. Objectives such as cost impose limits on how those results will be accomplished.

1.1.1 System elements

Figure 1.1 shows Christensen's (1985a) model of system composition and performance that identifies three classes of elements: hardware/software, personnel and procedures. Hardware/software includes machines and their controls, displays, mechanisms and software programs. Personnel are the humans who perform in roles as operators, maintainers and end users. Managers are personnel who supervise the performance of operators and maintainers.

The term personnel has traditionally referred to the operators and maintainers who are employed in a system. Both roles have defined responsibilities. Both roles can have a significant effect on performance, reliability and safety. Users can also significantly affect performance. Users can easily be under-emphasized in studies because they can be comparatively harder to recruit and less predictable in their behaviors.

Procedures are the actions that need to be performed by either personnel or software. Communication links allow for information to be transferred within a system and enable

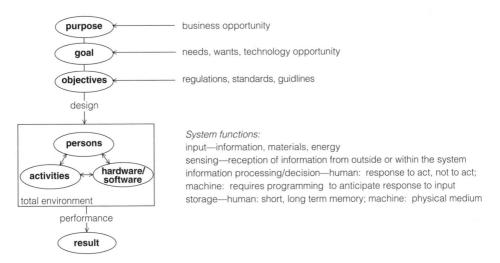

Figure 1.1 System elements and functions—the elements of a system are selected and configured to produce a result that is intended to satisfy its purpose and objectives. How each of the elements perform and interact with each other and with the total environment produces results at a certain level of performance. Evaluation is used to compare those results and the way that they are achieved with the system's goals and objectives. Changes can be made to elements and their relationships in order to bring results into line with goals and objectives.

Source: Adapted from Christensen, J. (1985a)

components to interact. All three classes of elements are considered to interact with each other. Their interaction produces a net result beyond the sum of the individual elements.

Each system exists in a context that Christensen refers to as the 'total environment.' For example, the total environment for a highway transportation system includes the city in which it exists. Even though the highway may be well designed, elements in its total environment can have a significant effect on its performance. An individual who drops a cinder block from an overpass onto autos that are driving on an expressway produces an adverse effect on an otherwise well-designed ground transportation system. This is how the total environment can intrude into a system and, in this example, degrade its performance.

1.1.2 System definition

Czaja (1997:17–40) accounts for six approaches to system development: traditional, socio-technical, participatory, user-centered, computer-supported, and ecological interface design (EID).

1.1.2.1 TRADITIONAL

In the traditional approach, system elements are developed according to a four phase process: concept, physical design, implementation and evaluation.

Sharit (1997:302–37) offers four ways to view such systems in order to better understand them: physical configuration, automation and workload, information flow, and partitioning techniques. Physical configuration considers the goals, agents and their connections, and timing constraints that affect information availability and reaction. The automation and workload view considers whether sufficient attention resources are available for the human to coordinate with an automated system. It also considers human attitudes toward automation and the consequences of error. Information flow considers the type, form, content, timing, rate and direction of information that is needed to realize system objectives, as well as channels through which it flows and the agents who must act on it. Partitioning techniques break a system apart into its logical subsystems, examine their relationships, and seek alternate ways to accomplish the same result more effectively. The human factors challenge in each of these instances is to account for the role of people in the system and the system's effect on people.

Levis (1999:427–8) finds that the traditional approach can be effective when the requirements are well defined and remain essentially constant during the development period. Complex system requirements typically change rapidly. The evolution of workplace studies over the 1990s has demonstrated that requirements can be complex and elusive.

Shortcomings in the traditional design model have led to difficulties. The traditional approach does not allow for requirements that change over time or new improvements in technology that become available. The traditional method's failure to take the human element into account can result in heavy emphasis on hardware and software and ignorance of broader social, organizational and political issues. As a result, five alternative approaches to system design have been developed to improve on the traditional model.

1.1.2.2 SOCIO-TECHNICAL

Socio-technical system design emphasizes the fit between social and technical systems and the environment. Macroergonomics, which focuses on the relationship among humans, organizations, the environment and machines, follows a socio-technical approach.

1.1.2.3 PARTICIPATORY

In participatory design, individuals who are part of work groups and those who benefit from their product apply ergonomic principles and concepts to the design of a system. Participatory design has been applied to the design of products, work environments, and jobs.

1.1.2.4 USER-CENTERED

User-centered design considers both the human and the technical subsystems in the broader context. Users are typically consulted throughout the design process. The user-centered approach has been applied to product production, particularly in human–computer interaction. Christensen's model shown in Figure 1.1 includes the characteristics of a user-centered design.

1.1.2.5 COMPUTER-SUPPORTED

Computer software programs can be used to provide information retrieval, management and transformation in the development of complex technical systems.

1.1.2.6 ECOLOGICAL INTERFACE DESIGN (EID)

EID is a cognitive systems engineering approach that is used to analyze the work domain (through means-end analysis) and individual behavior traits. Jens Rasmussen's (1983) skills-rules-knowledge hierarchy (which is described in Section 3.4.3) is used to determine how information is to be presented. The process offers system designers a set of maps to use in their analysis of operators and their work domain.

The way that a system is defined influences a research and development team's approach to a project assignment. More often than not, the client will set the bounds for a research and development project. Even so, the team may be able to redefine the level of approach. In fact, redefinition of the level of approach or the focus of the problem can be a significant way to influence a project's direction.

Abstraction hierarchy is one partitioning technique that is used to describe the system at several levels in order to separate the properties and functions of the system from its physical elements. Abstraction hierarchy can be used to perform functional analysis (Chapter 7). It can also be used to benefit the design ideation process. Instead of simply modifying an existing product or creating a new one, American design thinker Jay Doblin used abstraction hierarchy to create fewer artifacts through what he termed 'denovation.' Denovation is the practice of using the more abstract approach to a need to find a better way to fill it. The best solution might be no product at all. Such a shift in the level of approach sought a more effective solution opportunity by redefining the problem compared with its client's original intention. (See Problem definition, Section 4.3.2.3). Table 1.1 shows how abstraction hierarchy can be used to partition systems. Two approaches are shown in this example. In the first, navigation is considered a goal. In the second hierarchy, navigation ranks as merely a component of the larger air transportation system.

1.1.3 System types

Systems range both from simple to complex and in the percentage of functions that are allocated to either humans or hardware/software. Figure 1.2 compares a variety of solutions that have been created in order to meet ground transportation needs according to two variables: flexibility and complexity. Different ways to get around have both advantages and drawbacks. The simplicity and flexibility of walking meets many needs over a short distance. For needs beyond a short distance, significant numbers of people (markets) may be willing to forego the flexibility of walking for the benefits that more complex systems offer. Manual wheeled products are more convenient over longer distances although they are less flexible. One can't easily take skates, a skateboard, or a bike across a grassy field. The addition of power trades-off flexibility and low cost for convenience. Scheduled public transportation covers distance for less cost, but at a penalty in flexibility. One must take a bus or train when it is scheduled,

Table 1.1 System hierarchy—two examples illustrate how system definition specifies the hierarchical relationship among elements. Navigation is essential to long distance air transportation. When defined as a goal, four levels of elements are subordinate to it. When defined within the context of the air transportation system, though, navigation is a component.

navigation	goal
avionics	system
radar	product
display	subassembly
cathode ray tube	component
transportation	goal
air transportation	system
aircraft	product
cockpit	subassembly
navigation	component

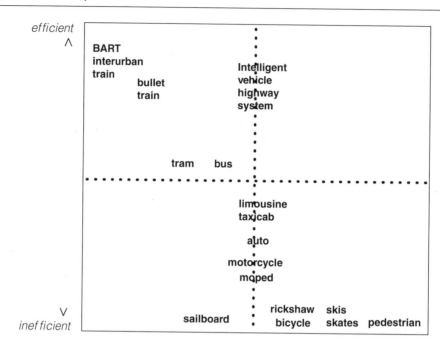

efficient

∧

BART
interurban
train
  bullet
  train

Intelligent
vehicle
highway
system

tram   bus

limousine
taxicab

auto

motorcycle

moped

rickshaw   skis

∨

inefficient

sailboard

bicycle   skates   pedestrian

inflexible <   > flexible

Figure 1.2 Types of systems—development teams choose and configure system elements to achieve a purposeful result. Recent developments in intelligent controls have made it possible to create systems which are both efficient and flexible. The number and type of tasks assigned to hardware/software is a factor which determines the degree of automation (See Function allocation, Section 7.4). In the set shown below, ground transportation systems can be configured to accomplish a variety of results. More desirable solutions would combine both flexibility and efficiency. That is the promise that the intelligent vehicle highway system (IVHS) offers.

following a prescribed route. The evolution of the intelligent vehicle highway system (IVHS) is intended to use autos in a more efficient way by guiding them along highway through the use of automatic control systems. Automated rail systems such as the Bay Area rapid transit (BART) system in San Francisco are intended to maximize efficiency by minimizing the number and salary cost of employees. Which solutions are best depends on the needs and resources of the marketplace (see Chapter 12: Cost-effectiveness). Maps can also be created to address a lower level of abstraction. At the product level, for example, automobiles can be plotted according to miles per gallon versus price. Reviewing clusters of products will show examples of how the market place serves certain needs (e.g. high mileage autos). It may also reveal gaps in coverage that could point the way to new product opportunities.

1.1.4 Performance

In Christensen's model, the interaction of people, hardware/software and procedures results in performance. This performance, which is shown in Figure 1.1, is the sum of all elements' activities and interactions in the context of the total environment. Performance and the results it produces can be compared to the system's original purpose. The comparison can determine whether the system is performing in a satisfactory, marginal or unsatisfactory manner.

All system elements have both limits and capabilities. As all things are finite, there are limits to what a product or service can or cannot do. These performance limits impose constraints on the way a system is designed. Those who perform research and development assume responsibility for understanding those constraints by using capabilities and respecting limits for each of the elements.

The tolerances for hardware and software are based on materials, the design specification (the way the materials are to be configured), wear and robustness (the ability to withstand change). Each aspect of hardware and software performance can be reduced to quite accurate models of when failure can be expected. However, human performance is far different. Humans vary widely in behavior such as perception and cognition, in physiology such as physical strength and endurance, and in tolerance for environmental stressors such as temperature, vibration, pressure, and acceleration. Chapter 2 provides an overview of these and other limits and abilities that affect human performance.

Materials performance can be quantified and predicted, which is a feature that engineers rely on. Human performance is far more complex and variable and can only be represented in terms of probability. Many engineers have shied away from accounting for human performance in their development work. Why? It may well be that the data on human performance are not as clean or reliable as those for materials. It may be that materials tend to perform in a linear fashion, whereas humans perform in a non-linear fashion. Whatever the reason, this partitioning of knowledge about human performance from engineering activity has led many engineers to disregard the role of people within the OE. It has also led to psychologists, physiologists, designers and other human-centered professions populating the field of human factors (also referred to as ergonomics).

1.1.5 Failure and reliability

Failure is the inability to meet system objectives perfectly, infinitely. Reliability is the ability to consistently perform in a way that satisfies system objectives. This can be true of elements such as personnel, hardware and software and the collection of all elements as a whole. System reliability can be increased by determining the probability of error or failure, then taking steps to counter it. Redundancy is one remedy that is used to improve reliability by providing back-up to essential elements.

Reliability is the condition of being available for the purpose the product was created to serve. Hardware and software reliability rests on the likelihood of failure and is sometimes expressed as mean time to failure (MTTF). While a product may have excellent performance, it also must be reliable, or available for use. For example, my colleague Dale Fahnstrom's convertible red Fiat was his pride and joy. As much as he liked the car, it unfortunately spent a great deal of time in the repair shop. So, Dale's Fiat performed very well qualitatively (e.g. satisfaction of sun and wind on a beautiful day, handling, well-considered form and vibrant orange-red finish) and quantitatively (e.g. acceleration, speed). Unfortunately, Dale rarely got to enjoy that performance because the Fiat's reliability was so low.

Error has been defined in a variety of ways, as Section 2.8 shows. The traditional engineering definition is that error is a human action that was unintended, not desired (with respect to a specified set of rules or an external observer) or resulted in a task or system going beyond desired limits (Jones 1999:689). Why does that error occur? Human-centered system design presumes that the cause is inherent in the system design and not in any single system element (including humans). The issue is when and how failure is likely to occur and what will be done prior to failure in order to protect people or assets from loss. For example, tire performance and failure affect automotive safety. For decades, a tire that was punctured would fail catastrophically (blow out) causing an auto to swerve dangerously. Training courses taught student drivers how to manage the course of the swerve to attempt to avoid an accident. This technique sought to manage the difficulty but it did not eliminate it. The driver who had a blow out accident may have lost control of the car, but the error was system-induced. Eventually, a product solution solved the problem: tires that deflated slowly. This graceful degradation made driving safer by enabling the driver to maintain control and get the car off of the road to relative safety. More recently, a tire was introduced to the market that would make it possible for a driver to continue to drive for 40 miles after deflation. As a result of such changes in tire design, highway driving was made more reliable and drivers are no longer put on the spot.

1.1.6 Evaluation

How system elements perform will affect how well system goals are achieved. Evaluation methods are used to collect performance to compare desired and actual performance (see Chapter 8: Evaluation methods). Unfortunately, cost and competitive pressure can force development teams to limit assessment testing to a bare minimum or skip it entirely. Effective system design and evaluation relies on performance assessment in the context of actual or near-actual conditions. A solution concept that has been tested by users under actual conditions is much more likely to succeed in actual use than a concept that has only been assessed in the limited confines of a lab or company headquarters.

Evaluation involves understanding performance, knowing how to assess it and using that knowledge in the context of the product life cycle. Evaluation can provide insight into the match between needs and performance. In a broader sense, it can also assess the match between assumptions about the solution and the needs that it is intended to serve. Performed well, evaluation can detect the potential for paradigm shifts that are conditions under which entire assumptions about a system and its value are radically changed by new thinking or technology.

I have never been forced to accept compromise, but I willingly accept constraints.
(*Charles Eames*)

1.2 Requirements

Requirements are established in order to assure that a product will perform to certain (usually minimum) standards. Understanding which requirements apply to a project and how to comply with them are an essential part of research and development.

Requirements are characteristics that are necessary or essential and set the stage for the creation of a product. Table 1.2 spells out the variety of requirements that can be brought to bear on a development project, including standards, regulations, guidelines, specifications, goals/objectives/criteria and needs and wants. Requirements that address critical issues such as safety, reliability and cost are normally codified. Yet what does necessary or essential mean, and who decides how influential each requirement will be? The source for a requirement influences its importance. International organizations, national governments, trade organizations, corporate staffs, project managers and customers may all have a say depending on the application area.

Table 1.2 Requirements—that which is necessary or essential. A variety of requirements are used in order to define the nature of development work.

Standards	Quantifiable performance levels for system elements that can be established and measured
Regulations	Standards that have been sanctioned through legislation to have the force of law
Guidelines	Procedures developed with a firm in order to ensure continuity of critical aspects (e.g. performance) among divisions, departments, and products
Specifications	Requirements that are written for a specific system
Goals	Desirable end states
Objectives	Aspects of the goal that can be quantified
Criteria	The ways in which goals are to be met
Needs	Necessities; that which one must have
Wants	That which is desired, but not necessarily needed

1.2.1 Standards

Standards are quantifiable performance levels for system elements that can be established and measured (e.g. speed, accuracy). Standards are developed in order to protect the natural environment and consumers, to improve the ability of firms and nations around the world to develop and sell and use products that are compatible with each other.

Codes are essentially the same as standards. The U.S. Life Safety Code, for example, is an ANSI (American National Standards Institute) standard that governs all of the requirements for facilities in order to protect residents from fire hazards.

Lowell (1999:413–26) defines standards as process-oriented technical documents that are used to establish uniform testing procedures, to define the engineering characteristics of materials and parts and to provide for physical and functional interchangeability and compatibility. Standards can also be used to set requirements for quality, reliability, maintainability and safety. Table 1.3 shows a selection of the organizations that draft standards that often pertain to research and development in human factors. Examples of the standards that the organizations create are also shown.

1.2.2 Regulations

Regulations are standards that have been given the force of law through legislative sanction. Most often, this step is taken in order to protect human life, safety and health. For example, regulations that are administered by the Occupational Safety and Health Administration (OSHA) are standards for workplace safety with which all qualifying firms in the United States must comply.

1.2.3 Guidelines

Companies develop guidelines in order to ensure the continuity of critical requirements such as performance among divisions, departments and product lines. For example, products from each division of a large corporation such as IBM should look and perform like IBM products. How IBM products are to perform can be communicated by company guidelines. Procedure manuals can also serve as guidelines. For example, the traditional General Motors 'Motor Vehicle Dimensions and Procedures Manual' described tools such as drafting templates,

Table 1.3 Standards—requirements to ensure safe and sufficient performance that may have the force of law (as regulations do). Standards are drafted by professional, trade and manufacturing organizations. A number of U.S. organizations that have authored standards of interest to development teams are shown below, along with examples of standards.

International Standards Organization (ISO)

ISO 9000	Quality management and quality assurance standards
ISO 9001	Quality systems—model for quality assurance in design, development, production, installation and servicing
ISO 9002	Quality systems—Model for quality assurance in production and Installation
ISO 9003	Quality systems—Model for quality assurance in final inspection and test
ISO 9004	Quality management and quality system elements
ISO/IEC12207	Information technology software life cycle process

The American Society for Quality Control (ASQC) standards parallel ISO standards.

Occupational Safety and Health Administration (OSHA)

Department of Defense

MIL STD 1472D	Human engineering design criteria
MIL H 46855	Human engineering requirements for military systems Equipment and facilities
MIL STD 1801	User computer interface
MIL STD 1800A	Human engineering performance requirements for systems
MIL STD 1478	Task performance analysis
MIL STD 882	System safety program requirements
MIL HDBK 763	Human engineering procedures guide

National Aeronautics and Space Administration (NASA)

Food and Drug Administration (USFDA)

American National Standards Institute (ANSI)
National Electrical and Safety Code
Life Safety Code

Society of Automotive Engineers (SAE)

Building Officials and Code Administrators International (BOCA)
National Building Code
National Mechanical Code

---

defined a uniform set of dimensions for GM vehicle interiors and exteriors and specified methods for their use.

Guidelines are also used in order to spread knowledge throughout the organization. New interface design knowledge that the firm's research labs develop can be spread through guidelines that the company's other divisions are encouraged to follow.

1.2.4 Specifications

Specifications (sometimes referred to by the slang abbreviation 'specs') are requirements that are written for a particular system. They may also set forth procedures that are to be used in order to verify whether those requirements have been met.

Specifications may articulate requirements at each of four levels in a hierarchy: as a whole, subsystems, sub-subsystems, and hardware/software items. There are two kinds of specifications: 'design to' and 'build to.' 'Design to' specifications state the capabilities and the performance of a system in order to solve a customer or operational problem. 'Build to' specifications give the details that are necessary for implementation (Chapanis 1996:58–78).

1.2.5 Best practice

Best practices are norms of professional conduct that are considered to be acceptable among those who lead the field. Each professional group develops its own standards of practice. In some professions, best practices form the basis for licensure such as in medicine and law. In other areas of professional practice such as human factors, they form the basis for certification. Human factors organizations and their technical and professional groups develop their own best practices and research.

1.2.6 Goals, objectives and criteria

Planning creates a logical flow of thought that connects the purpose of the system to the activities that the system is required to perform.

Briefs that are written to define research and development projects routinely specify project requirements. Briefs cite standards with which the solution must comply. In addition, the brief creates a structure for the problem through the definition of goals, objectives and criteria.

Goals are desirable end states, such as 'a healthy person,' 'an inventory of products ready for purchase,' and 'passengers and cargo in the desired location.'

Objectives are aspects of that goal that can be quantified. Making a successful product is a goal. 'Fifty percent market share in the United Kingdom after three years of sales' is an objective that supports the goal.

Criteria are the ways in which goals need to be met. Criteria regularly include cost, safety, efficiency, reliability, maintainability and comfort. Each system is unique in its goals and criteria. Those who develop systems strive to anticipate both criteria and changes in criteria that might occur throughout the system's projected life cycle.

Trade-offs are the decisions that system developers, operators and users make in order to negotiate conflicts among criteria and how they will be met. Such conflicts can threaten system goals if they are left unmanaged.

A plan accounts for the means, schedules and resources that will be used to employ the strategy. Table 1.4 shows the relationship of planning to the activities that occur within a system.

Chapter 14 provides further information on human factors communications including briefs and reports.

1.2.7 Needs and wants

Earlier, design was defined as the application of technology to serve human need. What exactly *are* needs? Needs are necessities; that which we must have. Want, on the other hand, is often held in contrast: that which is desired, but not necessarily needed. It is fruitless to try further

Table 1.4 System activities—what is done in a system can be sorted into a hierarchy of purposeful activity, from most general to most specific. Each is intended to fulfill requirements that were described in Table 1.2. Planning creates a logical flow of thought that connects purpose to functions. The system's purpose is its reason for being. The goal is an end state that the system strives to achieve. Objectives are aspects of a goal that can be quantified. A strategy is the intention to organize resources. A plan accounts for the means, schedules and resources that will be used to employ the strategy.

Function	Classes of events that are intended to use resources
Task	Elements of functions that can be scaled to the individual level
Subtask	Elements of tasks that are the smallest divisible portion of behavior; also known as 'activity'

definition between the two because one person's needs may be another's wants and vice versa. The distinction becomes evident only when making the trade-off decisions on how to proceed with a concept. Friedman (1996:16) contends that 'there are no needs, only wants.' Because 'we can never have enough of everything, we must accept trade-offs among the different things we value ... .' See Chapter 12 on Cost-effectiveness for further discussion.

The engineering community has distilled needs and wants into the voice of the customer. While it is valuable to listen to what customers say they want, customer preference is not the entire story. Hamel and Prahalad (1994b:64–70) contend that 'customers are notoriously lacking in foresight.' Many of today's businesses represent only the articulated needs of customers. The remaining unexploited opportunities include customers who are not yet served as well as the unarticulated needs of all current and prospective customers. As a result, unarticulated needs present the significant opportunity in development.

Design leaders often tap unexpressed needs or wants by presenting users with new concepts to consider. Paul Green (1995:165–208) provides an example through his report on three methods that were used to collect information on user needs and wants related to automotive products:

Clinics—The manufacturer recruits 50 to 200 potential customers to compare a future prototype vehicle with two contemporary competitors. Customers are asked to make pairwise comparisons of products by ranking them or rating them on general qualities (e.g. an auto model's interior space).

Focus groups—A skilled moderator leads a group of 10 potential users through a 2 hour session to identify common opinions on topics of interest to the manufacturer.

Surveys—Participants at shopping malls and driver license facilities are invited to rate or rank various alternatives such as warnings or symbols or to write down what alternatives mean.

In each case, participant locations are chosen based on attitudes and preferences. Clinics are often conducted in Los Angeles because fashion and product trends tend to appear first in California. On the other hand, participant responses in Minneapolis, MN and Columbus, OH 'are typical of the United States as a whole.' Surveys can also be made available via the Internet (see Chapter 9 for information on Web-based surveys).

Mitchell (1995:246–67) described the use of the rapid fire review technique to study how appearance influences ease of use perceptions among users. Subjects were shown 50 images for 2 seconds apiece and asked to rate each from 10 (easy to use or of high quality) to 1 (difficult to use or of low quality). Subjects were shown the images a second time and asked why they assigned each rating. Ratings per product were summarized.

As 'the product, in fact, is no more than artifact around which customers have experiences,' Prahalad and Ramaswamy (2000:79–87) contend that development 'is about the customer becoming the co-creator of the content of their experiences.' Enlisting the participation of user communities in product development will amount to active collaboration in order to truly personalize products. Nemeth (1996b, 1998) has presented a design for use model that features such a dynamic interaction and cites new roles and tools suited to it.

Rapid prototyping is a research and development method that has been successfully used to elicit such information that would otherwise go undetected by the use of traditional techniques such as individual interviews and focus groups. Chapter 4 provides more information on rapid prototyping.

Chapter 9 provides guidance on how to develop interviews and questionnaires that can be used to collect information on needs and wants.

A product heads toward obsolescence the moment its design is set.

1.3 Life cycle

Artifacts in the human-made environment are similar in some ways to living things in the natural environment. Both are considered to evolve through a sort of life cycle: creation, growth, optimization, decline and dissolution.

Systems and the environments in which they exist change continually. A product heads toward obsolescence the moment its design is set. How fast it progresses toward obsolescence depends on the amount and rate of change in technology. The rate also depends on changes in the needs that caused the system to be developed in the first place. While the rates of change vary, sooner or later a system's elements will need to be changed in order to accommodate new situations and criteria that need to be satisfied. Services can also be considered to have similar traits.

Figure 1.3[1] depicts the general pattern that systems tend to follow through what might be considered their life span. Negative cash flow represents the investment of funds to develop solution concepts. Cash flow implications for development can have significant results during the life cycle. The early commitment of significant resources to aggressively develop concepts can result in shortened development cycles, getting concepts right the first time, pre-empting competition and more immediate generation of revenue as market acceptance boosts sales. The sales that the new product generates recoup research and development costs and, through time, sales growth generates revenue up to the point where market demand reaches an upper limit. That limit can be the result of market size (only so many are willing or are available to purchase it), competition (another product has entered the market and stolen sales opportunities) or changes to market needs which make the product less satisfactory. Depending on these and other factors, there may come a point in the cycle in which sales begin to decline. Product planning seeks to determine when such a decline might be expected to occur. Research and development initiates the development of a product that would be

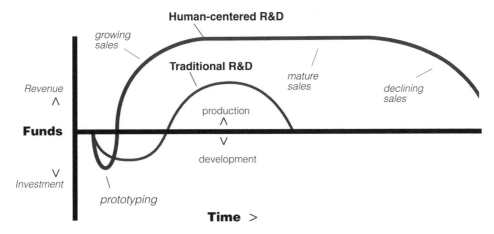

Figure 1.3 Life cycle—products, whether hardware, software, systems, or facilities, all originate, mature and obsolesce. This diagram, based on concepts developed by Patrick Whitney and the faculty of the Institute of Design (Illinois Institute of Technology, Chicago, IL) represents the comparative life cycle difference between traditional and human-centered research and development. In human-centered R&D, aggressive simulation and evaluation through rapid prototyping requires faster expenditure. The benefits of getting the solution right sooner are paid back by quicker time to market and a better fit with customer and client needs.

Source: Adapted from Institute of Design (1994) *The Strategic Value of Human-Centered Design*. Reproduced by permission of Illinois Institute of Technology.

expected to replace current product revenue by offering a better fit with current market needs. While convenient to diagram, decisions on whether to modify or retire products, when to retire products and which products to retire are difficult for firms to make.

Not all of the traits of artifacts are identical to those of living things. In contrast with living things, human-made systems do not typically show whether they are doing well or are having problems. It can be a challenge to know how well a system is performing. Are each of the elements that have been selected to be a part of the system performing at capacity? Is the configuration of those elements optimal? Can the system be optimized further, or is the system obsolete and actually living on borrowed time? The breadth, complexity and interdependence of systems can make these difficult questions to answer.

1.4 The development process

Firms vary in the approach that each takes to research and development. Even so, there are steps that account for the information and action that is typically necessary to accomplish design and human factors activity in the context of development.

1.4.1 Manufacturing

Figure 1.4 shows Robert Bateman's view of the way that research and development activities fit into the life cycle of a system. Portions of the diagram which are shown in gray indicate major steps: define problem, gather data, determine feasibility, plan (define requirements, identify alternatives), analyze (evaluate alternatives), design (build prototype, test, install), implement, use (evaluate) and retire. The design, prototype and test cycle may be repeated until the solution meets requirements. Evaluation includes means to either plan a product modification or to retire it. To create a follow-on product in time, it is necessary to complete planning before retiring its predecessor.

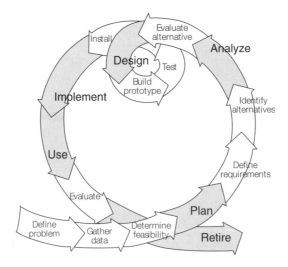

Figure 1.4 Development process—Robert Bateman describes the phases of product development within the context of product life cycles. To create a follow-on product in time, it is necessary to complete planning before retiring its predecessor. Major phases in the development cycle are shown as grey shaded arrows.

Source: Bateman, R. Reproduced by permission of the author.

Figure 1.5 shows Kreifeldt's (1992:1144–63) more traditional engineering view of the development process for consumer products that are well considered with regard to human factors: identify needs, information phase (patents, literature, morphological, ordering), consumer phase (questionnaires, interviews, focus groups), planned research, hazard analysis, specifications, creative design session, conceptual design, prototype design, verification and final solution. User response to the final solution and new technology opportunities compels the start of a new cycle. Funding, competitive strategy and other influences that are mentioned in Section 4.1.2 determine when a new development cycle will be started.

The development process in organizations such as manufacturing firms is a collaborative process among many departments. Factors such as size and business sector affect a firm's approach to research and development. Even so, there are procedures that firms follow so that its departments can work together.

Table 1.5 summarizes the development process that is typical for a product manufacturer. Acting in concurrent fashion, management and many departments including research and development (R&D), marketing, sales, and operations collaborate in order to develop and implement new products. The table also shows how human factors and design activities fit into the process.

Management develops the business plan in order to define the business opportunity and create a strategy to achieve it. The plan allocates capital and resources and sets production volumes over long (five-year) and short (one-year) terms. Plans by research and development, marketing, manufacturing, sales and other major departments spell out how the vision in the business plan will be achieved. The marketing plan identifies market segments of interest, which in turn creates traits profiles that are of interest to human factors. For example, a manufacturer that chooses to begin making products for use by elders will need to become familiar with the physical and behavioral traits of that group. Ideas that are not driven by objectives can still be brought to the attention of the organization by a need or opportunity proposal (see Section 4.1.3.2). If the idea is considered to have merit, resources such as research funding can be applied in order to look into it further.

Interest in bringing new knowledge into the organization leads to research. At this point, the organization knows neither the nature of the problem nor its solution. A research brief by R&D management identifies issues of interest, resources and a schedule. The investigator(s) gather the information that will be needed in order to define the problem. User segments, their traits, and their interests can be identified. Depending on the project, research can also define objectives and assess the market potential, the business opportunity, and may use diagrams and rough sketches to illustrate possible directions for the solution. Results are provided in a research report.

Human factors analysis methods (see Chapter 6) are performed in order to assess competitive products and to determine actual and potential dysfunction. Standards and guidelines are audited to collect information on requirements.

A product brief by R&D management, based on what was learned about the problem during research, sets the direction for the design phase. During design, a staff or consulting design group generates solution concepts in response to project objectives. Human factors design guidance methods (see Chapter 7) are used to analyze what needs to be done and to determine what humans should be expected to accomplish. Evaluation methods (see Chapter 8) can also be used in order to determine how well solution alternatives meet project objectives and fit human limits and abilities. The design phase results in solution prototypes, a vocabulary of components and their variations, and a report that provides the logic and references behind the artifacts.

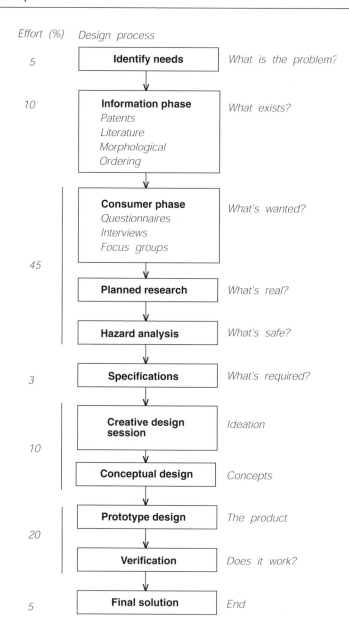

Effort (%)   Design process

5	**Identify needs**	*What is the problem?*
10	**Information phase** *Patents* *Literature* *Morphological* *Ordering*	*What exists?*
45	**Consumer phase** *Questionnaires* *Interviews* *Focus groups*	*What's wanted?*
	Planned research	*What's real?*
	Hazard analysis	*What's safe?*
3	**Specifications**	*What's required?*
10	**Creative design session**	*Ideation*
	Conceptual design	*Concepts*
20	**Prototype design**	*The product*
	Verification	*Does it work?*
5	**Final solution**	*End*

Figure 1.5 Ergonomic design process for (consumer) products—Kreifeldt (1992) presents a view of consumer product development, posing questions or results of interest in each phase, as well as estimating the relative amount of effort that is invested in each phase.

Source: *Ergonomics of Product Design*. Kreifeldt, J. (1992) in Salvendy, G. ed. *Handbook of Industrial Engineering*. Copyright © 1992 John Wiley and Sons. This material is used by permission of John Wiley & Sons, Inc.

Table 1.5 Product research, development and production (manufacturing)—whether it takes six months or five years, manufacturing product development follows this general series of steps. The roles that human factors and design perform are shown in parallel with the evolution of a product from business plan to use and then to retirement. While many departments collaborate in every step, one department is responsible for the completion of each, as shown by this key: M—Management, R—Research and Development, Ma—Marketing, S—Sales, O—Operations.

Development process	Human factors	Design
Business plan (M)—Define business opportunity and strategy	Identify market segment, and create traits profiles	
Product plan (M)—Define and pursue interest areas to articulate business plan		
Need/opportunity proposal (Any part of the firm can participate)— Bring idea, problem, opportunity to attention of organization		
Research (R)—Gather all information necessary to generate a problem definition, design objectives, and conceptual solution	Audit regulations, standards and guidelines Perform analysis methods to assess competitive products, determine actual and potential dysfunction Develop design constraints for product requirements brief	
Design (R)—Translate design objectivesinto product solutions. Define vocabulary	Perform design guidance methods to create basis for element selection and configuration Perform evaluation methods to determine how well solution fulfills requirements	Acceptance Analysis Problem definition Ideation Idea selection Implementation
Market analysis and financial model (Ma)—Determine concept potential		
Design closure (R)—Final statement of line and product vocabulary		Finish concept development, art and models for all parts/products to be made. Finalize materials and processes specifications
Marketing proposal (Ma)— Identify needs/benefits, customers, motivation, introduction plan		Articulate product benefits
Development engineering (R)– Develop specifications to make design concept into producible product. Specify materials, processes	Perform engineering analysis and testing to verify final specifications and prototype performance in actual environment, before production	Negotiate trade-offs to maintain concept integrity. Revise part visuals

Table 1.5 (continued)

Development process	Human factors	Design
Release for quotes (R)—Obtain tooling and piece parts cost estimates, quotations		
Business proposal (M)—Request manufacturing resource commitment		
Capital allocation request (Ma)— Request management approval for capital funding		
Release for tooling (M)—Authorize vendors to build tools and produce parts		Review first parts with development engineer and vendors
Product introduction (Ma)— Develop plans, programs to bring product to marketplace		
Part approval, release for production (R)—Put engineered products into production cycle		
Pre-sell (S)—Begin standard lead time order entry		
Product roll out (Ma)—Verify and implement business, marketing programs	Monitor customer and sales force reaction to product, accident information, customer service reports and recalls	
Production (O)—Produce, inventory, ship new product		
Sales (S)—Conduct sales training. Begin routine sales activity. Assist (if appropriate, train) customers		
Customer service (S)—Maintain product during use		
Management (M)—Monitor program objectives, assessing: process, objectives, profitability		

With a specific solution to review, marketing can determine both its potential through market research and its financial feasibility. Both can indicate the need for design solution changes, which culminate in design closure. This final version of the solution and its variations is used in order to produce a development brief. The development brief, which is also written by R&D management, describes how the product will be manufactured and sold.

Marketing refines its understanding of user needs and how the product satisfies them and develops a plan to introduce the product to the market. At the same time, R&D pursues the development engineering phase. Specifications are written for vendors and departments regarding materials and production processes. Tests are conducted on materials, components, and prototypes. A development report is written to collect development phase information for review with the team members who are responsible to plan manufacturing. Design attention ensures that changes that are made during the development phase protect the integrity of the solution that was agreed upon at design closure. Human factors attention turns to the evaluation of final prototypes in order to verify final specifications and to assess product performance in actual use prior to release.

Development engineering obtains quotations for the cost of producing parts and their machine tools (e.g. dies, molds, jigs). Quotes form the basis for a final business proposal that management relies on to approve funding for production. Upon quotation and funding approval, vendors and departments are given the go ahead to produce machine tools while marketing develops plans to introduce the product to the market. After finishing machine tools, the designer and development engineer verify that the tools produce parts according to specification and approve them for production.

With product production set, sales can begin to take advance orders and marketing can conduct a final check of the business and marketing plans. As the operations department produces the product, it is sent (usually briefly) to inventory and shipped. Sales trains staff members in product features and benefits, then conducts routine sales activity.

During use, customer service provides customers with maintenance and warranty assistance. Sales and customer service staffs may also train users. Management continues to track any of the significant variables in the business plan, including profitability (revenue and costs), materials and processes used to produce the product, and more. During customer use, accident reports, warranty issues and customer reaction provide valuable human factors information that can serve as the basis for improvements and the next generation of the product.

1.4.2 Software

Traditional software system development followed a seven step model: problem definition, requirements gathering, system analysis, coding (write software programs), testing, user acceptance testing and rollout. The ease of developing prototypes has changed the process significantly. Now, the initial product concept and refinements to it provide users and developers with a context to think about requirements (DiBiasi 2002).

Current approaches to software development vary widely. A recent description of the software development cycle in a small business consists of four phases that are shown below along with products that are produced within those phases (Carnahan 2002).

Inception—Product abstract defines the business case, market needs and client requirements for one application. Product feature specification defines the features, major concepts, rules, attributes, and user interfaces for one application.

Elaboration—Product development specification (PDS) is the functional specification that relies on the input of one or more feature specifications. The initial PDS describes software architecture, the early version of the design, cases in which the software will be used, diagrams that demonstrate use sequences, and all of the deliverables that will be produced. The final PDS also includes detail design. The prototype is an initial version of the product that is used to demonstrate how the product will operate. It can also be used to quantify high risk areas.

Construction—Software deliverables are the product components that, taken together, will be sold to the customer. 'Spec and tech' accounts for the finished product's specifications and the technical features.

Transition—Sales collateral describes the product benefits and features for prospective customers.

Table 1.6 summarizes a general model for software research and development.

Table 1.6 Product research, development and production (software)—software follows a faster development cycle than manufacturing and the divsions between phases are less distinct. Production is often done in frequent small releases, making cycle time faster and development cycle time minimal. Departments collaborate in every step, one department is responsible for the completion of each, as shown by this key: M—Management, R—Research and Development, Ma—Marketing, S—Sales, O—Operations.

Development process	Human factors	Design
Business plan (M)—Define business opportunity and strategy	Identify market segment, and create traits profiles	
Product plan (M)—Define and pursue interest areas to articulate business plan. Develop product abstract		
Inception		
Research (R)—Requirements elicitation and analysis	Elicit and analyze requirements. Summarize design constraints	Develop initial concept for software architecture, function and appearance
Gather all information necessary to generate a problem definition, design objectives, and conceptual solution	Perform analysis methods to understand user wants and needs, assess competitive products, determine actual and potential dysfunction	
Develop product feature specification		
Elaboration		
Design (R)—Translate design objectives into product solutions. Create product development specification. Perform design, construction, integration	Perform design guidance methods to create basis for feature selection and configuration	Create follow-on prototypes to visualize system, control functions
Market analysis and financial model (Ma)—Determine concept potential	Perform evaluation methods to determine how well initial solutions fulfill requirements	Create follow-on prototype versions based on initial review and exploratory test results
Construction		
Development (R)—Produce software deliverables to translate design concept into product ready for release	Perform alpha, beta site usability validation testing to verify final specifications and prototype performance in actual environment, before release for production	Finish concept development. Verify and optimize program performance
		Create specifications in order to document technical features
Marketing proposal (Ma)—Identify needs/benefits, customers, motivation, introduction plan		Articulate product benefits
Transition		
Product introduction (Ma)—Develop plans, programs to bring product to marketplace		
Release for production (R)—Put products into production cycle		

Table 1.6 (continued)

Development process	Human factors	Design
Pre-sell (S)—Begin standard lead time order entry		
Product roll out (Ma)—Verify and implement business, marketing programs		
Production (O)—Produce, inventory, ship new product	Monitor customer and sales force reaction to product and customer service reports	
Sales (S)—Conduct sales training. Begin routine sales activity. Assist (if appropriate, train) customers		
Customer service (S)—Assist customers while product is in use		
Management (M)—Monitor program objectives, assessing: process, objectives, profitability		

> It is better to ask some of the questions rather than to know all of the answers.
> (James Thurber)

1.5 System issues

Society has created and examined the value and role of the human-made environment since its inception. A range of issues affects the way we view the human-made environment and our role in it. Four issues have immediate implications for human-centered design: clumsy tools, interdependency, the role of humans, and the capabilities of intelligent products.

1.5.1 Clumsy tools

Failures of systems that were created to improve human performance have cast doubt on research and development models and methods. For example, the United Kingdom's National Health Service conversion to electronic patient records was intended to be an improvement over paper documents. Instead, the new system actually impeded the work processes that it was intended to improve (Heath and Luff 2000:31–57). Why? Suchman (2001b) suspects that 'there's a much more dynamic and contingent character to the way that human activity is organized than AI [artificial intelligence] models were assuming.' Initial notions of human behavior that were based on goals and plans were far too simple for what actually occurs among people. Luff *et al.* (2000:4) contend that the difficulties in failed intelligent systems 'may not be so much associated with poor design but more related to the general objectives underlying the system, particularly with respect to how designers are considering the activities they aim to support.'

In response to these shortcomings, sociologists have developed methods such as workplace studies and conversational analysis in order to improve understanding of how people collaborate. The use of observational methods such as those in Chapter 6 can also improve investigator awareness of actual human behavior.

1.5.2 Interdependency

More and more, complex systems are being designed to rely on each other. However, manufacturers, agencies and controlling authorities pursue separate agendas. Collaboration that must occur does not, leaving the possibility for one system's failure to affect many more. The performance or failure of a remote and seemingly unrelated system can have an immediate and drastic effect on others.

The experience of American Telephone and Telegraph (AT&T) during the early 1990s demonstrated how the failure of one system can have substantial ripple effects. In September 1991 an attempt to switch a New York AT&T facility to diesel power (in response to a local utility request) failed. No staff members were stationed in the engine room and back-up members were away at training. Despite a series of alarms, no staff members noticed that their equipment was being powered by batteries alone. After six hours, the batteries did not have enough power to make the switch back to city power. The facility's failure prevented long distance telecommunications traffic in and through New York, held up air traffic control communications, and forced the delay or cancellation of flights across the U.S. (Schwartz 1991:48).

Interdependency is not limited to existing conditions. New ideas can have unforeseen effects. Economist Joseph Schumpeter's notion of disruptive innovation allows that seemingly minor inventions can have unpredictable results. In essence, 'small innovations can have large consequences' and their long-term effects are difficult or impossible to know (Rothstein, 2001).

1.5.3 The role of humans

The role that humans play in semi-automated and automated systems remains a subject of significant controversy. The glistening promise of what new technology offers can divert attention from deeper consideration of issues. The rate and scope of systems development makes it difficult for people to truly understand the larger implications of new technology. Humans, Sheridan contends, have a weak grasp on the actual and prospective implications for intelligent systems and the human role in their operation. Sheridan (1998:20–25) finds that:

> We warm to the idea of 'human-centered automation' but when the alternative meanings of the phrase are examined, we find the substance thin and the real potential questionable. As humans become supervisors and as computers become mediators or intelligent agents, we realize that teaching some tasks to the machine isn't worth the trouble, and other tasks, those we can't define easily, we cannot program machines to do.

Jeremy Rifkin (1995) argues that advanced technology displaces or replaces workers who are unable to transition into other roles fast enough to support themselves. The pressure on individuals to find meaningful new roles in the workforce produces social change that will be disruptive unless it is mitigated by equally dynamic changes to social policy.

1.5.4 Product capabilities

Systems are being developed at a level of complexity and at a rate that can be hard to grasp. The prospects for what products will become (particularly those with artificial intelligence features) is both thrilling as well as a subject for concern.

Moore's Law is Intel Corporation founder Gordon Moore's observation in the 1970s that the semiconductor processing capacity doubles in complexity every two years. Until recently, it was thought that physical limits would end the trend by 2010. The development of nanotechnology (the construction of materials at the atomic level) may extend the trend through 2030. Continuing the exponential improvement creates the potential for computers to be a million times as powerful as computers in the year 2000. Many view this in a positive light. By contrast, Sun Microsystems Chief Scientist Bill Joy takes a sobering view of Moore's Law. Left unchecked, Joy warns that the confluence of commercial forces and the power that this growth in technology represents are 'propelling the development of systems that will ultimately bring calamity'(Stross 2000:44). Common wisdom has held that the rate of growth in processing capacity would diminish as semiconductor circuits reached physical limits of size reduction. Instead, recent developments in hard drives and optical fiber networking have made it possible for processing capacity to grow twice as fast as Moore's Law (Yang 2000:37–8).

Rather than attempt to predict the evolution of products and systems, this text is more concerned with implications for humans. As these and other issues show, the application of technology for human use requires attention well beyond how to run a project or whether a technology is feasible. Those who work in research and development are compelled to also speak to larger issues of whether or not products will actually serve human needs and, if they do, how.

1.6 Human factors practice

Everyone designs who creates a course of action that is intended to change existing situations into preferred ones. Even so, there are a number of professional roles that perform design activity according to specific types of applications. Table 1.7 accounts for many of the professions that create the operational environment.

Product ('industrial') designers develop new concepts in mass-produced human-scale products and visualize them through drawings, renderings and models. Visual communications ('graphic') designers create products to inform and train. Architects develop concepts for residential and commercial buildings and public spaces. Engineers ensure that product concepts are suited for manufacture, and that buildings are ready for construction. In particular, industrial engineers are '... concerned with the design, improvement, and installation

Table 1.7 Development practitioners—a variety of practitioners create the operational environment. Each has an opportunity to develop concepts which reflect an understanding of human abilities and limits.

Role	Practice—designs and develops:
Architect	structures
Product, or 'industrial,' designer	human-centered products, systems
Communications, or 'graphic,' designer	informational products, systems
Environmental designer	spaces
Apparel, or 'fashion,' designer	clothing
Human factors specialist/ergonomist	human-centered products, systems
Engineer	machines
Software designer	computers
Systems engineer	systems
Engineering manager	engineering teams

of integrated systems of people, material, equipment, and energy' in order to design work (Salvendy 1992). Software systems analysts develop the plan for a software program, including the interface and procedures that will be used to operate it.

Human factors is the development and application of knowledge about human physiology and behavior in the operational environment. Those who work in the field, referred to as human factors professionals or ergonomists, perform research and analysis on human performance, including both limits and abilities.

The term 'ergonomics' has come to be used interchangeably with human factors. Derived from the Greek *ergos* (work) and *nomos* (natural laws), it developed as the study of work industrial settings. Traditionally, ergonomics has been the term used more often on the European continent. Worker safety, health, and comfort have been a key concern in many European countries for years, and are often a cause taken up by labor unions on behalf of their members.

Terminology aside, the human-made, or 'operational,' environment is created to benefit people. Any development effort that fails to take human abilities or limits into account will not benefit humans and may put them in jeopardy. Table 1.8 portrays the knowledge and related fields that are within, and that relate to, human factors. Psychology, the science of mind and behavior, has developed theories of human cognition and perception that are crucial to activities such as learning and training. Physiology, the branch of biology that deals with living organisms and their functions, has produced theories that address the human response to stress in the physical world. Life support and protective equipment and apparel are based on physiological knowledge of human physical capabilities and limits.

Anthropology, the science of human social relations and response to environment, has contributed concepts such as culture and anthropometry. Observation methods are valuable tools that the social sciences and anthropology in particular have provided. The science of numbers, their properties, relationships and functions, known as mathematics, has developed probability theory, leading to statistics as a means to study the character of populations and relate cause and effect.

Industrial engineering, the study of manufacturing job and facility design, has contributed theories such as time study and motion study, and produced methods that include job design and the understanding of human error probability. Product ('industrial') design, the use of explorative visualization to make the possible real, has developed means to simulate possible solutions through the creation of diagrams, illustrations, mock-ups, and prototypes.

Table 1.8 Human factors practice—human factors draws on knowledge from a variety of fields and methods to develop systems that are user-centered. The examples that are shown are not intended to be exhaustive. Rather, they demonstrate the relationships among fields and between knowledge and practice.

Field	Knowledge	Aspects of practice
Psychology	Cognition, perception	Learning, training
Physiology	Stimulus-response, Stressor tolerance	Life support, protective equipment, apparel
Anthropology Sociology	Culture, society	Observation
Mathematics	Probability theory	Statistics
Industrial engineering	Time study, motion study	Job design, human error probability
Design	Visualization	Mock-ups, models, prototypes

1.6.1 Evolution

Efforts to improve human performance have been undertaken since early times (Figure 1.6). Evidence of early tools shows human ability to modify what exists in order to improve human effectiveness. It has only been since the 1400s that efforts have been made to systematically improve the design of procedures and tools to accomplish work. Thus, during the fifteenth century, the number and type of errors that occurred in early printed versions of the Bible created extensive manual corrections, then printed sheets of errata. Manual typesetting was one of the first documented, systematic efforts to improve human performance. Astronomy and navigation took substantial steps forward when consistent procedures and accurate timepieces (chronometers) were incorporated into the process of celestial observation (Bailey 1989:28–34).

During the Industrial Revolution, the Age of Machines (1750–1870) witnessed the evolution of cybernetics, the science of communication and control theory concerned in particular with the control of automated systems. For example, Jacquard's punched cards used to control weaving machines. The era also cultivated an awareness of the relationship between humans and machines. During the Age of Power (1870–1945) 'the emphasis ... was on adapting people to their work' through such techniques as selection, classification and work schedule management (Christensen 1987:4–15).

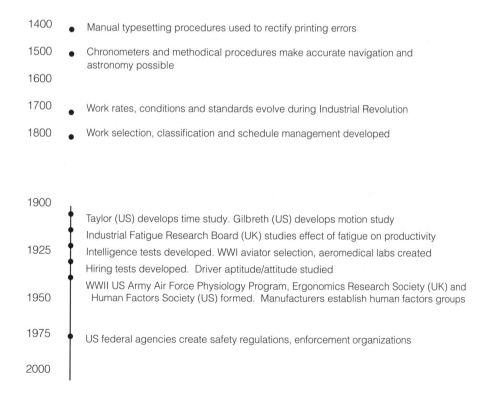

Figure 1.6 Human factors time line—efforts to extend human abilities through the use of machines and procedures were recorded in the 1400s. It wasn't until the Industrial Revolution that there was an awareness of the relationship between human and machine. The practice of human factors and ergonomics greatly expanded through the 1900s.

In the early 1900s, attention focused on altering the design of work in order to improve performance. Early developments resulted in what is known today as industrial engineering. Frederick Taylor developed what is now known as time study by examining the best way to do a job. His method was to discover the best employee's methods for performing work, and determine how long the tasks took. Those tasks were then taught to other less skilled employees who were paid according to how well they performed the specified job. His methods resulted in dramatic improvements to productivity at Bethlehem Steel, where a time study enabled 140 men to shovel the same amount of iron ore which had previously required 600 men.

In the same era, engineer Frank Gilbreth collaborated with his wife, psychologist Lillian Gilbreth, to develop motion study. Their motion study research method observed tasks, broke each into basic elements of motion (termed 'therbligs' in a puckish derivation of the family name) and eliminated all but the most essential motions.

Through the 1920s, behavioral scientists of the Industrial Fatigue Research Board in England used counters on looms and meters on motors to collect quantitative data to study the effect of fatigue on productivity.

During World War I, attention was paid to quantifying human traits. Psychologists focused on the development and administration of intelligence tests. A few, such as Yerkes and Dunlap, created prototype training aids and studied the anthropometry of gas masks. The need for aviator selection led to the creation of aeromedical laboratories, such as the lab at Wright Field in Ohio (later Wright-Patterson Air Force Base).

In 1927, studies on the way working conditions, specifically illumination intensity, affect production began at the Western Electric Hawthorne plant (which manufactured components for the telephone system) and lasted for twelve years. Heavy unemployment during the Depression years of the 1930s turned attention to techniques, primarily hiring tests, which could be used to choose the best candidates for work. Public enthusiasm for the automobile led to research into driver aptitudes, attitudes and traffic safety.

World War II created significant changes in science and the technology it produced. Sophisticated new systems outstripped the ability of operators to use them. 'Psychologists were engaged to work with engineers to produce machines that required less of their users while at the same time taking full advantage of people's special abilities' (Bailey, 1989:28–34). Fitts, Chapanis, Small, and Christensen brought their background in psychology to the development of new equipment. By applying the stimulus-organism-response (S-O-R) model to technological problems, they demonstrated how machines and systems could best account for human nature. Their efforts were not readily accepted by engineers, though. Initial resistance to their contributions stemmed from the fact that '… data usually were not expressed as functions that were related or relatable to important systems criteria' (Christensen 1987:5).

Active support of human factors research by the U.S. Department of Defense eventually resulted in the creation of the Army Air Force Physiology Program. It also spawned government centers of research which included the Psychology branch of the Aeromedical Laboratory, the Human Engineering Branch of the Naval Research Laboratory, and the Medical Research Council of the Applied Psychology Research Unit.

By war's end, studies had been conducted into equipment and its design. In 1947, Chapanis, Gardner, and Morgan published *Applied Experimental Psychology*, one of the first books in the field. The U.S. government encouraged the aerospace and electronics industries to hire members to their staffs to perform human engineering. Attention was focused on the development of 'man-machine' (later 'human-machine') systems. Methods such as Miller's development of task analysis in 1953 could now be developed to account for how humans would participate in systems. Woodson and Conover published the *Human Engineering Guide for Equipment Designers* in 1954 (later revised in 1960) that accounted for how products and environments needed to be configured in order to aid engineers in designing their equipment from a human–operator standpoint in order to improve 'man–machine' operation. Van Cott and Kinkade's 1963 edition (and 1972 revision) of the *Human Engineering Guide to Equipment Design* made human engineering knowledge available to a wide audience.

1.6.2 Objectives

Richard Pew (1985a) describes six objectives for human factors work: increased efficiency/productivity, increased reliability, minimum training manpower costs, improved safety and habitability, increased user appeal and flexibility.

Quantifiable results are an issue in any organization when it is necessary to justify the cost of human factors-related activity. The more directly such activity can be 'tied to the bottom line,' the easier it is to demonstrate a direct benefit to the organization. The objectives are ordered starting with results that are easiest to quantify (increased productivity) and ending with results that are most difficult to quantify (increased flexibility).

1.6.2.1 INCREASED EFFICIENCY/PRODUCTIVITY—MINIMUM TIME PER UNIT OF OPERATION, AND MINIMUM REJECTS

The amount of time it takes for each production line worker to perform an operation on a product produces a certain cost: the compensation that is paid for the worker's time. Even a small reduction, multiplied by many operations a day, can have a significant impact. It can also open the way for an increase in productivity, yielding more products per shift.

Errors in assembly require rework in order to correct the mistake. Fewer errors in production results in a greater percentage of products ready for further production and sale.

Some of the human factors actions that can improve the way the job is performed include motion study, time study, functional flow analysis, workload analysis, changes to workstation arrangement and the introduction of new equipment. Such improvements can cut the time it takes to perform each operation, or can reduce the frequency with which mistakes are made.

1.6.2.2 INCREASED RELIABILITY—MINIMUM HUMAN/SYSTEM MALFUNCTIONS, EFFECTIVE INSPECTION/MAINTENANCE PROMOTED, LOGISTIC SUPPORT ENABLED, LESSENED LIKELIHOOD OF FAILURE UNDER STRESS, LOAD, WEAR, AND MINIMUM MAINTENANCE COST

Section 1.1.5 on failure and reliability described the need for products to be available for use and to perform as expected.

Human factors activity that can improve product reliability include failure mode and effects analysis, use of an event tree, and fault tree analysis. Design alternatives can also be compared with regard to maintenance, supply and materials/components performance.

1.6.2.3 MINIMUM TRAINING MANPOWER COSTS—MINIMUM SPECIAL TRAINING REQUIREMENTS, REDUCED TRAINING TIME, MINIMUM NEED FOR SPECIAL SKILLS AND ABILITIES, AND EASE OF MAINTENANCE

The need to develop and remember unique skills to operate and maintain products imposes a burden on those who must use the products. The need for staff to take time from work to learn new skills is a cost to any organization. It costs a business to hire specialized workers, pay increased salaries for special skills and produce and maintain specialized tool inventories.

For example, office equipment in the 1970s provided a great boon to businesses. Electrostatic copiers and computers transformed the workplace. At the same time, early generations of these products required special skills and training. The last thing that an employee who was trying to operate a copier wanted to see was the 'call key operator' message. That signal indicated the need for a specially trained colleague to fix whatever problem was preventing the machine from operating correctly. Later generations of copiers and computers, which had the benefit of human factors attention, reduced the need for special skills. The advent of the bit-mapped cathode ray tube (CRT) display and the graphical user interface (GUI) made the creation of the Apple Macintosh personal computer possible. Iconic symbols that appeared on the display when the unit was turned on made it possible for its user to figure out how to operate the Mac, without taking time off for training. Saving the cost of training as well as the cost of lost productivity was a key feature that Apple used to sell the Macintosh upon its introduction.

Products that enable users to operate and maintain on their own can be made possible through activities such as field observation, decision-action analysis and flow analysis, and early user participation in the design process.

1.6.2.4 IMPROVED SAFETY AND HABITABILITY—FEWER INJURIES, OPTIMAL OPERATOR WORK LOAD, AND FEWER HAZARDS

Employees who are lost to the workplace due to injury incur a variety of direct and indirect costs to their firm, family and society. Their loss can result from accidents that can occur anywhere: on the job, while traveling, at home and more. Costs include the cost of replacement workers as well as reduced productivity if the replacement worker is less familiar with the job. There are costs of immediate and follow-on health care insurance and insurance administration. Family members often are pressed into service to care for the individual at home and to provide services for the family that the worker had performed.

The causes of accidents are not obvious. Products that do not appropriately take human considerations into account can induce behaviors that are contrary to safe operation. The effective use of human factors expertise, including development methods described in Part II of this work, can create products which lessen exposure to conditions which result in accidents. Methods such as critical incident study, failure mode and effects analysis, use of an event tree and fault tree analysis can be used to focus attention on what has gone wrong or might potentially go wrong.

1.6.2.5 INCREASED USER APPEAL—BETTER JOB SATISFACTION, BETTER SERVES USER NEEDS, AND ACCOMMODATES VARIOUS POPULATIONS OF USERS (E.G. CONSUMERS WITH DISABILITIES)

Changes to job design and job enrichment can create a new role for managers and staff, making better use of skills and abilities. Improvements to the way that people are expected to work can improve their satisfaction on the job and, by extension, their productivity. Human factors attention including task analysis and function allocation can make such improvements.

Better-designed products often sell better than those that are not as well designed. The relationship between a better product and increased sales can be blurred by many other factors that also figure into the mix. They include but are not limited to pricing, discounts, the competition, distribution and availability. Even though it is not as easy to quantify, it is safe to assume that products that serve user needs better and serve more users better are likely to sell well. Human factors participation in each stage of research and development will do much to ensure products are well considered from the standpoint of human needs and abilities.

1.6.2.6 FLEXIBILITY—EASE OF CHANGE TO PROVIDE NEW OUTPUT FOR REPROGRAMMING, ADAPTABLE TO NEW TECHNOLOGY, AND USE IN NEW ENVIRONMENTS

Attention to the extension and adaptation of products to new uses can increase market potential and extend product life. Both serve to better amortize the cost of new product development.

1.6.3 Organizations

Organizations have been formed to promote professional development. The Ergonomics Research Society was formed in 1949 at Oxford by a variety of professionals with an interest in human performance. The Human Factors Society was established in the U.S. in 1957 (becoming, in 1993, the Human Factors and Ergonomics Society). Conceived in 1957, the International Ergonomics Association (comprised of existing national professional organizations) has sought to expand awareness and use of human factors, particularly in developing countries. The Board of Certification in Professional Ergonomics was formed in 1993 to certify practitioners in the field.

In recent years, human factors groups have been created in the computing and telecommunications industries. A variety of U.S. federal agencies have sponsored research into human factors in civilian applications, including:

- Federal Highway Administration (highway and road sign design)
- National Transportation Safety Board (safe operation of all means of transportation)
- Federal Aviation Administration (airline, aircraft and facility design and operation)
- National Highway Traffic Safety Administration (improvements to car crash worthiness, lighting systems, and controls)
- National Aeronautics and Space Administration (space station design).

Other U.S. federal organizations have grown from the public interest in safety, including the Consumer Products Safety Commission (CPSC), Occupational Safety and Health Administration (OSHA), and the U.S. Food and Drug Administration (USFDA).

1.6.4 Issues

Christensen (1987:4–15) sees human factors devoting increased emphasis in the future to product safety, understanding individual differences and the performance of groups and their effects on society, the development of instructional systems which are more responsive to individual differences and the realization of computing systems' potential.

No other discipline is better suited to address the issue of reliability and accountability of complex organizations and systems. For example, the simple inability to get information from software firms on how their product operates, or how it relates to other software and hardware products, can prevent a customer from using what he or she has already purchased. The human factors practitioner (along with design and engineering counterparts) takes the user point of view to identify such issues in advance and to determine how well they are resolved prior to product launch.

David Meister (1998:4) has identified future challenges that are specific to test and evaluation, a field of human factors that affects development and is a major concern of this work. For Meister, the separation between the laboratory and practical application is an overriding concern. 'The fundamental problem of HF [human factors] is the transformation of behavioral principals into physical systems and the determination of the effects of design on human performance (in other words, crossed domain lines).'

The issue has much to do with how we use what is known. There is a substantial body of literature on human behavior. However, there is little in the way of a means to convert what is known into the reality of products, other than heuristic iterations. Is there a way to better translate what is done in the laboratory into products which are used daily life (i.e. in the operational environment)? Is there a way to better know how the use of products in daily life affects human performance?

Research activities, often in academic settings, are driven by different agendas than the commercial world. As a result, the data developed in the lab may be limited in practical utility. Research conclusions need to be validated in the reality of daily operations. The application of human factors information, though, is driven by needs that are unique to a particular environment or product. For example, new information may be developed on the eye and how it adapts to various levels of light. While that research may be done to answer one focused question regarding driving performance, there are many other application areas (e.g. displays) which might benefit from an accurate translation of that knowledge outside of its initial concerns. No means exists at this point to accomplish that useful and necessary role.

Meister also points out that no current quantitative database of human performance exists. That prevents the prediction of how people will operate or maintain equipment (as shown by the descriptions of methods in Part II that involve human reliability). It also makes it difficult to determine the value of design recommendations.

1.7 Summary

Much of the human experience each day relies on items that were made to extend human abilities and overcome human limits. The sum of all the products, services, environments, and systems which humans have produced comprises the 'operational environment (OE).' Understanding systems and how they work is a key to the creation of products that work well. A 'system' is any collection of elements that is organized to achieve a purposeful result. The performance of system elements yields results that are intended to fulfill its purpose, a desirable goal state that is to be achieved. Requirements (including standards/regulations, guidelines, needs and wants) set the type and level of performance that is necessary.

Products evolve through a sort of life cycle: creation, growth, optimization, decline, and dissolution. The cycle of research and development runs in parallel, including problem definition, data gathering, feasibility assessment, planning, analysis, design, implementation, evaluation and retirement.

Human factors is the development and application of knowledge about human physiology and behavior in the operational environment. Those who work in the field, referred to as human factors professionals or ergonomists, perform research and analysis on human performance, including both limits and abilities. The human factors practitioner (along with design and engineering counterparts) takes the user point of view to identify user-related issues in advance and to determine how well they are resolved prior to product launch. As a result, no discipline other than human factors is better suited to address the issue of reliability and accountability of complex organizations and systems.

2 Human abilities and limits

How can the human element be incorporated into products?
Human roles (2.1)

How does sensation and attention affect human performance?
Input (2.2)

How do perception, cognition, memory, thinking and learning affect human performance?
Mediation (2.3)

How do response and human physiology affect human performance?
Output (2.4)

How do external and internal environments influence human performance?
Response to environment (2.5)

What affects an individual's choice to perform?
Motivation (2.6)

How do humans change to accommodate the world around them?
Adaptation (2.7)

What causes unexpected behavior? What are the implications?
Human error (2.8)

How do group and individual considerations affect performance?
Group vs. individual (2.9)

In recent years, the term user-centered has been used to describe anything that is designed for operation and use by people. What exactly does it mean to be user-centered? It means that the research and development team must have both a knowledge and an understanding of human nature. Chapter 1 described the hardware/software, procedural and human elements that comprise a product. Each element is expected to perform a role in certain ways. Hardware and software performance can be specified and forecast. However, it is equally important to know what humans can and cannot do (and will or will not do). Two fields of knowledge, human behavior and physiology, provide the answer. Both serve as the basis for the development of products that are in fact human-centered.

Behavior is the actions or activities of the individual. Figure 2.1 shows a model of human information processing that describes behavior from the view of human factors. Aspects of behavior that pertain to research and development include attention, perception, memory, cognition, decision-making, learning, response and motivation.

Physiology is the study of biological functions of the human organism. The sensory, skeletal, muscular, nervous and limbic systems are the physical systems that are of particular importance to human performance. The aspects of physiology that are of interest to research and development include physiomotor (anthropometry and biomechanics), strength and fatigue (physical or muscular). Engineers may also be familiar with two related scientific

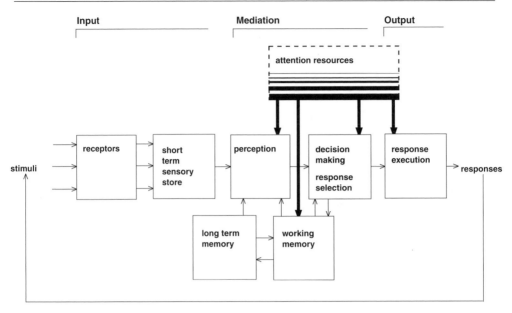

Figure 2.1 Model of human information processing—Wickens (1992) describes a qualitative model of human information processing that incorporates stages which are used to perceive sensations, transform data and choose action. Each stage takes time to perform. Additional considerations (e.g. uncertainty) can extend performance time. Attention resources are finite. Choices on how to allocate attention resources have significant implications for work design (Section 7.9) and workload assessment (Section 8.5).

Source: Adapted from *Engineering Psychology and Human Performance*, Wickens, C. (1992).

terms: kinetics and kinematics. Kinetics involves the effects of forces on the motions of material bodies or with changes in a physical or chemical system. Kinematics deals with motion separate and apart from considerations of mass and force.

These aspects of performance can be considered in three groups: input (attention and sensation), mediation (perception and cognition) and output (response). Four additional issues are of importance: response to environment, motivation, adaptation and error.

There is a rich literature that has been written on human limits and abilities that will not be duplicated here. Instead, this chapter provides an orientation to the roles that humans are expected to perform in the operational environment. It will also account for the nature of human abilities, limits and the implications for the performance of those roles based in part on work by Christopher Wickens (1992), Robert Bailey (1996) and K.R. Boff and J.E. Lincoln (1988). Understanding what humans can and cannot do (and will or will not do) will serve as a foundation for the chapters that follow.

2.1 Human roles

Chapter 1 described the nature of the operational environment that includes the human as an element of products, including systems. From early times through the 1800s, human work consisted largely of repetitive strength-based tasks. As culture, society and its products have evolved, the roles that people expect to fill and are expected to fill have also changed. That evolution has changed the behavior and physical abilities that are required of humans to operate, maintain and use products. In recent years, more developed countries have witnessed the growth of supervisory and trouble-shooting tasks that rely far more on behavioral rather than physical aptitude. There has been a great deal of interest in the engineering aspects of human behavior over the past two decades due primarily to proliferation of computing power into communications and product controls and displays.

Richard Pew (1985b) accounts for seven roles that humans perform in the operational environment and are summarized in Table 2.1: decision maker, monitor, information processor, closed-loop controller, information encoder and storer, discriminator and pattern recognizer and ingenious problem solver.

2.1.1 Decision maker

The decision-making role considers alternatives, estimates outcomes and elects a course of action. Jens Rasmussen's (1983) widely accepted skills-rules-knowledge model of how humans make decisions is described in Chapter 3. Artificial intelligence (AI) software programs have been developed to automate rule-based decisions.

2.1.2 Monitor

Two types of monitor roles exist in systems: vigilant and supervisory. As a vigilant monitor (e.g. a lifeguard at a beach or swimming pool) an operator is expected to detect information that can occur infrequently and at unpredictable times. As a supervisory monitor (e.g. airline pilot) the operator acts in the role of back up to a complex controller of a high stakes system.

The more work that is taken away from the operator and assigned to system software, the less qualified the operator will be to deal with what follows the current state. If such a transition occurs, the change can be abrupt and may have no advance warning. Such an abrupt transition can result in a quick shift from boredom to overload for the monitor, who is suddenly thrust into the role of a closed-loop controller (described in Section 2.1.4). Skill-based activities continue to be automated. As a result, more jobs are being created that have these characteristics. Functional allocation (Section 7.4) deals further with these issues.

2.1.3 Information processor

Data have shown that humans can perform as a multi-channel processor (see Section 2.2.1 Attention). However, humans typically behave as a single channel, limited-capacity processor.

Table 2.1 Human roles in systems—Richard Pew (1985b) accounts for seven roles that people are called upon to perform in systems, based on proven human abilities and limits. Role performance varies within populations, within individuals, and under differing conditions. Human limits and abilities form the basis for the allocation of functions to humans. See Function allocation (Section 7.4).

Role	Example roles
Decision-maker	Manager
Monitor	
Vigilant	Beach or pool lifeguard
Supervisory	Airline pilot
Information Processor	Data processing clerk
Closed loop controller	Machine operator
Information encoder and storer	Student, job trainee
Discriminator, pattern recognizer	Interpreter, inspector
Ingenious problem solver	Scientist, researcher, designer

Human response time is limited by the number and probability of choices, the capability to understand and use controls and displays and other stimulus factors such as errors, slips and mistakes. The range of choices that is possible affects response time (see Section 2.4.2.1 Reaction).

An individual can improve information processing speed and accuracy through practice although the rate of improvement will vary. People are versatile in handling many different input and output codes, yet are slow information processors. For that reason, many information-processing tasks have been automated.

2.1.4 Closed-loop controller (operator)

As a closed-loop controller, humans moderate the operation of other components. A vehicle operator is an example. As simpler inner loop tasks continue to be automated, people will continue to serve as an adaptive controller. Closed-loop controller performance is limited by the amount of information (bandwidth) that an individual can process. That limitation can delay the transmission of guidance information in tasks such as tracking (following a target at a distance).

2.1.5 Information encoder and storer

Most skilled tasks rely on memory capacity. The ability to encode information enables humans to store and recall information. Psychologist George Miller's pioneering work in the 1950s showed that human short-term memory is limited. Methods of organizing information into patterns such as the use of rhymes or acronyms can improve memory performance. To assist encoding and storing, products should offer memory aids whenever human memory plays a significant role. Section 2.3.2.1 provides further details on memory.

2.1.6 Discriminator and pattern recognizer

The ability to discriminate among items and to recognize patterns requires the ability to distinguish among objects and events and their change and to perceive selectively. The role of discriminator and pattern recognizer finds application in auditory patterns (e.g. speech communication) and visual patterns (e.g. reading print and performing visual inspections). People have excellent ability to locate and recognize patterns and can perform tasks that require this ability.

2.1.7 Ingenious problem solver

Problems often require the ability to think originally about how to develop and implement a solution. Humans are able to bring previous skills to bear on a new problem. Individuals can devise an approach to try out a solution and use successive approximation in order to attempt a solution. People are neither rapid nor always reliable in using deductive procedures. Individuals can have great ingenuity, can exercise judgment in situations where rules for action are incomplete, can make good use of approximate and heuristic methods and can estimate probabilities and likelihood ratios. Humans also have the ability to achieve insight, which is essential to ingenuity (skill or cleverness in planning or inventing). Chapter 3 provides further information on how humans solve problems.

2.2 Input

Attention and sensation enable the individual to be selectively aware of the environment outside of the self and to provide the information that is needed for the subsequent processes of mediation and output to occur.

2.2.1 Attention

Because the short term sensory store (STSS) is limited in its ability to process information, handling the amount of information that passes through it requires a selective focus of attention. Theories differ as to whether humans employ a single resource or multiple resources to manage the use of this limited capability. Wickens (1992:74) considers attention as 'one of the most formidable bottlenecks in human information processing.' Attention is normally discussed in terms of visual senses (Section 2.2.2.1) and auditory senses (Section 2.2.2.2) and is considered to be either selective, focused, or divided.

Selective attention—Selection is the choice of what to pay attention to. An error in selective attention is the unwise deliberate choice to pay attention to an environmental source of information that is less than optimal for the task at hand.

Focused attention—Focus in attention is the ability to withstand distractions. Failure to maintain focus results in the intrusion of external environmental information in spite of an individual's efforts to shut it out.

Divided attention—Divided attention is attention that is required by more than one activity at the same time. The performance of divided attention tasks relies on an individual's ability to perform two or more tasks using what Wickens calls time-sharing. Scheduling activities, switching activities and concurrent processing can all be employed in order to time-share. Three conditions may impede time-sharing: the confusion of task elements, cooperation between task processes and competition for task resources. How individuals perform time-sharing is open to debate between those who favor either the single resource theory or the multiple resource theory.

Supervisory control tasks such as piloting an aircraft focus on optimizing selective attention. This is done by dividing the environment into channels and monitoring the channels for critical events that may be expected to occur. In less structured search tasks such as inspector, a series of considerations influence visual search patterns: expectancy, display factors and salience, search coverage and useful field of view, fixation dwells and peripheral vision. Expectancy (where the individual expects to find significant information) appears to be influenced by expertise. For example, an experienced auto driver scans the road in a broader pattern and checks the rear view mirror more often. By contrast, a novice driver will focus on the road immediately ahead of the vehicle and the right side of the vehicle (Wickens 1992:76–89, 80–1, 89–94, 94–6).

2.2.2 Sensation

Sensation is the physical trait that captures sensory information and transforms it for use by the brain. Perception is the brain's awareness and processing of sensory information and is behavioral. For that reason, perception is discussed in this text as part of mediation (Section 2.3).

Bailey (1996:50–4) accounts for five traditional senses that act as sensors for stimuli: vision, hearing, smell, taste and touch. He also mentions five human senses that have been considered as senses more recently: cold, warmth, pain, kinesthetic and vestibular. Three of these senses (cold, warmth, pain) have been included within the category of touch in this review. While other senses such as hunger and thirst exist, they are considered to be outside of the scope of human performance.

Human roles in the operational environment rely on the ability to detect and process large amounts of information. Their ability to detect and convey large amounts of complex information make vision and hearing the two preferred senses for communication.

The eye is a sensor for radiation of electromagnetic energy in a portion of the spectrum (400 to 700 nanometers) that is considered to be visible. The cornea forms the exterior of the eyeball and is filled with vitreous humor, through which light passes from the lens to the retina. The retina is located on the back of the cornea and contains light-sensitive receptors: roughly 25 million rods that are sensitive to dim illumination and 6 million cones that are sensitive to bright light, shapes and colors. The lens at the front of the eye focuses images on the retina. The pupil controls the amount of light that enters the lens by constricting in bright light and widening in dim light.

The field of vision in which the eye can see detail (termed the fovea) covers about 2 degrees of visual angle. To place objects in foveal vision, the eye moves by either continuous pursuit movements or jumping saccadic movements. Saccadic movements either scan an area of view or fixate on the center and surrounding field of view at a specific location for a period of time (referred to as dwell time) (Boff and Lincoln 1988:8.101).

The ability to detect colors is limited to 60 degrees to either side of the center of vision. The ability to detect color within that field narrows from blue at the widest through yellow and red to green at the narrowest. Bailey (1996:54–62) identifies four aspects that affect visual performance: stimulus size, brightness and contrast, the area of the retina that is stimulated and an individual's physiological and psychological condition all affect visual sensitivity.

Size of stimulus—Visual angle can be used to determine the required size of an object at a given distance, or the required distance from an object (e.g. type characters) for a given object size. Visual angle is formed by two rays (imaginary lines) that are joined at the lens and extend to the top edge and bottom edge of the viewed object. Visual angle is useful to figure the size for objects that need to be seen at a distance (e.g. signage).

Brightness—Brightness is the subjective sensation of an individual viewing an object. Visual sensation is a factor in such considerations as legibility and depends on sufficient brightness. Variation in brightness affects visual sensation. For example, Edwin Land's research in the 1970s demonstrated that brightness and color judgment are relative, not absolute.

Visual field—Visual field is the area in which an individual can sense visual information and is important to such considerations as hazard detection (e.g. the ability of an auto driver to see overtaking vehicles). Considered in plan (also called bird's eye) view, the field can be described according to view by one eye, by both eyes, with eyes in a fixed gaze, with eyes able to move or by allowing the head to move.

Physiological and psychological condition—Visual acuity is optimal in young adults and comparatively poor among young children. Age brings about physical changes that can cause a decline in visual acuity in adults. Visual acuity may also be impaired by some medications.

Vision dysfunction can be caused by hyperopia (farsightedness), myopia (nearsightedness), humor impairments, glare sensitivity, blindness (visual acuity that cannot be corrected to better than 20/200), color weakness and color blindness (confusion among red, green, gray) and night blindness.

Implications for visual performance are addressed in Attention (Section 2.2.1).

Sound is waves of pressure that travel through a medium such as air or water. Waves strike the eardrum, transmit sound to receptors in the inner ear and nerve impulses are transmitted to the brain. Three characteristics are of interest: frequency, amplitude and phase. Frequency is the number of waves of pressure that occur in a second and is referred to in terms of cycles per second, or Hertz (Hz). Pitch is the auditory sensation of frequency that an individual experiences. Loudness is the auditory sensation of amplitude.

The audible range for humans is generally recognized to be 20 Hz to 20,000 Hz, with the greatest sensitivity in the range of 1,000 to 4,000 Hz. Sensitivity varies by as much as 20 decibels (Db) or more. After age 50, higher range frequency perception declines. After age 65 few individuals can hear tones over 10,000 Hz. Hearing dysfunction results from conduction and nerve deafness. Conduction deafness is the loss of hearing at all frequencies. Nerve deafness is the loss of hearing, primarily at higher frequencies due to nerve damage (Bailey 1996:63–7).

Humans have an excellent ability to identify the position of a sound source in space. The ability to localize sound is based on three considerations: differences between the sounds reaching the two ears, differences that vary with the sound direction (referred to as phase) and the distance of the sound source relative to the listener (Boff and Lincoln 1988:2.801).

Sound information differs from visual information in that sound does not have to be scanned, it can be detected from any direction and it is comprised of signals that are transient (have a beginning and end). Auditory messages vary according to the dimensions of pitch, location of the source, loudness and semantic content. The greater the difference between two messages along a given dimension and the greater the dimensions of difference, the easier it will be to focus on one message and ignore the other.

Auditory means are preferred over visual means to provide warnings because the ears cannot be shut. Humans can pay attention to more than one auditory signal, although which signal is preferred depends on a number of criteria. For instance, content can be used as a trigger to attract attention. Rather than using potentially startling loudness as a warning, it can be more effective to attract attention using pertinent material in the signal (Wickens 1992:106). For example, rather than a loud 'Danger!' warning, an assertive (yet calm) male voice cautions London Underground riders to 'Mind the gap' between train and platform when entering or leaving trains.

Auditory issues that bear on human performance include the ability to distinguish speech sounds and language, the ability to distinguish speech from background noise and the familiarity, size and context of vocabulary.

2.2.2.3 KINESTHETIC

Kinesthesia provides an individual's sense of the location and movement of the body and parts of the body in three-dimensional space. Kinesthetic sense provides information from the stimulation of sensors that are within the muscles and joints. Bailey (1996:71–2) considers kinesthetic sense the third most important sense behind vision and hearing. The performance of tasks that involve physical activity often requires an awareness of the location of the limbs before, during and after a movement. Kinesthetic sense makes it possible to control voluntary motion without the need to watch where one's limbs are.

Boff and Lincoln (1988:3.101) consider kinesthesia to be the second means of touch sensation. 'Haptic' perception refers to both cutaneous and kinesthetic sensation and forms most of an individual's daily tactual experience and activity. The sense of limb and body position can be either active or passive. Passive kinesthetic perception results from limb movement that is imposed on a subject by an experimenter or an apparatus such as a force feedback device. Active kinesthetic perception is based on intentional movement.

2.2.2.4 TOUCH

Touch is the sense that provides information about objects and events by the stimulation of receptors in the skin. Tactile sensation responds to pressure and vibration (e.g. objects pressed against the skin, patterns traced upon the skin) and to hot through cold temperatures.

Pain is the presence of uncomfortable touch stimulus that results from stimuli as varied as a pin prick, an electric shock, excessive pressure and blunt force impact. Pain can be sudden or chronic and prolonged. It can be experienced in the skin, in the muscles, tendons, or joints or viscera and 'can motivate and deteriorate human performance almost more than any other stimulus' (Bailey 1996:67–9).

2.2.2.5 VESTIBULAR

The vestibular system includes the utricle and the canals that are located in the temporal bone on either side of the head. These labyrinth organs sense signals that are created by head movements. The utricle is sensitive to the magnitude and direction of linear acceleration. The canals are sensitive to the magnitude and direction of rotary acceleration.

The vestibular sense coordinates with the visual and kinesthetic senses in order to maintain posture and balance. Failure to coordinate the three senses can result in spatial disorientation, dizziness and nausea. Human factors studies of slips, trips and falls deal with the causes and effects of failure to maintain balance. Certain professions such as aircraft pilots (and gymnasts) require exceptional vestibular sense. Spatial disorientation continues to be the most probable cause of 7 to 10 percent of fatal flight accidents.

2.2.2.6 SMELL AND TASTE

Specialized areas in the upper nostrils that are sensitive to vaporized odorous stimuli provide the sense of smell. Smell is influenced by combinations of chemicals in the environment. Some individuals can be blind to odors temporarily (e.g. as a result of congestion) or permanently. As a result, smell can only be used to convey fairly simple information. The presentation of simple information can be useful, though, in certain circumstances because humans are sensitive to specific odors (Bailey 1996:70–1). For example, methyl mercaptan is added to odorless natural gas in order to signal a leak. Prolonged exposure to odor will result in a loss of awareness of the sensation. Teenage athletes continue to demonstrate this at the expense of their parents' olfactory senses.

Taste has no significant role in human performance with the exception of professionals in cosmetics, food and beverage development.

2.3 Mediation

Mediation can be considered as the processes that occur following input (attention, sensation) and prior to output (response). Mediation processes include perception, cognition, memory, thinking and learning.

2.3.1 Perception

Perception can be thought of as bringing the past to bear on the present so that the present makes sense. Those who are experienced may be able to recognize patterns that others who are inexperienced cannot detect. Section 2.3.2 (Cognition) describes the nature and role of schemas (mental structures) in cognition. Training is used in order to provide schemas that are consistent with a product (Bailey 1996:110–1). Boff and Lincoln (1988:6.001) define perception as the integration of information that is provided by the senses in order to make sense of the world. Perceptual organization is 'the process by which we apprehend particular relationships among potentially separate stimulus elements (e.g. parts, features, dimensions).'

2.3.1.1 THE NATURE OF PERCEPTION

Bailey defines perception as an active process in which an individual constructs multiple schemas (schemata) in order to anticipate certain kinds of information. Expectations appear to direct perception, making it easier to identify a word or image when it matches what is expected. Those anticipations make it possible to accept the information that is provided by the senses as it becomes available.

2.3.1.2 HAZARD-RISK PERCEPTION

Individuals vary in their perception of danger and their ability to respond to it. The ability to perceive a hazard has significant implications for individual safety. The cost of injury and accident affects individuals, organizations and national productivity. Many causes of such loss can be prevented or at least minimized.

Loss ranges from minimal (such as time lost due to a minor injury) to moderate (such as the loss of sophisticated equipment and machines) to mortal (loss of life). A poorly crafted interface can cost an individual hours of work that is valued at hundreds of dollars when a data file is lost on a personal computer. Misreading a navigation display on an airliner in flight has cost the lives of its crew and passengers as well as the plane itself.

In 2000, US$512.4 billion in economic losses were incurred due to unintentional injuries, as Table 2.2 shows. Lost quality of life resulting from those injuries was valued at an additional US$1,158.7 billion. In 2000, there were 97,000 reported deaths due to unintentional injuries, of which 5,200 occurred in the workplace, 29,500 (and 7.1 million disabling injuries) occurred in the home and 2,200 occurred in public places such as parks. Of the 125 million days lost in 2000 because of work injuries, 80 million were caused by injuries in that year and 45 million were due to prior year injuries. Injuries that were incurred in 2000 were expected to result in 60 million days lost in future years (U.S. National Safety Council, 2001, National Safety Council <www.nsc.org>, 2002).

Table 2.2 Costs of unintentional injuries by class, 2000 ($ billions)

Cost	Total	Motor vehicle	Work	Home	Public non-motor vehicle
Total	512.4	201.5	131.2	111.9	82.6
Wage/productivity losses	259.8	71.5	67.6	70.9	53.6
Medical expenses	93.5	24.6	24.2	26.4	19.6
Administrative expenses	72.6	48.0	22.3	4.9	4.5
Motor vehicle damage	55.5	55.5	2.2	—	—
Employer cost	20.9	1.9	11.5	4.4	3.5
Fire loss	10.1	—	3.4	5.3	1.4

Source: Injury Facts 2001, National Safety Council

2.3.1.2.1 Definitions—A hazard is a set of circumstances ('often a combination of unsafe conditions and practices') that could cause loss such as damage to property, personal injury or death. Hazards can be either intrinsic (obvious) or latent (not obvious). Risk is the probability that damage will occur over a defined period of time (Freeman 1991:40–1). Danger is the product of the two.

2.3.1.2.2 Behavior—Perceptions of hazards are related to individuals' sense of the severity of consequences that may result from exposure to the hazards. Perceptions of risk for common items tend to be based on recent personal experiences and news reports that are available in memory. A product will be less likely to be perceived as hazardous the more familiar a person is with it and the more contacts the person has had with it (Sanders and McCormick 1993:675–9).

2.3.1.2.3 Dysfunction—Why is it that users remain unaware of hazards? Laughery (1993:8–13) points to five causes: hidden hazard, new technology, ambiguous information, accident infrequency and misperception.

- *Hidden hazard*—There is no obvious and open evidence. For example, a fractured support structure for a walkway cannot be seen by someone other than a qualified inspector
- *New technology*—Hazards are not generally known, or appreciated, by people
- *Ambiguous information*—Conditions allow for misunderstanding of conditions with no easy way to verify them
- *Accident infrequency*—People who have no recent experience in accidents are less aware and can assume they are unlikely to incur a loss
- *Misperception*—A user who has the wrong mental model of how the product is used can also misperceive what might go wrong.

Familiarity can also lead to poor hazard awareness.

2.3.1.2.4 Implications—Laughery (1993:8–13) finds that little or no attention is paid during the design process to the knowledge requirements of users and/or the knowledge states of users. Worse still, where knowledge requirements are addressed, inappropriate assumptions are often made regarding what people know or what they will do. Further, little effort is made to assess whether or not such assumptions are valid.

The usefulness of a product is frequently affected by a mismatch between the product and those who use it. Development teams typically do not think about how users perceive hazards or risk.

2.3.1.2.5 Assessment—Hazard-risk assessment, the evaluation of hazard and risk, is covered as a cognitive function in Section 2.3.2.4.

2.3.2 Cognition

Cognition is the set of mental processes that takes place between perception (pattern recognition, awareness of sensation) sensation and response (output). They include intellection and movement control. Each cognitive process is used to transform, reduce, elaborate, store, recover and use sensory input (Bailey 1996:91). Pew (1985c) describes perceptual motor and cognitive skills as 'the build-up of schemata; hierarchically structured prototypes or generic units of organization of features of a task or skill.' In Figure 2.1, these cognitive processes occur during the phase that is labeled decision and response selection. However, schemata are not updated until a mismatch has occurred (Rasmussen 1986:151).

Problem solving is the response to a situation for which an individual lacks a predetermined notion of what to do. The process of solving a problem includes problem definition, the choice between either algorithmic or heuristic solution strategies and the decision regarding whether one or many solutions are required.

Seven models have been proposed by various authors in order to describe how cognitive skills are developed and operate. Reason (1988:35–49) summarizes four traits that all seven share: cognitive control modes, a restricted workspace, automatic processors and activation.

Cognitive control modes—Processing is separated into conscious and unconscious modes and cognitive activity is guided by their interplay. Conscious attentional control is typified as 'serial, slow, laborious, resource-limited, analytic and computationally powerful, processes accessible to consciousness, essential for coping with novelty and cannot be sustained for any length of time.' Unconscious automatic control is typically 'parallel, rapid, effortless, no apparent limitations, intuitive, operates by simple heuristics, processes beyond the reach of awareness (only products available to consciousness), capable of handling routines and recurrences but often ineffective in the face of change.'

A restricted workspace—Attentional processing occurs within a working memory or global workspace. Powerful computational operators are brought to bear in working memory upon five to six discrete informational elements to identify goals, choose means to achieve them, monitor progress toward them and detect and recover from errors.

Automatic processors—Specialized processors are internalized useful regularities from the past. A processor is the recollection of what has worked successfully in the past. Processors serve as expert knowledge structures that can be employed over brief time spans in response to specific triggering conditions. These structures are held in a long-term knowledge base and control the bulk of cognitive activity. Human error (Section 2.8) describes slips as circumstances in which a processor is activated but is not intended.

Activation—Most authors imply that processors can be activated by more than one influence, including frequent and recent employment, current signals, features shared with other knowledge structures and emotional factors.

Cognition's most significant implication for performance is in skill assessment (which is needed for hiring) and skill development (which is needed for training). Task analysis (Section 7.6) spells out the level of skills that will be required for job performance and guides job design, hiring requirements and training materials and programs.

Eberts (1997:1328–74) recommends cognitive modeling as an alternative to experimentation using prototypes, even though its results are less valid. He contends that cognitive modeling makes it possible to save on time and expense by modeling products based on only a sketchy description for users with various levels of expertise.

Chapter 3 expands on reasoning, problem-solving, decision-making and creative thinking.

2.3.2.1 MEMORY

Memory is the ability to retain mental impressions. Current concepts of memory acknowledge the difference between a temporary, limited, volatile short-term memory and a more permanent long-term memory. How memory is structured is still under debate.

For Judith Olson (1985), memory does not act as a unitary whole. Memory is instead a series of three separate entities: sensory register, short-term memory and long-term memory. Each operates with different principles that are summarized in Table 2.3.

Table 2.3 Properties of memory—Judith Olson (1985) describes a series of memories that have unique and individual properties.

Feature	Sensory register	Short-term memory	Long-term memory
What it takes to get information in	Attended or un-attended stimulation	Requires attention or cognition	Repetition, rehearsal or encoding
How information is maintained	Not possible	Continued attention or rehearsal	Often not necessary
How long unmaintained trace lasts	0.25 to 2 seconds	Up to 30 seconds	Minutes to years
What format information is in	Literal copy of the input	Auditory	Semantic, auditory visual, abstract
How much information can be held	5 (auditory), 9 (visual) letters	Small: 7+/–2 chunks	No known limit
Source of forgetting	Rapid decay Overwriting from successive inputs	Decay, or overwriting from successive attended inputs	Possibly no loss

The short term sensory store (STSS) that is shown in Figure 2.1 serves as a momentary collection point for sensory input. The individual has almost no control over it except to pay closer attention to an environmental channel according to expectations.

Short-term memory is in auditory format. It has a small capacity and is able to manage up to roughly nine unordered items (Miller 1956). Short-term memory is poor for keeping track of the order of information and is more fragile than long-term memory. The individual can control and maintain information by grouping information, making items distinctive and rehearsing.

Long-term memory is in semantic, visual, auditory and abstract formats. It has a large capacity and is virtually permanent. The individual gains control over long-term memory by encoding information into rich meanings and linking items together and by being clever at finding items that appear to be lost (Olson 1985).

2.3.2.2 THINKING

Human directed thinking is the set of internal activities that include the handling of information and the ability to make decisions.

2.3.2.2.1 Human information processing—Abstract thought is comprised of two modes: mental images that resemble what they represent and more abstract mental structures such as concepts and propositions. Concepts are used to describe classes of events or objects, or relations between them. For example, the class 'tools' includes instances such as hammers, wrenches, screwdrivers, pliers, saws, routers, planers and milling machines. Propositions are used to relate concepts by making an assertion about them that can be proven to be true or false. 'Wrenches turn nuts' connects a subject (wrenches) with a predicate (nuts). Directed thought relies strongly on such hierarchical thinking, by making it possible to subsume many details into a larger category (Glietman 1986:267–73). For example, classical story tellers would break a long account into portions (referred to as chunking). Cues such as using each finger on a hand would help the story teller to keep track of segments and to recall each in a complete series.

2.3.2.2.2 Decision making—Decision making is the deliberation of information and the selection of one course of action from a number of possible alternatives. In decision making, an individual samples cues or information sources about the true state of the world, formulates a hypothesis about the true state of the world (making a diagnosis of current conditions or a forecast of future conditions), searches for further cues to confirm or refute the hypothesis, chooses an action and monitors the outcome by sampling cues (Wickens, 1992:258–61). This sampling and hypothesis formulation is referred to in the human factors literature as situational awareness. Situational awareness is a key aspect in the performance of many operator jobs such as driving.

Both uncertainty and risk can confound decision making. Uncertainty is the lack of complete knowledge about the future. Gary Klein (1997:276) found uncertainty can be due to missing, unreliable, ambiguous or conflicting, or complex information. That lack of complete knowledge reduces the probability that information can be known to be true to less than 1.0. Risk is the exposure to possible loss or injury. Risk needs to be evaluated if choices are to be made about future states.

The example of combat provides some insight into how uncertainty and risk affect decision making. Military commanders learn early in their careers that the combination of uncertainty and risk in combat inevitably clouds their decisions. Military author Carl von Clausewitz used 'friction' to refer to this phenomenon. Military writers subsequently dubbed this phenomenon the 'fog of war,' referring to the gap between perception and reality that is inherent in highly uncertain and risky circumstances.

The effort that is involved in making a decision can be made easier if there is a correlation between previous cues that indicated a certain condition and current cues. Decision making is also easier if the decision can be made by recalling a predetermined response from long-term memory. That saves having to make a knowledge-based analysis. Section 3.4 provides a more thorough description of the practical implications of decision making.

Gary Klein (1997:284–8) presents an alternative view of making decisions that does not fit accepted artificial intelligence or rational choice strategies. He describes a range of human abilities in naturalistic decision making which he refers to as sources of power. These abilities include intuition (pattern recognition), mental simulation (seeing both past and future), the use of leverage points to solve ill-defined problems, seeing the invisible (perceptual discrimination and expectancy), storytelling, reasoning using analogue and metaphor, communicating intent, rational analysis and drawing on the experience base of one's team. As a non-analytic approach, naturalistic decision making links channels decision making across opportunities and trades accuracy for speed (thereby allowing errors). They can be used in context to provide a way to build a base of experience, and enable a decision maker to redefine and search for ways to achieve goals.

Decisions can be assisted by the use of decision aids such as well-designed displays that make it possible for an individual to review and assess information while making a choice.

2.3.2.3 LEARNING

Learning is the practice of gaining knowledge, understanding or skill by study or by experience. Human factors pioneer Paul Fitts found in 1964 that learning progresses smoothly through three identifiable phases: early cognitive, intermediate association and late autonomous. During the early cognitive phase the individual organizes knowledge. During the intermediate association phase the individual works to eliminate error. In the late autonomous phase the individual demonstrates high levels of proficiency.

Knowledge that is learned and kept in the mind is considered to be declarative, conceptual, or procedural. Declarative knowledge is stable factual knowledge regarding the world and can be studied or rehearsed. Conceptual knowledge consists of core concepts for a specific domain and interrelations among those concepts. Procedural knowledge consists of production rules for the steps that an individual performs in order to complete a task, and can be practiced and performed. Expertise is determined according to an individual's amount of experience and amount of knowledge. Novices typically require a good deal of information in working memory in order to perform a task. Experts can recognize patterns and can determine the best problem-solving strategy to employ to use stored information most efficiently (Gleitman 1986:258–9, 272–3).

Training is a program to provide knowledge, often to prepare for performance of a skill. Training is task-oriented and can be accomplished through formal means (e.g. classroom lecture) or informal means (e.g. on the job). The challenge in human factors is to find the most efficient way to enable the learning process to occur. Transfer of training is the ability to learn a new skill or to learn a skill in a new environment, based on what has been learned previously. Positive transfer occurs when a training program and target task is similar. For instance, a good simulator will prepare an operator to successfully make the transition to using actual equipment. Negative transfer occurs when skills that are learned in the performance of one task inhibit the successful performance of another task (Wickens 1992:241–2). For example, keystroke patterns that are used to operate one software program may produce an undesired result when performed while using another program. This is termed habit interference.

2.3.2.4 HAZARD-RISK ASSESSMENT

Hazard-risk assessment is the trade-off decision that an individual makes by accepting potential exposure to harm in order to enjoy a perceived benefit. A pedestrian who crosses the street against a red traffic light trades-off protection from being hit by a car for the benefit of arriving at a destination sooner.

2.3.2.4.1 Expectations—Individuals have expectations of the products they use while coping with daily life. The consumer expects there to be no design defects, no manufacturing defects, that the product has been safety tested and certified by someone, to be appraised of residual risk and that some misuse is tolerable. Individuals tend to use technical and psychosocial factors in order to assess the acceptability of hazard and risk.

Technical—Technical factors include product knowledge, probability assessment, generalization from sample to population and the necessity of the product.

Psychosocial—Psychosocial factors include the value to individual, value to society, voluntary versus mandatory activities, foreseeability, personal experience, overconfidence (i.e. feeling lucky), the influence of friends and media, leveling, obvious hazards and risks and blaming the victim (Christensen 1985b).

2.3.2.4.2 Risk reduction—Development team members are responsible for exercising due diligence in order to reduce the risk to users. Due diligence requires an understanding of the behaviors and physiology of known as well as potential users. To prevent accidents, designers must anticipate the consequences of unsafe practice and integrate state of the art protection. To do that, the research and development team will carefully analyze possible hazardous situations by following a 'what if?' approach early in the development cycle. The team must compensate for normal human risk-taking behavior, incorporate controls to the full extent of current state of the art and augment the normal adult desire to remain injury-free by taking reasonable and prudent preventative action (Freeman 1991:17–18).

There are a variety of ways to reduce the risk of exposure to hazard. Table 2.4 shows five remedies that are available to the development team that range from greatest to least impact on potential hazards. Two primary means are available to prevent exposure to the hazard: make a design change to the product or environment or take the hazard away entirely. Three secondary approaches are available in case the primary means cannot be used: guard, warn, or educate. Guarding in order to protect from the hazard can be done at the hazard (e.g. placing a muffler on a noise hazard). A guard can be placed at an intermediate point. For example, barriers are installed between highways and residential areas in order to protect residents from traffic noise. Guards such as earplugs or hearing protectors can be placed on the resource (i.e. the person). Warnings can be used in order to notify of the hazard. Individuals can be educated through training programs.

The correct method to use is the one that most effectively protects from the hazard, not the one that is most convenient or costs the least. For example, if a guard can be installed, then a warning alone would not be sufficient (Freeman 1991:21).

Five human factors research methods are of import to hazard-risk assessment: critical incident study (Section 6.6), hazard analysis (Section 6.6), failure modes and effects analysis (Section 8.3), fault tree analysis (Section 8.2) and cost-effectiveness analysis (Chapter 12).

> ... Designers ignore the human element of the system at their peril.
> (Judith Orasanu and Michael Shafto)

2.4 Output

Behavioral response and aspects of human physiology make it possible for the human to influence the environment that is external to each individual.

2.4.1 Response

Response is the action that an individual may take as a result of the preceding input and mediation processes. An individual can voluntarily respond by using the hands, feet or vocal organs. Response can also occur involuntarily through changes to the autonomic nervous system that controls the heart, lungs and adrenal glands.

Interaction with products involves some kind of response. Responses are often limited to the kind that can be recognized by hardware and software or conform with procedures. Such predetermined responses can include actions such as speaking, depressing keys on a keyboard or keypad or operating a rotary dial. Requiring many kinds of responses from the same individual may induce errors. Some kinds of responses may be more prone to errors than others. For these reasons Bailey (1996:310) recommends careful consideration of the responses that a user will be required to perform.

Table 2.4 Accident prevention techniques—Freeman (1991) lists five steps that can be used to prevent people from being exposed to hazards. Development teams are compelled to take the most aggressive possible action to protect the user.

Primary	
Design	Change the product or environment
Remove	Take the hazard away
Secondary	
Guard	Protect from the hazard
Warn	Notify of the hazard
Train	Educate the user

2.4.2 Physiology

Physiology is the science of the functions of living organisms and their parts. Reaction, response, physiomotor considerations (i.e. anthropometry and biomechanics), strength and fatigue are aspects of physiology that play a significant role in human performance.

2.4.2.1 REACTION

Vigilant monitor tasks require sustained alertness. The performance of such work requires the ability to react in a timely, accurate manner. Reaction spans the range from onset of stimulus, to the start of a response, to the completion of a response. The more complex the perceptual and cognitive processes, the longer the time required to react (Boff and Lincoln 1988:9.101).

Two types of time are of concern to reaction: response time and reaction time.

Response time—Response time is the duration between stimulus onset (e.g. a red traffic light turns to green) and the start of response (e.g. the driver's foot begins to move toward the accelerator).

Reaction time—Reaction time begins at the onset of response yet does not end until the response is completed (e.g. the foot depresses the accelerator). The difference between the two is that reaction time includes decision making. In the case of an auto driver, the decision whether the intersection is clear of other vehicles plays a role. A driver in rural Saskatchewan might react to a green light in response time. A driver in Naples, Italy had better respond to a green traffic light in reaction time. The pavement there is littered with the remnants of autos whose drivers have not.

Wickens (1992:312–4) refers to reaction as selection of action. The time that is required for that selection can range from short to long. One of the foremost influences on the time that is taken to react is 'the degree of uncertainty about what stimulus will occur and therefore the degree of choice in the action to make.' Low uncertainty leads to short reaction time. Any other circumstance leads to longer time to react.

2.4.2.2 PHYSIOMOTOR

Anthropometry and biomechanics account for aspects of human stature, structure and physical function.

2.4.2.2.1 Anthropometry—Anthropometry is the measurement of the body's static physical and dynamic functional features. Static skeletal or contour dimensions are taken when the body is in a fixed position. Skeletal measurements account for the distance between prominent landmarks such as the pupils, or joints such as the length from the hip to the knee. Contour measurements record the distance around body parts such as the waist. Dynamic functional dimensions are taken while subjects perform a physical activity such as reaching up, reaching forward or reaching down from a standing or a seated position. Data are available for hundreds of dimensions.

Each dimension occurs in normal distributions among 20 to 60 year old male and female populations and are routinely reported by percentile. Male height can serve as an example, as Figure 2.2 illustrates. A 5th percentile male standing height of 63 inches (162 cm) means that just five percent of males in that population are smaller. The 50th percentile height of 68.3 inches (173 cm) means that half the male population is shorter and the other half is taller. The 95th percentile height of 72.8 inches (185 cm) means that 95 percent of males are smaller (Sanders and McCormick 1993:415–20).

Males who fall below the 5th or above the 95th percentiles can find dealing with the operational environment to be a challenge because product dimensions routinely accommodate 90 percent of the population. Those who fall below the 5th percentile may have difficulty seeing over an auto dashboard. Those who fall above the 95th percentile may hit their head on hanging signs, lighting fixtures and doorways. Issues such as reach and clearance are compounded by the overlap that occurs between male and female populations. For height, the upper percentile of women is taller than the lower percentile of men. Because many products have to accommodate both genders, reach considerations typically need to accommodate the 5th percentile female. Clearance is typically based on the 95th percentile male.

Anthropometric data can be used to design for extremes. Use of the maximum expected height and shoulder width makes it possible to specify clearance for passageways, hatches and doorways. Use of the minimum expected dimension such as arm and leg length can be used to specify reach to hand- and foot-operated vehicle controls. Types of adjustment and adjustment ranges are also based on anthropometric data.

Products such as clothing are based quite closely on anthropometric data although this is not true of all products. Transaction counters in such places as retail stores can be designed to mean dimensions because, even if a customer is significantly shorter or taller, their adjustment to use the counter will likely be for only a short time. Anthropometric data can be of particular benefit in applications such as workplace design and layout, clothing, postural support, displays and controls. In each instance, anthropometric data needs to reflect the physical stature of the intended user population.

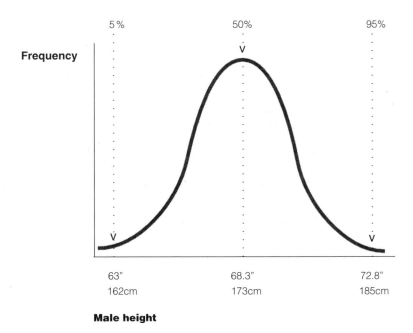

Male height

Figure 2.2 Anthropometry: population percentiles—anthropometry embraces a range of human measurements and manages them as percentiles of a normally distributed population. The example shows male height percentiles. The physical dimensions of architectural and product features are most often based on the 5th to 95th percentile. There are proportionately so few men taller than 72.8" or shorter than 63" that it is not considered economically feasible to design for them. The issue is the same for women and for other types of measurements.

The series of nine *Humanscale* templates (Diffrient *et al.* 1974, Diffrient *et al.* 1981a, 1981b) has served for years as a popular reference for human anthropometric information. Human modeling software programs are now used in combination with computer-based product models to evaluate considerations such as visibility, reach and clearance. Example programs are described in the Appendix. Anthropometric data are available from the National Institute of Science and Technology (NIST).

2.4.2.2.2 Biomechanics—Biomechanics is the study of the human body according to mechanical principles in order to understand its structure and movement. Kinesiology is the study of human motion according to the structure of the skeletal and the muscular systems. Through kinesiology, it is possible to understand the motions of body parts including flexion, extension, abduction, adduction, rotation and circumduction and how the body produces those motions through muscular exertion.

Skill in motor activity is the ability to use the correct muscles with the correct amount of force in order to consistently perform the desired response in the correct sequence. There are five major classes of movements: static positioning, discrete, repetitive, sequential and continuous. Static positioning is used to hold an item. Discrete movement is used to reach a stationary target. Repetitive movement performs a series of discrete movements on the same object such as hammering a nail. Sequential movement reaches toward a number of stationary targets such as keys on a keyboard. Continuous movement such as operating the steering wheel of an auto requires muscular control adjustments (Sanders and McCormick 1993:276–83).

Section 2.4.2.1 described response time as the difference between onset of a stimulus to the onset of a response. Movement time covers onset of response to completion of the movement. Movement accuracy, distance and speed influence an individual's ability to perform tasks according to performance requirements.

2.4.2.3 STRENGTH

Muscles stabilize the body and transmit energy outside of it. Muscle contraction is used to maintain postural stability and to effect movement. Muscular contraction produces lactic acid. If recovery is insufficient the build-up of lactic acid produces local discomfort and eventual lethargy. In terms of human performance, strength is defined as the maximal ability of a muscle or group of muscles to exert force against an object. Personal factors that affect strength include genetics, body dimensions, physical condition and motivation. Both age and gender may have an effect on strength and may be a consideration in hiring.

Age—Maximum strength tends to peak around 25 to 35 years of age. Even though physical exercise can increase muscle strength in adults by 30 to 40 percent, overall strength declines with advancing years. Certain kinds of strength such as knee extension and hand-grip are more affected than others such as trunk flexion and elbow flexion.

Gender—Females' mean strength is 67 percent of males' mean strength for all muscle groups. Female strength for specific groups can be as low as 35 or as high as 85 percent of male mean strength. Females in the 95th percentile may be stronger than males in the 5th percentile. Females tend to compare favorably with males in lower-extremity lifting, pushing and pulling tasks. Females compare less favorably with males in tasks that involve bending and rotating the arm about the shoulder.

Because individual abilities vary, performance-related strength tests should be used to determine an individual's suitability for hiring independent of gender. No job should require individuals to exert maximum strength because the ability to exert control decreases as exertion increases. Designing for maximum strength can, therefore, create a hazard.

Improvements to job design, work selection and work training can all be used to minimize the potential for overexertion and possible injury in strength-related tasks such as lifting, carrying and pushing (Sanders and McCormick 1993:245–59).

2.4.2.4 FATIGUE

Fatigue is weariness from physical labor or stress. Rodgers (1997:269–88) defines local muscle fatigue as the result of active muscle effort and relaxation during task performance. Whole body fatigue is the result of task performance over a shift.

Muscle contraction requires recovery in order to offset fatigue. The greater the effort that is required, the longer the time that will be needed for recovery. This applies to high strength tasks as well as tasks that are performed over a long period of time. Heavy efforts in localized muscles that are sustained from 6 to 40 seconds will require very long recovery times. Even light efforts such as maintaining posture can become fatiguing if they exceed eight percent of maximum voluntary muscle strength and last longer than 20 seconds.

An individual who performs a self-paced job will monitor the personal sense of fatigue. To reduce the discomfort the individual will change to a less-fatigued muscle group (e.g. switch a tool from one hand to another), use larger muscle groups (e.g. shift from using the hand to using the whole arm), or increase secondary work (e.g. check with a supervisor). An individual who performs an externally-paced job will often speed up in order to get time to recover during the task cycle (Rodgers 1997:269–88).

Mental fatigue is an issue in the performance of many human work roles. The performance of repetitive and simple tasks such as continuous vigilance tends to deteriorate over time. The effects of prolonged work on more complex tasks are less direct, less uniform, less general and less certain. Boff and Lincoln find that 'As fatigue increases, skilled subjects seem to lose the smooth coordination of the parts of the task and attention is allocated less efficiently' (Boff and Lincoln 1988:10.801).

Rodgers (1997:269–88) makes four design recommendations to lessen the affects of fatigue for tasks that require muscular effort: make tasks that impose heavy stress on a muscle group as short as possible; minimize the effort that is needed to perform tasks that require the static loading of a muscle group (e.g. maintaining posture) using such approaches as changing posture, or providing supports; avoid creating tasks that require high effort frequently; and moderate or high muscular efforts should not last longer than 10 to 15 seconds unless they are followed by recovery periods that last 5 minutes or longer.

When possible, use machines to perform tasks that require power and assign responsibility for supervisory control to humans. A visit to a construction site can provide multiple examples of how this trade-off has been accomplished.

2.5 Response to environment

The environment can be considered from two viewpoints: external and internal to the individual. How an individual responds to the environment is unique to each person. What may be intolerable for one person is acceptable for another and delightful for yet another. The challenge in research and development is to accommodate the range of human responses to environment.

2.5.1 External environment

The external environment (both natural and operational) comprises the world outside of the individual. External factors (zeitgebers) are all of the factors outside of the body including

Table 2.5 Stressors and strains—stressors are aspects of the environment outside of and inside of the individual. Strains can result from exposure to stressors. Either are acute (temporary) or prolonged. Every individual responds differently to stressors and strains. For that reason, research and development teams need to design for the range of responses that are possible. Examples of external stressors and strains and the internal stressors and strains that result are shown below.

External—outside of the individual	
Stressors	
Acute	Heat, cold, darkness, noise, vibration, sleep loss
Prolonged	Tropical climate, arctic climate, repeated motion
Strains	
Acute	Eye strain, hearing loss, heatstroke, frostbite, performance errors
Chronic	Physical degenerative condition (e.g. arthritis, repetitive motion injury, tenosynovitis)
Internal—within the individual	
Stressors	
Acute	Fear, anxiety
Prolonged	Phobias (e.g. fear of heights), life events (e.g. loss of spouse)
Strains	
Acute	Inhibition, avoidance
Chronic	Withdrawal, depression, ulcers, somatic symptoms (e.g. headache), accidents

regular activities. Aspects of the external environment that are of interest to performance include stressors, and strains that may occur as a result of stressors. Table 2.5 summarizes external and internal stressors and strains.

2.5.1.1 EXTERNAL STRESSORS

External stressors are conditions that can affect learning, performance and decision making. They include heat, cold, hunger, darkness and sleep loss. Stressors can be either acute (short in duration) or prolonged (long in duration). Acute stressors occur within a range that most humans encounter. Prolonged stressors are conditions that comparatively few individuals experience and require exceptional preparation.

2.5.1.1.1 Acute—Acute stressors that individuals often experience include noise, illumination, temperature and vibration.

* *Noise*—Noise is unwanted sound. More particular to human performance, it is sound that bears no informational relationship to the task at hand. Noise can be either continuous or periodic. Continuous broadband noise at about 100 dB affects performance on vigilance, serial response and memory tasks. A brief burst of noise can cause a startle reaction that induces autonomic nervous system activity such as rapid pulse and blood vessel dilation (Boff and Lincoln 1988:10.301).

  Loudness is the psychological experience of sound frequency and amplitude. While noise is considered a stressor it does not always impair performance. Noise can improve performance on certain tasks or on parts of a task. It may cause impairment on one aspect of a task and improvement on another performed at the same time. Depending on the task, the momentary distraction that a burst of noise creates can either impair performance or improve certain aspects of performance by acting as an arouser. Task performance after exposure to noise may also be affected. Background noise (such as white noise) can be used to mask distractions.

- *Vibration*—Vibration affects the ability to see detail quickly and easily, to manipulate controls precisely and can induce motion sickness. The extent to which a visual task or a manual control task is degraded by vibration depends on the nature of the vibration, the type of task and other variables. Vibration can be an issue in the operation of vehicles such as aircraft, maritime vessels, on- and off-road vehicles. Options that are available to accommodate the effects of vibration include changing vibration magnitude or frequency and altering posture or seating to modify the transmission of vibration to the body (Boff and Lincoln 1988:10.401).

- *Temperature*—The body strives to maintain a core temperature in the vicinity of 98.6°F (37°C). Four factors affect the body's ability to regulate temperature: dry-bulb temperature, humidity, air movement velocity and radiant temperature. Reaction to thermal stress can be affected by acclimatization (or lack of it), by psychological factors of expectancy and will and by diet.

  Heat begins to affect performance when it exceeds 86°F (30°C) and exposure exceeds 3 hours. The effects of high temperatures depend on the body's ability to maintain core temperature by sweating and radiation (Boff and Lincoln 1988:10.601, 10.602). Safe heat limits are the subject of much debate. Authors such as Ramsey and Morrissey have developed isodecrement curves (indicating constant decreases in performance) that demonstrate the effects of time and exposure to heat. Their research revealed that two classes of performance are significantly affected: reaction time performance and vigilance and complex task performance. In an attempt to maintain consistent core temperature in cold environments, the body shivers and restricts the flow of blood to peripheral tissues. Cool temperatures that are within the lower end of the human comfort range can improve performance on some tasks. Manual dexterity, a common measure of performance, declines when hand skin temperature falls below 60°F (16°C). Core temperatures below 95°F (35°C) are dangerous. Temporary amnesia can develop at 93°F (34°C), with cardiac irregularities and possible unconsciousness at 86–90°F (30–32°C). Extreme cold has less effect on body temperature and on performance than cold temperatures that are accompanied by high wind velocity. Additional clothing can sometimes offset the effects of a cold environment (Kantowitz and Sorkin 1983:607–17).

- *Illumination*—Illuminance is the amount of light that reaches a surface. Light that reflects off a surface makes it visible (able to be seen). Visibility is a key element of task performance. Visibility-related task performance suffers in the presence of illumination-related problems such as insufficient illumination, reflectance, shadows, low brightness contrast, flickering and glare. Glare is produced by brightness within the field of vision that is greater than that for which the eyes are adapted (Sanders and McCormick 1993:533–9). Reduction in visibility becomes more pronounced with the luminance of the glare. Reduction in visibility is greatest when the source of the glare is in the line of sight. As ambient illumination in the vicinity of the display increases, the visibility loss due to glare decreases (Boff and Lincoln 1988:10.501).

  The Illumination Engineering Society (IES) has established standards for lighting that are appropriate for the performance of different types of tasks.

- *Sleep loss*—Most adults need eight hours of sleep a night although some require ten and others do well with six. People will tend to sleep 10.3 hours out of 24 if all cues to the time of day are removed and they are allowed to sleep as long as they choose. Those who are sleep-deprived are less efficient and more irritable. Individuals who accumulate a large sleep debt experience attentional lapses, reduced short-term memory capacity, impaired judgment and the occurrence of 10 to 60 second microsleeps. Drivers who experience microsleeps put themselves at substantial risk for a serious and possibly fatal accident. Drowsiness can also result in industrial accidents, decreased productivity and interpersonal problems (Brody 1998).

Recovery from sleep deprivation can be fast. An undisturbed full night's sleep nearly completely restores performance efficiency.

The steady state of balanced control over physiological variables is referred to as homeostasis. That steady state still includes a variety of rhythmic variations in body functions. Circadian (diurnal) rhythms have a cycle length of 24 hours (e.g. peak body temperature). Ultradian rhythms oscillate faster than 24 hours (e.g. heart beat). Infradian rhythms are slower than 24 hours (e.g. woman's menstrual cycle). Diurnal rhythms result from a balance of internal and external influences and their disturbance may result in health or performance consequences. Most research has concluded that it is easier to advance an individual's diurnal rhythm (also referred to as one's internal clock). Such rhythms have implications for travel through multiple time zones (e.g. jet lag) and shift work (Kroemer *et al.* 1990:153–200).

2.5.1.1.2 Prolonged—Prolonged external stressors are elements to which an individual is subjected over a period of time, often referred to as a climate. Exposure to any stressor such as heat, cold, noise, vibration or illumination can be prolonged. Design for extreme environments prepares individuals to confront prolonged stressors in arctic, tropical, high altitude and underwater conditions.

2.5.1.2 EXTERNAL STRAINS

Strains can result from exposure to stresses. Strains that affect the primary means of communication, vision and hearing, can have significant implications for performance that are either short term (acute) or chronic (occurring over an extended period).

2.5.1.2.1 Acute—Acute strains induce a significant burden on the individual but they do not necessarily last for a long time. An injury can induce substantial physical pain. In both cases, therapeutic intervention and the passage of time may be able to restore the individual's performance ability as it was prior to the strain.

- *Vision*—Extended efforts to focus the unaided eye or to restrict the retina to occlude bright light can cause eyestrain. Exposure to bright light can bleach eye pigments. Certain types of job are more likely to experience vision strain. For example, welders may experience flash damage that can scar retinal tissue.
- *Hearing*—Normal hearing loss that is not related to work is the result of either the normal processes of aging or noise sources such as household noise or traffic. Occupational hearing loss occurs from either occasional or continuous exposure. Loss can be temporary or permanent. Even with temporary loss, the ability to recover diminishes over time and the loss can become permanent. Conduction deafness can be caused by obstructions or tissue scarring in the outer ear and usually results in partial hearing loss. Nerve deafness can result from damage to the internal ear, usually due to aging and typically results in the loss of hearing in the higher frequencies.

2.5.1.2.2 Chronic—Chronic strains are resistant to therapeutic intervention. Their persistence may require exceptional skills on the part of the individual in order to continue performance. Mental strain may benefit from consistent use of techniques such as meditation. Chronic physical strain may require medication and therapy. Some (such as arthritis) may require an individual to make a life change such as relocating to a more favorable climate.

The effects of stressors over time depend on the degree of exposure and mitigation efforts.

- *Vibration, noise*—Vehicles such as ships, aircraft, busses, trucks and automobiles need to be insulated in order to manage the transmission of vibration that can be manifest as tactile and/or acoustic stressors. Failure to take such action can have a significant performance effect on operators, crews and passengers.
- *Cold*—Protective clothing mitigates continuous exposure to cold. Without sufficient protection, cold strain can ensue. Manual dexterity declines as skin temperature approaches 59–68°F (15–20°C). Tactile sensitivity is substantially reduced below 46°F (8°C). Temperatures approaching freezing can result in the formation of ice crystals in skins tissue (frostbite). Vigilance drops when body core temperature drops below 97°F (36°C). Severe cooling of the skin and body can eventually result in nervous block that prevents the performance of activities, apathy and hypothermia.
- *Heat*—Exposure to heat also has a cumulative effect over time. An individual who is not acclimated to a hot environment may experience transient heat fatigue resulting in a decrease in productivity. If sweat glands become inflamed a rash (prickly heat) can result in discomfort and temporary disability. Loss of consciousness (fainting) can ensue if oxygen supply to the brain is insufficient. Imbalances in minerals and water can cause cramps in skeletal muscles. Dehydration places a strain on the circulatory system that can cause a variety of heat exhaustion symptoms including weakness, fatigue and possibly vomiting. If sweating stops, heatstroke can cause mental confusion, possible unconsciousness and brain damage and death. Treatment for heat stress disorders focuses on acclimation for transient fatigue, rehydration and removal to a cool area for heat exhaustion and soaking the body in cold water and medical attention for heatstroke (Kroemer *et al.* 1990:153–200).

2.5.2 Internal environment

The internal environment refers to conditions that arise within the individual in response to the external environment. The reader who is curious to know the role of emotions in human performance will find that this section addresses the topic. When performance is considered, emotion is normally referred to as internal stressors. Recent attention to the role of affect in design explores the influence of emotion on cognition. Affect 'regulates how we solve problems and perform tasks by broadening exploration of alternatives or concentrating one's focus of attention' (Norman 2002:39–40).

Folkman *et al.* (1979) contend that the way individuals think about a stressful situation affects their emotional response and how they cope. The traditional view holds that there are three functions of emotion: signal, interrupt and interference. The signal function is a cue to set coping or defensive processes in motion. The interrupt function makes it possible to attend to a more salient danger than one is currently doing. Interference can obstruct cognitive processing such as in the case of anxiety while taking a test.

Humans perform evaluation ('Am I okay or is there a problem?'), then secondary appraisal ('What can I do about it?') to confirm, enhance or reduce the threat, then employ coping strategies. Coping is actions or internal thoughts that are brought to bear to manage external and internal demands (and conflicts among them) that tax or exceed a person's resources. Ability to cope with circumstances relies on variable resources such as health/energy/morale, problem solving skills, social networks, utilitarian resources (money and social agencies) and both general and specific beliefs. Lack of such resources or the failure to employ them can result in strains.

2.5.2.1 INTERNAL STRESSORS

Anxiety, apathy, depression, loss of self-esteem and withdrawal have been cited as internal stressors by various human factors authors. As with external stressors, internal stressors can occur over either an acute (short) or a prolonged (extended) time.

2.5.2.1.1 Acute—Transient emotions can be caused by external events or internal states of mind. For example, performance anxiety is a familiar acute internal stressor for many individuals. Whether preparing for a speech or the performance of a free-fall parachute jump, anxiety flows from awareness that one's performance will have consequences (e.g. gain or loss of esteem). Some individuals seek occasions that will expose them to such stressors. Others will avoid such circumstances. In either case, coping strategies are psychological conditioning techniques that can be employed to manage such conditions.

2.5.2.1.2 Prolonged—Individuals can experience internal stressors over a long period of time. Protracted life stressors such as ill health, conflicts at work or the loss of a family member can have consequences that last for significant periods of time.

Phobias are prolonged internal stressors that result from what others perceive as acute stresses. While phobias are very real to the individual, they result from events or objects that are not perceived as stressors by the general population. How individuals deal with fear of heights provides an example. An aware individual who traverses an unprotected walkway hundreds of feet in the air would be right to be concerned for one's safety. An individual who suffers from aerophobia (fear of heights) might not even approach the walkway. By contrast, members of the American Indian Mohawk tribe who work as ironworkers have a reputation for their ability to walk unaided on the upper reaches of skyscrapers that are under construction.

2.5.2.2 INTERNAL STRAINS

Internal strains may result from exposure to internal stresses. Like external strains, they can have significant implications for performance that are either acute (short term) or chronic (occurring over an extended period).

2.5.2.2.1 Acute—Any individual who has experienced performance anxiety knows the acute strains of sweating and increased heartbeat that can result. In response to such situations, individuals may be inhibited or avoid the situation altogether.

2.5.2.2.2 Chronic—Internal strain over a long term can produce debilitating effects. Individuals who experience chronic internal strains can withdraw, experience depression and manifest physical symptoms of distress such as headaches and ulcers. Such individuals are also at increased risk for accidents.

Workload assessment (Section 8.5) is one means to determine the effect of emotions on performance. The NASA task load index (Section 8.5.3) is a method to measure subjective workload. One portion of the evaluation asks about the stress due to frustration that a situation imposes: 'How insecure, discouraged, irritated, stressed and annoyed versus secure, gratified, content, relaxed and complacent did you feel during the task?' (Bateman 2000a).

2.6 *Motivation*

Motives are the needs that cause a person to act. The reader may be most familiar with psychologist Abraham Maslow's mid-1950s proposal regarding a hierarchy of human needs. Maslow contended that humans attempt to satisfy basic through to advanced needs. Needs range from basic psychological (survival) and safety (security) needs to belongingness and love (affiliation), esteem (achievement) and self actualization. In the realm of human performance, motivation issues revolve around performance and satisfaction. Both would place higher in Maslow's hierarchy because motivation to perform well has more to do with affiliation, achievement and self-actualization than with survival or security.

In 1967, Galbraith and Cummings applied expectancy theory to the organizational context, extended the job and valence models that Vroom developed in 1964. Their valence-instrumentality-expectancy (VIE) model linked behavior (effort) to organizational outcome (e.g. wages, promotion, security). The means were three elements: the valence (anticipated satisfaction associated with an outcome), the instrumentality of performance for the attainment of second level outcomes and the expectancy that effort leads to performance. The model made it possible to predict job effort from the expectancy that a given level of effort led to a given level of performance (weighted by the valence of that performance level) (Mitchell 1982:293–312).

Whether a worker is motivated to perform well appears to depend on whether the individual is motivated by internal or by external influences. An individual who is sensitive to internal influence responds to the nature of the work itself and seeks the opportunity to be personally responsible for worthwhile work. 'Worthwhile' means that it is possible for a person to complete a whole piece of work, to enjoy variety in performance in order to exercise many skills and to receive feedback. Increased motivation leads to improved job performance. Whether an individual is motivated to perform well appears to depend on whether the work was rewarding in the past.

External influences have to do with job satisfaction. Such influences include influences outside of the work itself such as money and difficulty of commute. Workers who have not found work to be internally motivating will shift their focus of attention to external influences. As a result, they may not be motivated to perform well but may be satisfied with their job (Bailey 1996:154–71). Bureaucratic organizations can be a haven for this set of circumstances.

Bailey (1996:154–71) makes the case that meaningful work modules have a significant influence on worker motivation. Medsker and Campion (1997:451–89) describe a motivational approach to job design that succeeds well beyond the traditional mechanistic approach by enhancing job meaningfulness. Under the motivational approach jobs are designed to provide:

- Autonomy to make decisions about how and when tasks are to be done
- A link between the individual's work to the organization's mission
- Feedback from the task itself, a supervisor and others
- The opportunity to use a variety of skills and to grow in ability on the job
- Opportunities for participation, communication and recognition
- Adequate pay, promotion and job security.

Medsker and Campion report that jobs designed using a motivational approach 'have more satisfied, motivated, and involved employees who tend to have higher performance and lower absenteeism.'

Work design (Section 7.9) and workload assessment (Section 8.5) are two methods that have a substantial affect on motivation of those who perform the jobs that are created using these two methods.

2.7 Adaptation

Adaptation is the set of changes that take place in an individual that make it possible to deal with an adverse condition. A person who reaches the limits of adaptation experiences stress and may experience degraded performance. Computer scientist and psychologist Herbert Simon (1988:83) considers that adaptive limits are not a matter of behavior. Rather 'it is to physiology that we must turn for an explanation of the limits of adaptation.'

2.7.1 To environment

As the most aware and flexible element in any system, it is often people who change, sometimes to their own detriment. For example, employees who are motivated to perform well on the job will adapt to poorly designed workstations. The adjustments they make to their posture, reach, etc. can cause overuse syndromes such as carpal tunnel syndrome.

Adaptation can be positive. For example, many who live at sea level will experience the headache, shortness of breath and lethargy known as 'altitude sickness' when exposed to life at 20,000 feet. However, Tibetans, Peruvians and other peoples have learned to thrive in such conditions. Military screening and training for those who perform rescue or commando work includes regular positive adaptation to harsh conditions.

Individuals who press psychological and physiological limitations may experience stress from the negative effects of adaptation. An individual may use a variety of mechanisms and strategies to avoid breakdown under overload conditions: skip (allow events to pass unrecognized), partially process (adjust the procedure), queue (set priorities), or walk away (refuse participation).

2.7.2 To people

Adaptation to people affects an individual's ability to perform in the social context either in one-on-one interpersonal relationships or as a member of a team.

Most people are strongly influenced by competitive forces, social pressure and whether others are present in the workplace. The pressure that an individual feels to avoid rejection by others can have a significant effect on individual decision making.

Individuals maintain a personal sense of space and will make efforts to regulate it. Edward Hall's work in the 1960s showed that individuals regulate the proximity of others based on familiarity and social roles. Space designers need to manage trade-offs between privacy and proximity and allow workspace of at least 70 square feet per individual (Bailey 1996:451–2). Diffrient et al. (1981b) provide further guidance on the essentials of space planning and human use of public space.

Chapter 13 provides further information on the process of effecting change in an organization and includes guidance on building relationships.

> If an error is possible, someone will make it.
> (Donald Norman)

2.8 Human erroneous actions, or error

Among the elements described in Chapter 1, the human is the most complex and least pre-dictable. Accounting for the wide range of human behaviors (including error) is one of the more significant challenges to the design of human-oriented products.

The traditional engineering view of human error is that it is an action that violates system tolerances. Recent approaches treat error as human–system mismatch. The reasons for error can be attributed to task complexity, error-likely situations and behavioral characteristics.

Task complexity—Complex task sequences can overburden short-term memory limited capacity and long-term memory recall.

Error-likely situations—Situations that are a poor match for human abilities, limits, experi-ence or expectations are likely to induce errors. Such circumstances can occur because of inadequate attention to human factors research and engineering.

Behavioral characteristics—Demographic traits, skills, training, experience, emotional state and stress factors are a few of the characteristics that can induce error.

The result of such causes can range from minor inconvenience to the loss of 583 lives in the collision of two jumbo jets on a foggy runway in Tenerife, Canary Islands in 1977 (Park 1987:79–100).

Error has been classified by various authors according to task, system, or behavior. Swain and Guttmann (1980:2.4–2.8) describe five kinds of error according to behaviors of omission and commission as Table 2.6 shows. In errors of omission a portion of a task is skipped.

Errors of commission are the incorrect performance of a task. Three errors are consid-ered to fall within errors of commission: extraneous act, sequential error and time error. An extraneous act is a task that should not have been performed because it diverts atten-tion from the system. A sequential error is a task that is performed out of sequence. A time error is a task that is performed too early, too late, or outside of the time that is allowed. James Reason (1990:9, 213–16) considers error as 'a generic term to encompass all those occasions in which a planned sequence of mental or physical activities fails to achieve its intended outcome, and when these failures cannot be attributed to the intervention of some chance agency.' Reason correlates types of error with the three types of behavior that are inherent in Rasmussen's skills–rules–knowledge model, as shown in Table 2.7. Slips, result-ing from a failure of skill by inattention or overattention, occur at the skill level and the cognitive stage of execution. Lapses result from a failure of good rule or the application of knowledge and occur at the skill level and the cognitive stage of storage. Mistakes emanate from a failure or lack of expertise and occur at the knowledge level and the cognitive stage of planning. The generic error modeling system shown in Figure 2.3 describes error behaviors in the context of the skills–rules–knowledge model.

An individual seeks to achieve a goal state by applying the easiest means possible. Other means are employed when attentional checks indicate that progress may not achieve the

Table 2.6 Human error: Swain and Guttmann—Swain and Guttmann (1980) described five kinds of human error, three of which can be considered to fall within errors of commission.

Omission	Portion of a task is skipped
Commission	Task is performed incorrectly
Extraneous act	Task that should not have been performed, as it diverts attention from the system
Sequential error	Task is performed out of sequence
Time error	Task performed too early, too late, or outside of time allowed

Table 2.7 Human error: Reason—Reason (1990) classifies behavioral impediments to human performance according to influences on human behavior that can compromise performance, based on Jens Rasmussen's skills–rules–knowledge model.

Level	Error type	Cognitive stage	Failure modes
Knowledge	Mistakes	Planning	Failure of expertise, lack of expertise
Rules	Lapses	Storage	Failure of good rule application or knowledge
Skill	Slips	Execution	Failure of skill by inattention or overattention

goal state. However, error may also disrupt progress. Skills and lapses may occur at the skill level. Rule-based mistakes may occur at the rule level. Knowledge-based mistakes may occur at the knowledge level.

Reason's (1997) 'Swiss cheese model of defences' extends error from the individual to the organization scale. Organizations erect defenses in order to protect from exposure to hazards. Those

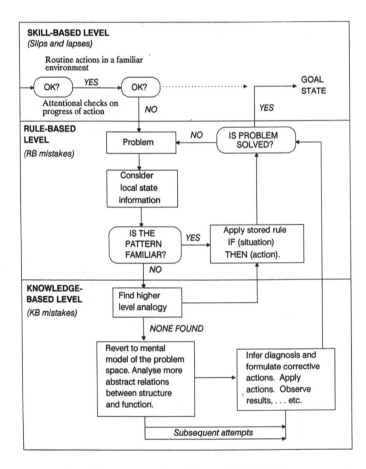

Figure 2.3 Generic error modeling system—James Reason (1991:64) demonstrates the application of Rasmussen's skills–rules–knowledge model to decision making.

Source: *Human Error*. Reason, J. Copyright © 1991 Cambridge University Press. Reproduced with the permission of Cambridge University Press.

defenses can be envisioned as series parallel planes separating hazards and people. Active failures and latent conditions at successive levels (including latent failures in organizational influences, unsafe supervision, preconditions and active failures of unsafe acts) can create a breakdown in the barriers. Enough holes and the necessary alignment among them (like a series of Swiss cheese slices) can result in exposure to a hazard. Shappell and Wiegmann (2000) have applied Reason's theoretical model to practical use in the human factors accident classification system (HFACS). HFACS has been used to identify and classify the human causes of accidents at all (organization, supervision, preconditions and unsafe acts) levels, primarily in aviation.

While mistakes are errors due to conscious deliberation, slips are errors that are caused by lack of attention. Donald Norman (1988:105–14) accounts for five kinds of slips that can occur due to lack of attention, which are summarized in Table 2.8. Capture errors occur when one frequently done activity supplants another. Description errors are the performance of a correct action on wrong object, triggering an incorrect action. Associative activation errors are the occurrence of an internal thought or association that triggers an incorrect action. Loss of activation errors occur when an individual forgets to do something. Mode errors occur when a changed meaning in the same context leads to incorrect action.

Individuals who do not understand a situation will create their own description of it that can vary widely from reality. Norman (1988:38) describes how humans insist on making sense of the world in order to understand experiences, predict outcomes of actions and handle unexpected occurrences. 'Mental models are often constructed from fragmentary evidence, but with a poor understanding of what is happening and with a kind of naive psychology that postulates causes, mechanisms and relationships even when there are none.' Such naive behaviors can lead to conditions that put humans in unsafe situations, without their necessarily being aware of it.

Figure 2.4 summarizes the nature and varieties of unsafe acts.

Why do errors occur? Rasmussen (1986:151) recounts that in 1905, Ernst Mach argued that 'knowledge and error flow from the same source, only successes can tell one from the other.' Rasmussen (1986:4, 150) considers that '... human errors are the manifestation of human variability, which is basic to adaptation and learning' which can benefit from a more fruitful point of view such as considering 'human errors as instances of human-machine or human-task mismatches.' In this light, psychologist Eric Hollnagel prefers 'erroneous actions' to the term 'error'.

2.8.1 Reliability

Human reliability is the obverse of human error. Rather than focusing on what may go wrong, reliability focuses on how likely it will be that an individual will perform a task error-free. Human reliability is a function of the task that a person is required to perform, the definition of failure that is specified and the environment in which the task is performed. Information on human reliability is useful during the performance of functional allocation (Section 7.4).

Table 2.8 Human error: Norman—Donald Norman (1988) partitions errors into those that are due to automatic behavior (slips) and to conscious deliberation (mistakes).

Slips (automatic behavior)
    Capture errors—Frequently done activity supplants another
    Description errors—Performing correct action on wrong object triggering incorrect action
    Associative activation errors—Internal thought/association triggers incorrect action
    Loss of activation errors—Forgetting to do something
    Mode errors—Changed meaning in the same context leads to incorrect action
Mistakes (conscious deliberation)

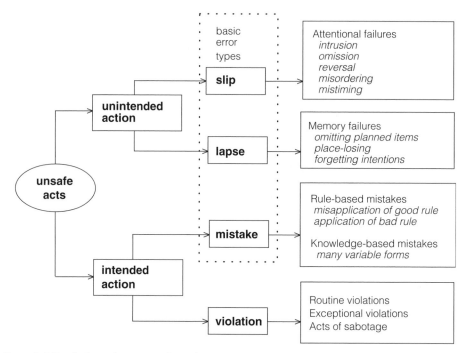

Figure 2.4 Psychological varieties of unsafe acts—Reason (1991) sorts unsafe acts according to whether they are intentional, separates errors from violations, and provides the basis for each. The design of products can anticipate the potential for such events and prevent or mitigate loss.

Source: *Human Error*. Reason, J. Copyright © 1991 Cambridge University Press. Reproduced with the permission of Cambridge University Press.

Variability in human performance can be quantified in probabilistic terms for repetitive discrete tasks or time-continuous tasks.

Repetitive discrete tasks—When dealing with well-defined tasks that have a definite beginning and ending, human error probability (HEP) is estimated from the ratio of errors that are committed to the total number of error opportunities using the formula:

HEP = number of errors/number of opportunities for error

The higher the stress in the subject, the higher the HEP is likely to be. Stress and inexperience can increase an operator's error probability by a factor of as much as ten.

Time-continuous tasks—Tasks that include vigilance such as monitoring a display and tracking, such as driving an auto can be modeled according to time to first human error during a defined period (Park 1987:79–100).

2.8.2 Protection from error

Table 2.9 summarizes five categories of strategies that can be employed in order to prevent human error as well as hardware and software failure. In the case of human error, performance models, summaries of error rates and HEPs can be used to predict error. Task analysis (Section 7.6) and Workload assessment (Section 8.5) can be used to specify performance as well as potential violations of requirements. Hiring and job assignment can be used to screen

Table 2.9 Error/failure prevention strategies—unrecoverable loss is unacceptable in most systems, especially those involving human life or valuable property. Christensen (1985b) accounts for the more frequently used ways to avert loss that can be caused by either materials failure or human error. Christensen had shown a question mark in the preparation/strategy cell for materials. Design and testing could serve as a useful materials preparation strategy.

Role	Strategy for materials	Strategy for people
Prediction	Materials and component testing	Performance models, Error rate summaries, HEPs
Specification	Standards	Task analysis, workload assessment
Screening	Quality assurance Quality control	Hiring Job assignment
Preparation	Design and testing	Training
Notification	Inspection, preventive maintenance	Displays, alarms, controls

candidates for hiring. Training is used to prepare individuals for task performance and error avoidance. Displays, alarms and controls can be designed to provide notification that an error has been made.

Two remedies are available to protect from the inevitable effects of error: reduce the likelihood of error or lessen the consequences of error when it occurs.

2.8.2.1 REDUCE FREQUENCY

Options to reduce error frequency include reducing the degrees of freedom, reducing time stress, redundancy, affordances and training and instruction.

- *Reducing the degrees of freedom*—Limiting the options that are available to the individual lessens the opportunities to violate system parameters. Providing a menu of predetermined options for an individual to select is an example.
- *Reducing time stress*—Relaxation of production quotas or deadlines makes it possible to slow the rate of work and lessen the number of errors.
- *Redundancy*—Redundancy can also lessen the likelihood for error. Assigning two capable persons to perform a crucial task makes it possible for one to monitor the other. Airline flight crews are an example.
- *Affordances*—Affordances are features that make the method of product use self-evident. Affordances invite users to operate a product successfully by making their function evident to users. While hidden tough latch hardware may present a clean appearance it can be difficult to understand how to open doors or drawers. Cabinet hardware such as wire pulls that announces what it does by the way it looks and its position lessens the likelihood of error.
- *Training and instruction*—Training and instruction provide the guidance that individuals need in order to perform procedures successfully. However, both training and instruction need to be designed with regard to user perceptive and cognitive abilities, their motivation and the use environment (Norman 1988:131–40, 187–205).

2.8.2.2 REDUCE CONSEQUENCES

Consequences can be reduced by lessening the degree of loss, or by making it possible to recover from loss. For example, an automatic software back-up utility program makes it possible for a person who has erroneously deleted a file to find it and restore it.

2.9 *Group versus individual*

Human characteristics in this chapter can be considered from two points of view: as humans in groups and as individuals.

2.9.1 Groups

Basic human factors research has discovered the range of abilities and limits that pertain to all humans, regardless of demographic details such as age and gender. Descriptions of the range of abilities among all humans are of limited value, though, because products are developed for groups, or subsets, of the human population.

In research and development, people are not treated as unique individuals, but rather as groups of individuals who share certain traits. Marketing creates and uses models that aggregate people into groups, termed market segments, that share similar characteristics. Manufacturing produces large quantities of artifacts for market segments because of the limitations that manufacturing processes, economies of scale and the cost of tooling impose. Even though new approaches to software production and flexible manufacturing are changing traditional assumptions about mass production, firms continue to cluster people into segments or populations.

Research and development uses the traits of market segments in order to determine the requirements for a product. To accommodate variations, designers will provide different versions of products and different ranges of adjustment. Individuals who do not fit the assumptions about the cluster find it necessary to make adjustments. For example, clothing sizes are based on clusters of traits that are related to physical stature. Individuals who do not match those traits must make adjustments, such as tailoring clothing to fit.

2.9.1.1 INFLUENCES

Groups evolve as a result of influences such as the economy, the public good, and regulations.

Economy—Commercial products rely on the willingness of businesses and consumers to pay money for finished goods (e.g. products that are developed for purchase by businesses and consumers). Requirements are developed in order to suit the needs that market research indicates matter to prospective purchasers.

Public good—Products that are created for the government serve the needs of the public sector (e.g. products for government agencies such as the military). Requirements are expressed by specifications.

Regulation—Society chooses to accommodate the needs of those who cannot obtain the good or service on their own (e.g. products for people who have disabilities). Requirements are articulated by regulations (e.g. the Americans with Disabilities Act).

Chapter 4 provides more information on other influences on the development process.

2.9.1.2 ACTUAL VERSUS INTENDED USERS

Products may be developed for one group yet may be sold to and/or used by another. The traits of the ultimate user may not necessarily coincide with those for whom the product was developed. This finds specific application in product liability. The implications also apply beyond national boundaries. For example, a product that is created for use in one country may be sold in another. The increasing globalization of markets makes the potential for such a situation more and more likely.

Development relies on correctly identifying who is the operator, the maintainer, or the user. Products of development are drawn from cultural values, usually that of the country in which the development is being done. A team in a U.S. firm can develop a product that rests on U.S. cultural presumptions. However, many products are exported to other countries. Unless sufficient research has been done to understand the implications for the culture in the country of sale, problems can occur. This area of interest is often referred to as cultural human factors.

Cultural human factors can influence any aspect of a product from the need for it to its interface design, physical dimensions and labels. This can be an issue for products that are sent from one culture to another. It can also affect the use of products by one who comes from another culture; a thought that occured to the author while once using a fax machine with Norwegian labels in Norway and a PC with Flemish labels in northern Belgium.

2.9.2 Individuals

Individuals vary in their ability to perform tasks. Human factors practice has developed areas of study (e.g. individual differences, aging) to better understand them.

Variation in individual performance can be the result of permanent, temporary or progressive causes.

Permanent—Permanent conditions can affect a career. Physical abilities or disabilities range from individuals with exceptional agility and strength to those who have impairments due to infirmities since birth or severe accidents. Psychological abilities or disabilities range from those who can perceive, reason and remember exceptionally well to those who have emotional and rational impairments.

Temporary—Temporary conditions (e.g. fatigue, injury, response to conditions such as illness or stress) can require adjustments to performance expectations.

Progressive—Progressive conditions (e.g. age, chronic debilitating afflictions such as arthritis) can cause a gradual change to an individual's ability to perform.

2.10 Summary

Human behavior and physiology serve as the basis for the development of products that are, in fact, human-centered. Humans perform seven roles in the operational environment: decision maker, monitor, information processor, closed-loop controller, information encoder and storer, discrimination and pattern recognizer and ingenious problem-solver. Individual ability to perform in these roles relies on human limits and abilities.

Attention and sensation enable the individual to be aware of the environment outside of the self and provide the information that is needed for the subsequent processes of mediation and output to occur. Mediation can be considered as the processes occur after input and prior to output and include perception, cognition, memory, thinking and learning. Means of output that include behavioral response and aspects of human physiology enable the individual to influence the external environment.

The environment can be considered from two viewpoints: external and internal to the individual. Both environments include stressors that are acute (short term) or protracted (long term) as well as strains that occur in reaction to stressors that are acute (temporary) or chronic (protracted).

Adaptation is the set of changes that take place in an individual that make it possible to deal with an adverse condition. Adaptation beyond an individual's limits can cause syndromes such as overuse injuries.

Human error has traditionally been viewed as an action that violates system tolerances but has more recently been considered a human–system mismatch. The reason why humans err can be attributed to task complexity, error-likely situations and behavioral characteristics. Remedies that are available to protect from the inevitable effects of error include reduction in the likelihood of error or lessening the consequences of error when it occurs.

Human behavior influences the way that individuals understand and solve problems. Chapter 3 explores these internal effects on human factors research. Chapter 4 examines external influences on research and development that include individual and team behavior, organizations, processes and events.

3   How we think about development problems

What internal influences affect how I solve problems?
Reasoning (3.1)

How does reasoning affect problem solution?
Philosophical thought (3.2)

What influence does intellect have on solving problems?
Intelligence (3.3)

How do people conceptualize and solve problems? Why are some better at it than others?
Problem solving behavior (3.4)

What gets in the way of successfully solving problems?
Impediments to problem solving (3.5)

Chapter 1 described how a variety of elements are assembled into a system in order to achieve a goal. In its broadest sense, a problem is a situation for which the human does not have a ready response. In the context of research and development, a problem can be considered as an obstacle to the achievement of intended system performance. An opportunity is a chance for advancement, that is, to take advantage of a new material or process, or to provide for a need that is under- or poorly-served.

Both human factors and the practice of design strive to create solutions to benefit those who use products, services and environments. As an applied art, design practice relies on intuition to visualize potential solutions and manifest them in physical form. This is the practice of *synthesis*. As an engineering discipline, human factors applies scientific knowledge to practice. Problems are analyzed and taken apart to examine their nature and to discover opportunities for solutions. Concepts are examined to determine their suitability. This is the practice of *analysis*. The practice of synthesis and analysis both rely on and complement each other throughout the development process. Dorst (1996:25) similarly describes the design process as 'a process of continuous alternation between generation and evaluation.' Chapter 5 will apply this alternation between analysis and synthesis to the process of research and development.

This chapter emphasizes the need for a link between leaps of imagination and thorough evaluation during the development process. In recent decades, it has become fashionable to view synthesis and analysis as separate practices. Technical disciplines, such as engineering, favor 'hard data' that is collected through empirical analysis. Creative disciplines, such as design, favor intuitive visualization and 'soft' data that is collected through methods adapted from the social sciences. In reality, the ability to use both individually or together provides those who work in development with the best means to solve problems. To be most effective, the individual or team performing development work will combine both approaches.

The purpose for development research is to learn what is needed in order to create a solution and to evaluate how well a proposed solution performs. Learning and decision making form the basis for solving problems. Success in development research relies on the cultivation of insight into what is significant about the need for a solution and why. Asking the right question is the essence of a successful solution.

This chapter provides a basic discussion of internal effects (what goes on within the individual) on problem solution. It discusses how humans use philosophy and rational thought to develop the right questions to ask, or how to seek answers. It will cover philosophical affects on how mankind views existence and identifies and acts on problems. It will also serve to introduce Chapter 4, which discusses how what is learned can best be applied to the creation of solutions.

> *The natural sciences are concerned with how things are. Design, on the other hand, is concerned with how things ought to be, with devising artifacts to attain goals.*
>
> *(Herbert Simon)*

3.1 Reasoning

Philosophical thought forms the foundation on which much of the rest of our knowledge is built. *Reasoning* is the practice of discovery or formulation through the exercise of logical thought. Established models of what is accepted to be good decision making, or reasoning, have been developed through the use of logic. A brief description and example of four different kinds of reasoning (deduction, induction, abduction, innoduction) will show how humans put logical thought into practice.

3.1.1 Deduction

Deduction is the practice of reasoning from a general rule or rules to make a statement about a particular case. This can be expressed in structural form as 'All A are B, all B are C, therefore all A are C,' and practically as 'All sedans are automobiles, all automobiles are vehicles, therefore all sedans are vehicles.'

3.1.2 Induction

Induction is the practice of reasoning from a number of instances to a general rule that covers them all. Induction is the foundation of empirical scientific reasoning, as the goal is to determine what different events have in common. An investigator will form a tentative assumption (hypothesis) about a general rule, then collect evidence through study to confirm, modify, or reject it. This is the scientific instance of synthesis (hypothesis) and analysis (experiment). For example, heat the same type of water in 100 identical pots in the same conditions over the same type of heat source that is set at various levels of intensity. Noting the temperature of the water in each pot will show that boiling begins when water reaches a temperature of 212°F (100°C). Inductive reasoning leads to the conclusion that water boils at 212°F.

N.F.M. Roosenburg and J. Eeckels (1995:68–81) propose two further types of reasoning, *abduction* and *innoduction*, and relate them to the design process.

3.1.3 Abduction

Abduction reasons from singular fact to singular fact. One concludes from consequence to antecedent, from observed symptoms to presumed cause, based on a general statement that is assumed to result from the case at hand. This form of reasoning occurs in legal, historical and medical science, as well as technology (e.g. troubleshooting). Roosenberg and Eeckels express abduction practically as 'If X steals a ring, X will leave fingerprints on the display case. The only fingerprints on the display case from which a ring has been stolen belong to X. X has stolen the ring.' Design practice sets a conclusive performance requirement and then seeks a solution that will make that requirement's fulfillment possible.

3.1.4 Innoduction

Innoduction reasons from general to general, from a class of possible intended states to a larger class of possible realizations. '... the kernel of the design process is the reasoning from function to form *and* its actualization' (Roosenberg and Eeckles 1995:76, emphasis in original). Designers conceive a solution and produce an artifact of it. In the process, they relate its form and its actualization to its function. In practice, properties (e.g. the physical laws of levers) can be used to create a new device that can be expected to perform in ways that can be anticipated.

3.2 Philosophical thought

The reader knows intuitively that research and problem solving is a much richer and chaotic experience than logic alone affords. Problem solvers do not proceed step-by-step. In actuality, the patterns of reasoning just described are helpful in 'making good' on a problem solver's insight that foresees the correct path to follow.

Aristotle, in his *Nicomachean Ethics*, speculated that two types of rationality exist in dealing with practical problems (such as, in his day, politics). The first type of rationality was *nous* (Gr.) (*intellectus* (Rom.)). *Nous* sought a vision of the end toward which humans strive. The other type of rationality was *dianoia* (Gr.) (*ratiocinare* (Rom.)), which works back from the end to the current state through practical wisdom (known by the term *phronesis*) by pursuing means to reach ends in elegant, yet not devious, ways.

Both *nous* and *dianoia* find their practical application in two kinds of decision making that can be recognized today: effective and non-effective. Effective decision making leads from precedents to consequences. Working through all possible permutations provides a mechanical check to verify that a solution is, in fact, valid.

By contrast with effective decision making, non-effective decision making leads from consequence to precedents, leaping first to a conclusion by intuitively constructing a proof and verifying it later through reconstruction using the effective method. This aptitude for intuiting the right goal, then mapping the well-reasoned steps that connect the current state to the desired goal, is 'elegant' decision making.

Both Aristotle's *nous* and *dianoia*, as well as effective and non-effective decision making, are historical examples of the interaction of synthesis and analysis; the leap forward to envision desirable ends and the attention to particulars that ensure the leap is valid.

Four schools of thought in Western civilization have sought to account for the elegant non-effective decision making process: rationalism, empiricism, pragmatism and phenomenology. Among these four philosophical schools of thought, pragmatism and phenomenology provide the best fit with research and development activity. Both pragmatism and phenomenology focus on the objective reality of existence, as evidenced by man's direct experience with phenomena. This awareness of objective reality informs our efforts to better know and live in the world.

3.2.1 Rationalism

Seventeenth-century rationalists held that the nature of man and the real world are both rational in nature. In rationalists' view, there is a fundamental harmony between subjective self and the objective reality, particularly if the end goal is worthy (ethical). The person who thinks is able to take subjective leaps by the process of seeking harmonious, or resonant, patterns.

French essayist Michel Montaigne (1533–92) contended that sense-experience (experience based on information that is provided by the senses) is relative and unreliable. The mind cannot obtain ultimate truth due to its reliance on sense-experience and mankind cannot resolve conflicts between claims of the senses and reason. Montaigne's arguments in favor of skepticism anticipated the eventual development of empiricism in the 1700s that would view reality based solely on experience.

Renaissance (fourteenth to seventeenth centuries) philosophical thought was typified by French philosopher René Descartes (1596–1650), creator of the modern scientific method. The development of mathematical principles during this time presented a new opportunity to describe the world and the human role in it. Subjective feelings and interpretations would be eliminated in favor of the clarity and order of mathematical reasoning.

German philosopher Gottfried Leibnitz (1646–1716) favored a universal science (beyond mathematics and logic) that could be used to describe the universe as a harmonious system (Coppleston 1960:13–36).

3.2.2 Empiricism

According to the empiricists of the eighteenth century, there is no harmony between subjective and objective rationality. In fact, there is little rationality in either. Skepticism brings about caution. Experience is simply a contingent series of events. The focus of attention is on outcome, or the data, of existence. Laws are nothing but custom and habit.

For Scottish philosopher David Hume (1711–76), the most prominent empiricist, all contents of the mind come from experience. Close observation of psychological processes and moral behavior through the experimental method and inductive assessment makes a science of man possible. As man seeks to extend the methods of science to human nature, he does not clearly differentiate between physical sciences and those of the mind or 'spirit' (Coppleston 1964:63–96).

3.2.3 Pragmatism

For pragmatists in the early to mid-1900s, a difference exists between the individual and the objective world. Pragmatists contended that existence consists of the interaction between the individual and the environment, in which the standard of success is whatever works. American philosopher Charles Peirce's (1839–1914) arguments stressed the fallibility and uncertainty of human knowledge and held that a theory is true insofar as it is useful. When internal or external experience does not agree with belief, one inquires in order to overcome doubt and either restores confidence in the original belief or substitutes a newer, better one. American philosopher John Dewey (1859–1952) viewed thought as the ultimate evolution of the interaction between the mind and the environment. Reality is the result of the interplay between thought and experience, resulting in potential actions that will be instrumental in changing the problem situation (Coppleston 1967:60–85).

3.2.4 *Phenomenology*

In contrast with preceding schools of thought, present day phenomenologists consider that the interaction between subject and object is a kind of harmony. Phenomenologists seek to eliminate the gap between the two, so that there is no distinction between subjective and objective worlds. For Edmund Husserl (1859–1938), mankind is a product of culture and experience. That formation influences one's self view and interpretation of object and event. Humans are to take the phenomena of the world as they appear, without making any decision to accept or reject them. Through this attention to the reality of existence, mankind can become entirely certain of it. Jacques Derrida's deconstruction seeks to strip away elements of language and other adornments that influence the authentic perception of phenomena.

3.3 Intelligence

Intelligence is the ability to learn, to understand, or to deal with new or trying situations. Notions about what intelligence is have evolved significantly in the past 150 years.

Educational psychologist Howard Gardner recounts that the early stages of scientific psychology began in the latter half of the nineteenth century and were led by Wilhelm Wundt (1832–1920) in Germany and William James (1842–1910) in the United States. Early practitioners theorized that the powers of intellect could be estimated by success in sensory discrimination tasks. Frenchman Alfred Binet (1857–1911) and his followers favored the measurement of more complex abilities such as language and abstraction and thereby created the beginnings of intelligence testing.

Swiss psychologist Jean Piaget (1896–1980) strove to counter the strong empiricism of intelligence testing by exploring how individuals go about solving problems. His approach suggested that the individual generates hypotheses and knowledge about the world in order to make sense of it. Piaget's concept of the evolution of intellect through time established him as the theorist of cognitive development.

Gardner (1983:12–30, 54–61) points to the recent change in both philosophy and psychology '… from external behavior to the activities and products of the human mind and, specifically, to the various symbolic vehicles in which human individuals traffic.' According to Gardner, such symbol systems are what makes human cognition and information processing unique.

> To my mind, human intellectual competence must entail a set of skills of problem solving—enabling the individual *to resolve genuine problems or difficulties* that he or she encounters and, when appropriate, to create an effective product—and must also entail the potential *for finding or creating problems*—thereby laying the groundwork for the acquisition of new knowledge. (Emphasis in original)

Gardner theorizes that individuals think differently according to five unique aptitudes, which he describes as intelligences: linguistic, logical-mathematical, spatial, musical and bodily-kinesthetic. In his view, each of these five intelligences varies in its worldview, perception, learning and decision making. The hallmark of each intelligence is an inclination toward and facility with, certain media of expression (as their names suggest). Each individual appears to have some element of each intelligence and strength in one.

- *Linguistic*—The conveyance of thought by the use of words, as expressed through a command of phonology, syntax, semantics, morphology and pragmatics.
- *Logical-mathematical*—The patterning of objects in numerical arrays. The creation of abstractions through the skillful handling of long chains of reasoning and the expression of those concepts through notation.
- *Spatial*—The ability to transform objects within one's environment and to make one's way amid a world of objects in space. The ability to conceive and manipulate images in multiple dimensions and to depict them graphically.
- *Musical*—The ability to create the controlled movement of sound in time, as expressed through pitch and rhythm.
- *Bodily-kinesthetic*—Inwardly, the focus of the exercise of one's own body. Outwardly, the focus of physical actions on objects in the world.

Those who practice roles that are related to research and development frequently use the first three intelligences (linguistic, logical-mathematical, spatial). Gardner's theory has further implications for development teamwork. Efforts that are made to build consensus among groups need to account for this variety in the ways individuals think. Traditionally, engineering presentations have been logical-mathematical, comprised of formulae and some connecting words. A more compelling approach might demonstrate key principles through the addition of verbal scenarios, illustrations and three-dimensional models to depict spatial relationships and so on. The challenge is to appeal through the use of many approaches, as no single approach can be completely informative. Chapter 14 provides further material on effective communication.

> *There is always an easy solution to every human problem—neat, plausible, and wrong.*
> (H.L. Mencken)

3.4 Problem solving behavior

Cognitive theory is the concept of learning in which humans and animals acquire and relate items of knowledge (cognitions) in order to form maps of reality and expectancies (Gleitman 1986:262–94). The process is based on Ulric Neisser's (1976:110–13) perceptual cycle of schemata that are embedded in cognitive maps to describe the process of human interaction with the actual world. Neisser's model identifies three internal schemata (schema of present environment, perceptual exploration, and actual present environment) and three external cognitive maps (cognitive map of the world and its possibilities, locomotion and action, and actual world) that flow in parallel in a kind of rotary cycle. Humans develop an internal schema of the present environment that provides a cognitive map of the world and its possibilities. The schema and cognitive map direct internal perceptual exploration that evokes locomotion and action. Both perceptual exploration and locomotion and action sample the available information of the internal actual present environment as well as the potentially available information of the actual world. Changes in the actual present environment and the actual world modify schema of the present environment and the cognitive map and the process continues.

Chapter 2 described the model of human information processing. As a brief review, perception, memory and thinking are the three domains of cognition. Sanders and McCormick (1993:61–8) offer definitions of the first two terms. Sensing enables the individual to collect visual, auditory, tactile, orientation and olfactory information. Perception is used to organize these stimuli into meaningful assemblies through detection, identification and recognition. Perception depends on the stimuli that are sensed, the task a person is required to perform, prior experiences and learned associations. Memory is the ability to retain information. It is most often considered to take the form of short-term sensory storage, working memory and long term memory. Gleitman (1986:262) defines directed thinking as 'a set of internal activities that are aimed at the solution of a problem' that incorporate decision making, learning and response.

How we use perception, memory and particularly thinking has a bearing on the ability to conduct research and to solve problems. While the exact nature of thought is a matter for continuing study, there are a number of factors that appear to influence it: situation, strategies, experience and behavior.

3.4.1 Situation

A situation is the set of conditions or circumstances in a given instance. Understanding the events that led up to the current circumstance and being aware of the conditions that comprise the circumstance both have a bearing on how an individual understands and attempts to solve a problem (Gleitman 1986:262–94). Endsley (1996:159–80) defines situational awareness (often referred to as 'SA') as 'a person's mental model of the current state of a dynamic environment.' That model relies on three elements: perceiving critical factors in the environment, understanding what the factors mean and understanding what will happen in the system in the near future.

3.4.2 Strategies

Two strategies are typically employed in solving problems: algorithms and heuristics. Algorithms, in which each step to achieve a result is specified, are thorough yet may be slow. Software programming has traditionally employed algorithms to solve problems. Heuristics are strategies that have been developed through the trial and error of experience. Heuristics (also referred to as rules of thumb) have worked in the past and may do so in the future. While heuristics are not as sure as algorithms they are much more efficient. This division between algorithms and heuristics parallels Aristotle's *nous* (searching toward the end vision) and *dianoia* (filling in from the end to the current state). In fact, heuristics are the modern day incarnation of Aristotle's *nous* and are a good fit with the nature of development problems. Part II of this text provides a selection of heuristics for use in human factors research.

3.4.3 Experience

Psychologist Karl Duncker demonstrated in 1945 that problem solving behavior relies on a hierarchical structure of activity, in which each major step is accomplished by the completion of its own series of smaller incremental steps. While newcomers need to follow each small step, experts can perform them in 'chunks' without thinking each through. This enables the expert to take fewer, more productive steps to reach a goal (Gleitman 1986:268–73). Duncker's model would later find a more practical application in Jens Rasmussen's model of problem solving behavior.

Figure 3.1 illustrates the widely accepted skills–rules–knowledge (S-R-K) model of human performance that was developed by Jens Rasmussen from 1979 to 1983. The S-R-K model identifies behaviors that are used to solve problems according familiarity with an environment or task. Behaviors range from the least effort to greatest effort that is needed to perform them.

3.4.3.1 SKILL-BASED BEHAVIOR

The individual employs stored patterns of well-learned (almost automatic, or prepro-grammed) instructions. Errors such as inattention or overattention can cause failure. Neuromotor abilities such as sports skills are often reduced to skill-based behavior.

Hazardous activities such as emergency procedures are taught to the point that they are performed automatically. For example, military parachute canopies descend at a faster rate than sport parachutes. That is because the military canopy is a tool that enables the jumper to quickly get to the ground and to perform a job. Sport jumpers can maneuver their more sophisticated chutes to maximize hang time and make a tiptoe landing. A 'smoke jumper' firefighter who uses a standard military canopy must perform a parachute landing fall (PLF) upon contact with the ground or risk injury. If the jumper is either inattentive (carelessly daydreaming) or overly attentive (anxiously watching the ground and thinking hard about what to do) injury can result. The best result comes from watching the horizon, keeping feet and knees together and doing an automatic PLF the instant that contact is made with the ground.

In a recreational example, golf swing can suffer from inattention if the golfer is distracted. On the other hand, if the same golfer pays such intense attention that he or she tightens-up and misses the ball, overattention has cost that golfer a stroke.

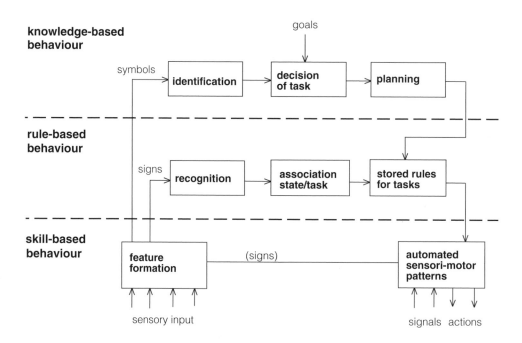

Figure 3.1 The skills–rules–knowledge formulation of cognitive control—Goodstein, Andersen and Olsen (1988) consider Jens Rasmussen's skills, rules, knowledge model as the 'market stan-dard within the systems reliability community.' Chapter 3 provides further information on Rasmussen's model and the practical implications for how humans make decisions.

Source: *Skills, Rules and Knowledge: Signals, Signs and Symbols, and Other Distinctions in Human Performance Models.* Rasmussen, J. IEEE Transactions on Systems, Man and Cybernetics. Vol. SMC-13. No.3. May. 257–66. Copyright © 1983 IEEE.

3.4.3.2 RULE-BASED BEHAVIOR

Actions with respect to familiar problems are governed by stored 'if-then' rules. The misapplication of good rules, or application of bad rules can cause failure. The auto driver who is familiar with his or her own car may find that the design of controls on the steering column of a rental vehicle does not follow the same set of rules. If he or she pulls lightly on the right steering wheel lever thinking that it will squirt washing fluid into the windshield, the rear windshield wiper may start instead. Such a result is the misapplication of a good rule.

A pipefitter who checks for the type of gas coming from a factory utility drop by opening the valve and simply sniffing will be unable to distinguish among carbon monoxide, argon and air. All three gasses are odorless, colorless and tasteless. If a gas connection that carries either of the first two is mistakenly chosen for a procedure that requires air, the results could cause a hazard. The use of a gas monitor should be implemented as a good rule to replace the pipefitter's bad rule.

3.4.3.3 KNOWLEDGE-BASED BEHAVIOR

The individual uses conscious analytical processes and stored knowledge in order to deal with novel situations for which actions must be planned. Failure can result from selectivity, workspace limitations, lack of consideration ('out of sight out of mind'), confirmation bias, overconfidence, biased reviewing, illusory correlation, halo effects, problems with causality or complexity. (The reader is invited to become more familiar with these phenomena in Reason 1990:86–94.)

Rasmussen's three types of behavior can be understood as a sort of stepladder model. However, individuals do not neatly follow each step. In order to conserve effort, an individual will typically seek the least demanding behavior. Figure 3.2 shows how individuals will take shortcuts. An individual will make 'leaps' across the ladder to an action that seems to make sense. This has significant implications for research and development. Even though it is considered to be knowledge work, those who do development work are subject to the same behaviors. For example, well-considered research work will include data collection in order to obtain observations. Yet thorough data collection can be a burden to efforts, time and budget. Skipping 'up the ladder' from observation and data collection to assumptions about the state of the problem will save effort, time and budget. However, it also truncates observations with the risk that knowledge about the problem will be too superficial to adequately interpret the situation at the knowledge level.

The skills–rules–knowledge model also has significant implications for understanding human behavior in systems. For example, adults typically apply personal experience in order to solve problems rather than thoroughly read directions. Using past experience instead of following written instructions is how a parent who assembles a bicycle on Christmas Eve can wind up with a selection of parts left over after completing its assembly. Those who work in research and development and who assume that an adult will read instructions from start to finish fly in the face of Rasmussen's pragmatic understanding of human behavior.

Arrogance begets ignorance.

3.5 Impediments to problem solving

In his work that describes the nature of human error, James Reason (1997:12–13, 53–96) reflects Rasmussen's skills–rules–knowledge model when he identifies three stages in the

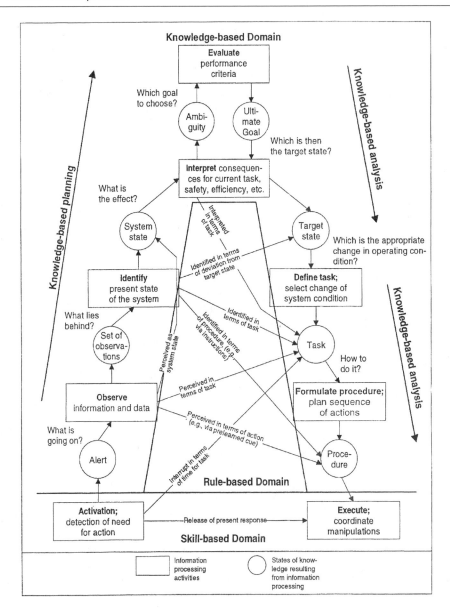

Figure 3.2 Problem solution model using skills–rules–knowledge behaviors—according to psychologist Jens Rasmussen, problem solution relies on three types of behaviors: skill-, rule-, and knowledge-based. Knowledge-based behaviors, which ascend from activation through interpretation, to evaluation and procedure execution, take the most effort. Rule-based behaviors (shown crossing the interior of the figure) seek to conserve effort by 'short-circuiting' the process. Skill-based behaviors expend the least effort, crossing directly from activation to execution.

Source: 'Task analysis'. Luczak, H. in Salvendy, G. (ed.) *Handbook of Human Factors and Ergonomics*. 2nd edn. Copyright © 1997. John Wiley and Sons. This material is used by permission of John Wiley & Sons, Inc.

human thought process: planning, storage and execution. Planning 'refers to the processes concerned with identifying a goal and deciding upon the means to achieve it.' Storage accounts for the delay between formulating intended actions and their execution. Execution 'covers the processes involved in actually implementing the stored plan.' Of the three, planning is most directly related to research and development.

Expertise is the skill of an expert, or one who has special knowledge. According to Reason, errors in planning which are called mistakes are due to either failure of expertise or lack of expertise.

3.5.1 Failure of expertise

A failure of expertise results from the inappropriate application of a pre-established plan or solution. For example, the leadership of a manufacturing firm may presume that making a few cost-driven improvements can make an increasingly obsolete product current. While that might be true, such a presumption may also ignore new requirements that have evolved since the product was originally developed. The product must observe those requirements in order to perform acceptably. The exclusive use of a cost-driven plan when new information must be learned and accounted for is a failure of expertise.

3.5.2 Lack of expertise

Ignorance of what needs to be known at the time it is needed is lack of expertise. If there is no solution thought-out ahead of time, a plan of action must be conceived solely on what is already known. For example, resource constraints may force a research and development team to make decisions 'on the fly' without gathering sufficient information. Skipping necessary research forces the team to assume that current knowledge is adequate. It may be adequate, but it may not. Rapidly changing product areas such as software and consumer electronics place significant pressure on research and development teams to maintain knowledge that is adequate to meet the competitive challenge.

Skills–rules–knowledge behavior is yet another instance of synthesis and analysis. Not content to plod through steps methodically, a problem solver will take an intuitive leap to a desirable state. Only sufficient attention to obtaining adequate and thorough knowledge about the problem will avert a lack of expertise mistake. Lack of expertise mistakes amount to the sole act of synthesis without the benefit of analysis to complement it.

Gary Klein (1999:3, 274–5) reviewed over 600 decision points in different domains and found three causes of error: lack of expertise, lack of information and mental simulation (the ability to imagine how a course of action might be carried out). When error occurs during mental simulation, the decision maker notices signs of a problem but explains it away. This dismissal of an early warning sign foregoes an opportunity to act in time to avert further difficulty.

Even when necessary knowledge is known, Gleitman (1986:287–91) finds that problem solvers can still fail due to a variety of counterproductive behaviors. They include fixation, set and motivation, confirmation bias, availability and conjunction.

3.5.3 Fixation

In this context, fixation is an obsessive or unhealthy attachment to a preferred way of thinking. Habits of ingrained, stereotypical thinking can result in rigidity and an unwavering, exclusive focus of attention on one approach to a task.

3.5.4 Set and motivation

As motivation to succeed increases, flexibility in behavior decreases. Increased motivation to succeed can induce reliance on proven strategies, which may be inappropriate for the task at hand.

Breaking a misleading mind set through a radically different approach to it (restructuring) can result in dramatic new insights. Riddles called brainteasers and creativity puzzles are popular techniques to loosen over-reliance on such habits of thinking. Allowing one's unconscious to continue working on a problem (incubate) by stepping back from deliberate attempts to solve the problem can break mindsets that block new thinking. A writer who suffers from writer's block will often take a break in order to relax and allow the free flow of thought to resume. Designers working on a project can often get an inspirational idea when on vacation.

3.5.5 Confirmation bias

Individuals tend to seek confirmation of hypotheses rather than seek information that may prove them to be false. Among researchers and designers this is also known as hindsight bias: the habit of viewing human factors from the viewpoint of the all-knowing observer.

3.5.6 Availability

Memorable or recent events that are more available to an individual's recollection lead subjects to overestimate how often events occur. Folk wisdom about the occurrence of events in clusters of three is an example of availability. The effect of availability can be detected using human factors methods such as workload assessment (Section 8.5), interviews (Section 9.2) and questionnaires (Section 9.1).

3.5.7 Conjunction

It is not unusual for two events to occur at the same time. The meaning an individual ascribes to that joint occurrence may be erroneous. Through conjunction, an individual erroneously estimates that the likelihood that two events will occur at the same time will be greater than the likelihood that either event will occur by itself.

Error and counterproductive behaviors are not insurmountable. Awareness and avoidance will prevent each of them from becoming an obstacle to individual problem solving. That opens the door to participating in the development process that is described in Chapter 4 and the development of effective relationships to effect change, which is described in Chapter 13.

3.6 Summary

In the context of research and development, a problem can be considered as an obstacle to the achievement of intended system performance. An opportunity is a chance for advancement, that is, to take advantage of a new material or process, or provide for a need that is under- or poorly served.

Reasoning is the practice of discovery or formulation through the exercise of logical thought. Established models of what is accepted to be 'good' decision making, or reasoning, have been developed through the use of logic including deduction, induction, abduction and innoduction.

Aristotle speculated that two types of rationality exist in dealing with practical problems (*nous* (Gr.) (*intellectus* (Rom.))) and *dianoia* (Gr.) (*ratiocinare* (Rom.)). While *nous* sought a vision of the end toward which humans strive, *dianoia* works back from the end to the current state through practical wisdom (known by the term *phronesis*) by pursuing means to reach ends in elegant, yet not devious, ways. Both *nous* and *dianoia* find their practical application in two kinds of decision making that can be recognized today: effective and non-effective, decision making. Effective decision making leads from precedents to consequences. Working through all possible permutations provides a mechanical check to verify that a solution is, in fact, valid. Non-effective decision making leads from consequence to precedents, leaping first to a conclusion by intuitively constructing a proof and verifying it later through reconstruction using the effective method.

Psychologist Horward Gardner theorizes that individuals think differently according to five aptitudes: linguistic, logical-mathematical, spatial, musical and bodily-kinesthetic. Gardner finds that each individual appears to have some element of each intelligence and a strength in one.

Rasmussen's widely accepted skills–rules–knowledge model of human performance identifies behaviors that are used to solve problems according familiarity with an environment or task. Behaviors range from the least effort to the greatest effort that is needed to perform them. In order to conserve effort, an individual will typically seek the least demanding behavior.

Problem solvers can fail due to a lack of expertise or failure of expertise, as well as of counterproductive behaviors including fixation, set and motivation, confirmation bias, availability and conjunction.

We live life through both our own experience and the knowledge of others. The person who asks the right questions is already on the way to successfully solving a problem.

This chapter has examined how individuals think about problems, from the viewpoint of logic, philosophy, intellect and behavior. Chapter 4 explores how issues and organizations affect user-oriented research and development.

4 What influences development?

By definition, research and development always occurs in the context of an organization. As a result, there are many additional influences that are external to the individual that affect research and development. Chapter 3 described what goes on inside the individual (internal influences) in order to solve problems. This chapter accounts for the influences that are external to the individual which also affect problem solving. Those influences include issues that pull and tug on research and development teams, the processes that organizations follow and the manner in which teams interact with others who have a say in the development process. This sets the stage for Chapter 5's introduction of human factors research methods.

4.1 Problems and opportunities

Work in research and development can come about either in response to a problem or to seize an opportunity.

In research, a problem can be considered from two differing viewpoints: as a question to be considered or as a source of trouble or annoyance. To pose a question, for example, an investigator would look into how individuals understand navigation through topics in a multi-dimensional information environment. The result would inform a team that is developing a web-based information product. To deal with a cause of trouble, an investigator might be called in to identify the reason for an accident or series of accidents that have caused injury or death. The result of the investigation and recommendations could be a new design that would remove, provide protection from or warn about the hazard that caused the accidents.

An opportunity is a chance for advancement. New technical abilities, new kinds of work, new materials or equipment and new skills and abilities all pose opportunities to create new products. Research and development can seek to exploit the benefits such new opportunities offer in order to benefit customers and gain a competitive edge.

Characteristics, causes and strategies influence how teams go about addressing a problem or seizing an opportunity.

4.1.1 Characteristics of development problems

Development problems are typically described as 'ill-structured' and 'poorly defined.' That may be true when compared with well-defined problems, such as algebraic proofs. However, development teams typically launch into a vigorous effort to engage a problem in any way that seems to hold promise for better understanding it. Kees Dorst (1996:25) notes that participants make an informed guess at a solution approach, then see how well it works. The process echoes the synthesis-analysis theme described in Chapter 3. By this 'process of continuous alternation between generation and evaluation,' the research and development team engages a problem using what is commonly referred to as a 'satisficing' strategy—a strategy to satisfy as many criteria as possible in as efficient a manner as possible. Satisficing does not necessarily produce a solution that is theoretically the best. It does, however, provide a solution that works within the practical limitations of the organization.

Good problem definition, solution criteria and robust shared project activities make it possible for a team to determine how well solutions meet performance requirements.

Development projects also vary in complexity from simple, narrow assignments to those that are complex or broad in scope. As with strategies, development organizations routinely pursue development activities on more than one level at a time. Table 4.1 illustrates how development work on a facsimile machine can fit each of the levels. Each type of development project offers an opportunity to improve the solution in terms of human factors. The larger the scope of the project, the more changes will be made to the current product. The greater the amount of change that is made, the greater the opportunity there will be for the development team to add value to the product.

The example in Table 4.1 shows five different opportunities to improve a product through human factors knowledge. In this case the example is a facsimile machine (fax). The simplest improvement would be to replace the keypad with a newer unit having tactile features that indicate activation, thus saving keying errors. The original display might be replaced with one that is more easily legible, making it easier for the increasing population of older users to use.

Conversion from a roll paper supply to a plain paper supply can improve maintainability and reliability. Plain bond paper has fewer chemical properties that may degrade through time and compromise its performance. Plain paper is more readily available. No special supplies are needed for the product to be ready to use.

Off-the-shelf components made by an original equipment manufacturer (OEM) can be assembled and presented in more insightful ways that can add value for users. For example, a new housing for the inner workings of the facsimile can include features that enable the user to more easily understand and program the unit (e.g. user interface) and diagnose and remedy

Table 4.1 Development complexity—each type of development project provides an opportunity to improve on the result, adding value through more current or thorough human factors research.

	Development project	Example: facsimile machine
Simple	Modification of existing product	Replace original keypad with new keypad
	Adaptation of existing product	Convert from roll paper to plain paper
	Enclosure for existing OEM components	Create housing for purchased fax engine
	Hybrid of existing components	Combine phone and fax into one product
Complex	New product	Develop personal 'pocket fax'

difficulties (e.g. simplified paper path). The combination of phone and fax into a single product can simplify communication options for the user and save correspondents the confusion of dealing with multiple numbers for phone and fax.

A new generation of fax can provide a variety of improvements. A 'personal fax' could allow those who work in remote sites (e.g. surveyors, claims adjusters) to have increased mobility and access to better quality image-based correspondence.

4.1.2 Causes to initiate development

Why launch a development project? Companies launch development projects to generate revenue from sales, to develop a competitive advantage by meeting needs of those with an interest in the result and to expand market share and improve margin. Urban *et al.* (1986:1–17) account for ten factors that can create the need for a new product. Each is a compelling reason for an organization or individual to commit resources to develop a solution.

- *Financial goals*—Satisfy the need for sound earnings growth.
- *Sales growth*—Seek lower per-unit costs of a product through economies of scale and thereby maintain profitability.
- *Competitive position*—Maintain or increase the standing of one's firm in the marketplace.
- *Product life cycle*—Replace a product that is experiencing a decline in sales with a newer, more profitable one.
- *Technology*—Exploit the opportunity to use prospective materials, processes and capabilities.
- *Invention*—Exploit the potential of a new concept by translating it into a design that can be built, manufactured, or packaged.
- *Regulation*—Respond to the creation or deletion of statutory requirements.
- *Material costs and availability*—Replace current materials with others that are more easily available or more cost-effective.
- *Customer requests*—Cultivate ideas a firm's customers provide into producible products directly related to their needs.
- *Supplier initiatives*—Review and act on supplier recommendations to consider a new approach, or material.

4.1.3 Competitive strategies

Urban *et al.* (1986:1–17) outline the reactive and proactive competitive strategies that firms employ in the pursuit of their goals. These strategies are summarized in Table 4.2. The strategy or strategy combinations the firm chooses depends on its culture and on the competitive market situation.

4.1.3.1 Reactive

Reactive strategies are employed to respond to other market influences when they occur and are typically short-term in their effect. They include defensive, imitative, second-but-better and responsive.

Making changes to existing products is defensive. Defensive changes such as adding features can keep costs at a minimum while attempting to add enough value to attract new interest.

Imitators will copy a new product and bring it to market before its originator can succeed. This method of 'knocking-off' a better-recognized and more costly product tends to occur in higher volume lower technology sectors such as furniture.

Table 4.2 Competitive strategies—Urban, Hauser and Dholakia (1986) outline two types of strategies that firms use to pursue their goals: reactive and proactive. Reactive strategies are employed to respond to market influences when they occur, and are typically short-term in their effect. Proactive strategies are used to take the initiative to bring about change and typically have longer-term effects. Human factors expertise can be used to support either.

Reactive	Defensive	Make changes to existing products
	Imitative	Copy a new product and bring it to market before its originator can succeed
	Second-but-better	Copy and improve on another product in order to save development costs
	Responsive	React to customer requests
Proactive	Research and development	Maintain an ongoing program to build base knowledge that can be used to generate a series of new concepts
	Marketing	Find a customer need and develop a product to fill it
	Entepreneurial	Support and develop one person's new product idea into a producible product
	Acquisition	Purchase another product or company

Firms that follow a second-but-better approach will copy and improve on another product in order to save development costs. The second-but-better strategy assumes that the first firm to develop and introduce a product will overlook opportunities that can be incorporated in its successor.

Reacting to customer requests is a responsive strategy. Some business sectors such as medicine and science have well-qualified customers who can provide sales staffs with informed guidance on what products to make (e.g. laboratory instrumentation).

4.1.3.2 PROACTIVE

Proactive strategies are used to take the initiative to bring about change. They include research and development, marketing, entrepreneurial and acquisition strategies. Effects from the use of proactive strategies are typically longer-term than the effects from reactive strategies.

Firms that rely on a research and development strategy maintain an ongoing program to build a base of knowledge that can be used to generate a series of new concepts. The cost of development can be offset by demand for product concepts that are new to the marketplace.

Firms that use research and development need to keep ahead of other firms that use imitative and second-but-better reactive strategies.

Finding a customer need and developing a product to fill it is a marketing strategy. Much of the development work in the United States relies on this approach. While it does not necessarily result in the best products, it does tend to generate sales.

An entrepreneurial strategy supports and develops one person's concept into a producible product. 3M Corporation is a U.S. firm that has made entrepreneurial strategy its main development method. The approach allows research scientists much leeway to follow their own instincts. Scientists who demonstrate product concept feasibility are awarded increasing allotments of development funding.

Through acquisition, a firm purchases another product or company that is new to the firm's market. Acquisition can be a quick way to flesh out a limited product offering or to seize the initiative in a new market.

Companies often choose more than one strategy to cover short- and long-term needs at the same time. In the mid-1980s Herman Miller, Inc. created its Milcare subsidiary to diversify beyond the firm's traditional office furniture business into the expanding market for health care products. Milcare's first year development efforts included four of the above-mentioned strategies used together:

- *Defensive*—Made changes to the existing Co-Struc materials handling system to add needed components, expand system offerings and solve some issues with manufacturability;
- *Responsive*—Responded to customer requests for certain product offerings by initiating research projects;
- *Research and development*—Launched product initiatives in materials management and related areas that the firm had chosen as significant opportunities for future business growth. Some required significant research, while others could be moved quickly to design concept development;
- *Acquisition*—Reviewed furniture made in northern European countries for possible sale to alternative care facilities in the United States. Milcare also identified a proprietary air/gas/vacuum distribution product that could be adapted for use with Co-Struc in order to complement its structural and storage features. Adding the gas distribution components made it possible to create products appropriate for advanced applications, such as pediatric and neonatal intensive care stations.

Herman Miller's development process already allowed for new ideas to be brought to the attention of management through what was called a 'need or opportunity statement.' In effect, the need or opportunity statement offered a means for the company to employ the proactive entrepreneurial strategy discussed above. The method made it possible for any interested party, including outside health care professionals such as physicians and pharmacists to bring new product ideas to the company's attention.

> *Nothing is more difficult to carry out, or more doubtful of success, nor more dangerous to handle than to initiate a new order of things.*
>
> (Machiavelli)

4.2 Influences on research and development

Psychologist Donald Norman (1988:151) notes that adverse influences on those who work in research and development come from a number of directions, including peers, professional near-sightedness and clients.

> Designers go astray for several reasons. First, the reward structure of the design community tends to put aesthetics first. Design collections feature prize-winning clocks that are unreadable, alarms that cannot easily be set, can openers that mystify. Second, designers are not typical users. They become so expert in using the object they have designed that they cannot believe that anyone else might have problems; only interaction and testing with actual users throughout the design process can forestall that. Third, designers must please their clients, and the clients may not be the users.

Further factors influence how research and development team members think about problems. They include: role in the project, culture and the characteristics of problems. Events, the existing base of products and resources also play a part.

4.2.1 Role in the project

Any participant with a stake in a project's outcome can be thought of as a 'stakeholder.' Each participant in the process views the nature of work to be done through his or her particular set of considerations. For example, senior management has the responsibility to both create the vision to anticipate the future and to take the initiative to lead the organization to realize that vision.

4.2.2 Culture

An organization's culture is the collective set of values by which its members define their organization's character. The mission (what the organization seeks to accomplish) and agenda (when those accomplishments are expected to be realized) impose a point of view on its members. That point of view focuses the efforts of those in an organization. David Meister's observations in Section 1.6 address the effects that point of view can have on human factors professional practice.

4.2.2.1 OPERATING CULTURE

Donald Norman (1995:35–6) finds that two cultures appear to have evolved in the realm of human factors professional practice: commercial and laboratory. Commercial firms are results-oriented and seek success in the marketplace. The practitioners who work there, however, often fail to reflect on broader questions or on how past experience might benefit future work. On the other hand, academic laboratories, which are more methods-oriented, seek elegance and precision in research techniques. In Norman's view, the academic research community's emphasis on methods does not adequately appreciate the practical importance of issues that are raised. He writes: 'Thus, in a university laboratory, there are three questions that are (surprisingly) seldom asked: Are the right questions being asked? Do the results make any practical difference? Will the results enhance our understanding or welfare?'

David Woods (1988:128–48) finds that researchers typically cope with the complexity of real world problems by defining 'system' in a narrow enough manner to limit its complexity and thereby make it more manageable. Unfortunately, the results from such closely limited models do not translate well into real world situations. Portions of the problem solving process may be missed or their importance may be underestimated. Some aspects of problem solving may only emerge when more complex situations are directly examined. He favors 'better understanding human behavior in complex situations' as the path to resolution of the issue.

4.2.2.2 CORPORATE CULTURE

Depending on a firm's culture, certain departments may significantly influence the research and development effort. Donald Schön (1983:336) describes the organizational setting for technical professionals. 'Whenever professionals operate within the context of an established bureaucracy, they are embedded in an organizational knowledge structure and a related network of institutional systems of control, authority, information, maintenance and reward which are tied to prevailing images of technical expertise.'

Four types of department-led cultures are described here: design, manufacturing, marketing and engineering.

4.2.2.2.1 Design—In a design-led company, the collaboration of senior management and a few senior designers forms a vision of offerings which the firm will produce. For example, D.J. DePree relied on Gilbert Rhode's concepts to lead his firm, Michigan Star, from a period furniture manufacturer during the American Depression of the 1930s into a nationally recognized producer of modern residential furniture: Herman Miller, Inc. In the 1960s DePree established Herman Miller Research Corporation to enable Robert Propst to create 'Action Office,' the first fully integrated open landscape office furniture system made in the U.S. Intended as a replacement for drywall partitions, Action Office complemented the growth of the American white-collar workforce during the 1960s and 1970s. Its success changed Herman Miller into a systems furniture manufacturer.

4.2.2.2.2 Manufacturing—Manufacturing-led firms base their vision for the future on the optimal use of their installed base of tooling and manufacturing processes. General Motors has leveraged its expertise in manufacturing to create autos that have similar functional specifications. Variations in auto model appearance and features are made available to appeal to the preferences of various market segments.

4.2.2.2.3 Marketing—Marketing-led companies rely on their marketing department to frame and exploit the opportunity to serve the needs of segments of potential purchasers. Procter & Gamble conducts extensive consumer research to tap into needs and wants that can lead to new household products.

4.2.2.2.4 Engineering—Engineering-led firms develop concepts that exploit the opportunity to use the most advanced technology available. General Electric develops a wide range of well-engineered high performance products from electric locomotives through jet engines.

4.2.2.3 PROFESSIONAL PRACTICE

Each professional group develops its own standards of practice. In some professions, standards form the basis for licensure (e.g. medicine, law). In others, they form the basis for certification (e.g. human factors). Best practices are norms of professional conduct that are considered to be the state of the art among those who lead the field. In human factors organizations, professional and technical special interest groups develop such best practices.

4.2.3 Characteristics of problems

According to David Woods (1988:128–48), the effect that the real world has on attempts to solve problems can be described according to four dimensions: dynamic, interconnected, uncertain and risky. Each affects the ability of an investigator to truly understand what is being studied.

4.2.3.1 DYNAMIC

Problems reveal themselves over time. Problems can change significantly over the course of a project. Frequent attention to the situation (situation assessment) can help to keep the investigator's understanding of the problem aligned with actual conditions.

4.2.3.2 INTERCONNECTED

Problem situations are made up of many interconnected parts. The failure of part of a system or changes that are made to one part of a system can affect a wide range of other parts in ways that cannot be predicted.

4.2.3.3 UNCERTAIN

David Woods (1988:138–42) finds that lack of certainty can cause available data to be 'ambiguous, incomplete, [have a] low signal to noise ratio, or [be] imprecise with respect to the state of the world.' Failures of attention can cause an investigator to overlook information which is available, unless prompted by some external means (e.g. an alarm message or signal) or knowledge-driven activation (e.g. self-direction).

4.2.3.4 RISKY

Outcomes that result from choices can have large costs. Development research, prototype development and evaluation prior to production are some of the means that firms use to limit the potential downside effects of risk.

4.2.4 Events

Technologies that are available for consideration and adoption vary depending on what is discovered, when it is discovered and decisions which are made with respect to that new information. Those decisions create paths that guide subsequent research. Passell (1996:60–1) defines this as *path dependence*, which is a pattern of '… small, random events at critical moments [which] can determine choices in technology that are extremely expensive and difficult to change … .' Perceived as slightly superior by producers or consumers, a particular technology is chosen over others, then replicated by other producers. Increases in sales make it possible to implement more cost-effective production methods. The efficiencies (economies of scale) that those methods afford can lower production costs. Market acceptance then grows with familiarity.

The prevalence of Matsushita's VHS format video recording over the technically superior Sony Betamax is a familiar example. While Matsushita entered the market a year after Sony, it liberally licensed its technology and offered units capable of playing two-hour tapes —long enough for movies. A half dozen other competitors entered the market within months. Even though Sony later countered with its own model that accommodated a two-hour-long tape, its fate was sealed.

On a larger scale, the cumulative effect of events produces paradigms. Fogler and LeBlanc (1995:11–24) describe a paradigm as 'a model or pattern based on a set of rules that define boundaries and specifies how to be successful within these boundaries.' Changes in need, opportunity or technology can change the basis for such patterns and present a new paradigm. Those changes can happen slowly or (in some cases) quickly. When the changes occur, the benefit of the new paradigm supplants the old. Those who had skills under the old paradigm lose their value, as the old rules no longer apply. Whether firms detect such paradigm shifts and how they comprehend and respond to them can make the difference between an industry leader and follower.

The creation of the Internet provides an example of a paradigm shift. For decades, immediate image transmittal relied on the use of facsimile (also known as a telecopier) machines. The United States' Defense Advanced Research Projects Agency (DARPA) made its high bandwidth computer communications network available for public use in the early 1990s. That made it possible within just a few years to reach most countries around the world by electronic mail and, eventually, the Worldwide Web. New work roles such as web publishing developed. New web-based applications such as collaborative work environments and retail commercial transactions are also evolving at a rapid rate. Internet start-up firms that have been created to serve this expanding market are creating a new business sector. Firms that fail to understand the value that the Internet offers (and to develop expertise in using it in order to achieve a competitive edge) may lose out to others that understand this paradigm shift.

4.2.5 Installed base of product

What already exists exerts a significant influence on research and development. For example, manufacturing tools and jigs, off-the-shelf (OTS) software programs, existing parts and subassemblies and much more are already in production. They create a sort of context for what is to follow. As long as an item serves some purpose, efforts will be made to use it in order to leverage the investment of time and funds. In service industries, procedures will be modified in order to minimize the cost of retraining.

4.2.6 Resources

Philosopher Michel Foucault's 'radius of addressability' contends that resources limit the ability to engage a problem and man's capability to understand it. Funding, time, staff and access to information can vary widely. In particular, the time that is needed in order to perform research can often suffer under the pressures of short development schedules. How much time is enough to adequately learn about a new situation? For anthropologist Harry Wolcott (1995:15–18) three months is the minimum time one should expect to spend in order to observe another culture and to be accepted as a participant observer. To be sure, the team that develops a new product faces a different challenge from an anthropologist who studies a culture. Even so, there are similar considerations. One needs to observe in order to learn. To observe thoroughly, one needs to be aware. Awareness is based on sufficient sensitivity to what is being observed and enough time to account for what actually happens. The newer an area is, the more care an investigator needs to pay to it.

> *Success, as I see it, is a result, not a goal.*
> *(Gustav Flaubert)*

4.3 Process

Research and development can fail under too much, or too little, control. A staid, pre-determined path prevents the versatility that is necessary to understand and solve the wide variety of development challenges. It also precludes the spontaneity that inspiration brings. On the other hand, defining a process makes it possible to understand it. Lending structure makes it possible to learn and to develop and realize expectations.

Problems and opportunities vary widely. As a result, there is no perfect model for how to perform research. Instead, there are considerations that need to be taken into account. A procedure for how to go about research and development ensures those considerations are accounted for. Organizations, such as the United States Department of Defense, have developed procedures that lend structure to what can be a very complex process.

4.3.1 Phases

Charlton and O'Brien (1996:13–26) note that development projects are necessarily more formal than individual creative activity for three reasons: to ensure reliability, to manage substantial amounts of information and to communicate among many participants who are often located in many different sites.

Defense systems have long been a development area that has relied on human factors work. Such systems are technologically sophisticated and must perform well under demanding conditions. Charlton and O'Brien describe the management of defense system development in four phases. Even though commercial projects are not the same as defense systems, the process used to develop them follows a similar pattern (see Table 1.5—Product development and production and Figure 1.4—Development process).

4.3.1.1 CONCEPT EXPLORATION AND DEFINITION (PLANNING)

This first phase identifies user needs and system requirements by learning about shortfalls in current capabilities and existing systems, learning about new technologies to exploit and finding out about the expressed preferences of intended users.

4.3.1.2 CONCEPT DEMONSTRATION AND VALIDATION (ANALYSIS)

Functions, or what the system is expected to do, are allocated to systems and system components. Trade-off analyses are performed in order to balance cost, performance and reliability.

4.3.1.3 ENGINEERING MANUFACTURING AND DEVELOPMENT (DESIGN)

Functional prototypes are evaluated with a sample of the user population. Either of two approaches can be taken in order to accomplish this task.

- Development test and evaluation (DT&E)—Problems are detected and corrected through an iterative process of design-test-design-redesign
- Operational test and evaluation (OT&E)—An independent tester determines the system's effectiveness and stability and performance to specification under realistic operational conditions.

4.3.1.4 PRODUCTION AND DEPLOYMENT (IMPLEMENTATION)

Follow-on tests are conducted in order to determine user satisfaction, training and personnel requirements are evaluated and information is gathered for the next version of the product.

Figure 1.4 adds a further step (retirement) that also has implications for research and development projects.

4.3.2 Activities

There is no single best way to come up with a good solution to a problem. There are many approaches to solving problems, including being methodical, opportunistic, inspired and even lucky. Through time, authors have developed models of how the human process of creative problem solving is currently understood, as Table 4.3 shows.

In *The Art of Thought* (1926), Wallas accounted for the steps that an individual follows in order to solve problems. They include: preparation (clarify and identify problem, gather pertinent information), incubation (unconscious mental activity while doing something else), inspiration and verification (check the solution). Wallas theorized that changing one's environment makes it possible for the unconscious mind to discover the true solution through incubation once the cues to previously unproductive mind sets have been removed (Gleitman 1986:280–1).

In 1968, Shulman, Loupe and Piper offered a model that included four steps: problem solving (discomfort moves individual to act), problem formulation (define and develop

Table 4.3 Problem solving process—Bailey (1989) and Fogler and LeBlanc (1995) account for authors who have developed models of how individuals go about solving problems. Each author identifies steps in the creative process. Jones (1992) describes three processes. While our understanding of human behavior continues to evolve, De Bono's seven-step model seems to be enjoying a fairly wide acceptance.

Wallas (1926)	*Shulman, Loupe, Piper (1968)*
Preparation—Clarify and identify problem; gather pertinent information	Problem solving—Discomfort moves individual to act
Incubation—Unconscious mental activity while doing something else	Problem formulation—Define and develop anticipated form of solution
Inspiration	Searching—Gather information
Verification—Check the solution	Problem resolution—Satisfaction that the problem is solved; discomfort subsides

Jones (1992)	*Fogler and LeBlanc (1995)*	*DeBono (1968–76)*
Black box—mystical internalized thinking	Define the problem	Acceptance
Glass box—Rational externalized thinking that incorporates three phases of activity:	Generate solutions	Analysis
	Decide course of action	Problem definition
Analysis	Implement the solution	Ideation
Synthesis		
Evaluation	Evaluate the solution	Idea selection
		Implementation
Self-organizing system— That which carries out the search for a suitable design. That which controls and evaluates the pattern of search (strategy control)		Evaluation

anticipated form of solution), searching (gather information) and problem resolution (satisfaction that the problem is solved; discomfort subsides) (Bailey 1989:115–18).

Jones (1992:45–56) accounts for three models of design problem solving. The first, 'designer as black box,' represents design problem solving as mystical internalized thinking which cannot be entirely accounted for. The second, 'designer as glass box,' models design problem solving as rational externalized thinking which responds to information by performing the steps of analysis, synthesis and evaluation in the interest of producing an optimum result. For Jones, the glass box model finds more success in design problems that can be decomposed, or broken apart, to be solved. The third, 'designer as self-organizing system,' relies on the design effort being split into two parts: 'that which carries out the search for a suitable design' and 'that which *controls* and *evaluates* the pattern of search (strategy control).' He finds the benefit in this third approach lies in the combination of an 'intelligent search that uses both external criteria and the results of partial search to find short cuts across unknown territory' (emphasis in original).

Edward De Bono and others have proposed a seven-step model: acceptance, analysis, definition, ideation, idea selection, implementation and evaluation. In their elementary introduction to problem solving, design authors Don Koberg and Jim Bagnall (1991:9–35) provide interesting insights into the nature and implications of the seven-step model. The descriptions that follow are based on their work.

4.3.2.1 ACCEPTANCE

Acceptance is the step that is used to identify the resources that will be used. Resources can include funding, people, information, equipment and materials. A written or oral brief will often be used to begin a phase by stating what the problem is. Careful review of the client's problem statement provides the team with the problem as given. The problem as given often identifies a symptom of conditions, not necessarily the cause. It is up to the investigator to translate the problem as given into 'problem as understood' by probing for the problem's true nature.

Fear of failure (or success), inadequate resources and inadequate motivation can impede acceptance. Among these seven steps, project acceptance is largely relegated to the project's start. However, each time the terms of acceptance change (e.g. new members join the team) acceptance needs to be reconfirmed.

4.3.2.2 ANALYSIS

Analysis is the chance to learn what needs to be known about an opportunity or problem. Every available resource needs to be employed. Information sources can include first-hand observation, interviews, primary resources (scientific journals, proceedings, technical reports, graduate theses), secondary resources (technical books, handbooks, textbooks, literature reviews, annotated bibliographies, abstracts, newsletters), regulations, standards, guidelines, anthropometric (human body dimensions) and biomechanical (human strength and lifting abilities) data, checklists and electronic resources such as web sites that are available through the Internet.

Where is the best place to start? One way to determine where to begin is to use the method of Pareto analysis (Section 6.1): compare each aspect of the problem to determine their relative importance. The aspect that presents the greatest threat to success or the greatest opportunity to affect performance deserves emphasis.

How much information is enough? First, survey the information sources that are listed above. Follow bibliographic citations and interview references to other key resources. Eventually, those references will lead to resources the investigator has already encountered. At that point the investigator can be confident that he or she has covered the topic. Results that are derived from the evaluation of solution concepts (see Chapter 8) will show whether further information is needed.

Good analysis brings new knowledge into the team or the firm. It also increases both the depth and the breadth of knowledge about the problem. However, analysis is not foolproof. Analysis can run afoul of inadequate acceptance. Information that is unavailable or overlooked can also impede effective analysis.

4.3.2.3 PROBLEM DEFINITION

Problem definition is the process of clearly stating the desired goal state that the project activity is intended to achieve. Based on insights that were gained during the analysis step, problem definition guides two steps that follow: ideation and idea selection. One premise of this work is that asking the right question heads a person in the direction of the right answer. For that reason, problem definition is the most crucial and challenging of the seven steps described here.

A 'ladder of abstraction' model can serve as a way to shift the ground from a problem as it is given to a new area which better matches the problem's true nature. By abstracting the 'problem as given' at least one level, it is possible to distill necessary functions rather than to simply revise what already exists. Look for opportunities to apply a new technology or new way of thinking to a solution area in a way which will better serve its goals. In a food preparation project, the goal state might be 'small bits of garlic, ready for use in cooking.' This changes a mundane project to design a new garlic press into a much broader exploration which opens the way for a different solution, such as a jar of minced garlic. Also, use verbal definitions that speak to what function is desired, free from embedded solutions. To explore new car design, for example, abstracting one step higher to 'transportation' can lead to the sort of change in mind set which brings new insights that a review of all other car types would not. Changing 'design a car,' to 'design a terrestrial translocator' frees the design team from the presumptions about 'car' and opens a wealth of opportunities.

Criteria are parameters that the solution needs to meet in order for the goal state to be achieved acceptably. Information that is gathered during analysis ('Customers are not willing to pay more than $50') can be translated into criteria ('Retail price must not exceed $50'). Other frequent criteria include requirements regarding time ('Task should take no longer than three minutes to complete'), errors ('User error rate should not exceed one percent') and training ('Instruction in use will rely on web-based media'). The way criteria that are phrased can be used to reflect their degree of importance to the project. Charles Owen found in his development of design factors that constraints on solutions can best be expressed as 'must,' 'should' and 'ought to.' The basis for the difference lies in requirements that are essential to system performance, or preferable, or nice to have if possible. This sort of refinement among project requirements becomes useful during trade-off decisions and solution alternative comparison.

The human factor is a significant criterion. Those who are expected to be operators, maintainers and users have a substantial bearing on the solution. Methods to account for human abilities and limitations include flow analysis (Section 7.1), function allocation (Section 7.4), task description (Section 7.5) and task analysis (Section 7.6).

Human factors requirements must be quantifiable, noting performance which must occur and how it will be measured. The time that is required to perform tasks and error rates in the performance of those tasks are two variables that are often used in performance specification. Vague, poorly defined requirements (e.g. 'the system must be user friendly') will lead to a result that is neither effective nor suitable. In the end, a product that enables a user to get a desired outcome under all expected conditions will result in the perception that the system actually *is* user friendly.

Charlton and O'Brien (1996:19) describe two approaches that are typically used for human factors requirements definition: characteristic-based and function-based. Both approaches can be of benefit to development work.

4.3.2.3.1 Characteristic-based requirements definition—Characteristic-based requirements take a least common denominator approach by relying on accepted ergonomic standards to specify solution requirements such as equipment size, controls location and so on. Using this approach ensures that the solution that results complies with generally accepted norms for safety, usability and comfort.

The standards need to specify the user population. Requirements that are set without regard to the prospective population of users may find that those who actually use the solution cannot operate it successfully. For example, health care facilities in the 1980s installed storage cabinets on a standard that reflected the stature of a nursing population that had descended primarily from western European families. During that decade, the nursing population witnessed the influx of many nurses from Pacific Rim countries. Their shorter stature made storage unit height a significant problem in the performance of their work.

4.3.2.3.2 Function-based requirements definition—In a function-based approach, requirements are defined in terms of their contribution to overall product or system performance. While the approach is more difficult to implement, function-based standards are potentially more open-ended. The challenge is to develop and reach agreement on human performance standards for the project at hand. An example of a function-based requirement would be 'make it possible for a user to operate an interface within elapsed time (t), committing no more than (n) errors.' Individuals or teams with limited resources may have difficulty allocating enough time to pursue this approach. It can also be a challenge to agree on performance levels that are acceptable.

There are a number of obstacles to successful problem definition. Confusion between process and result can result in a team performing research and development activity without a focus on an outcome. Even though a great deal of information is generated, the problem is not solved. Stating the problem in terms of a solution can also derail a project. For example, the problem definition that calls for 'a new way to grip a stack of up to 25 sheets of paper' is more open than 'design a new paper clip.' 'Clip' may be the wrong approach. The term also brings the burden of assumptions about what a clip is. Finally, teams that mistake the inconsequential for the essential may waste effort working on only a portion of the actual problem. For example, a team that has not done field observation may invest effort in equipment redesign and not be aware that user interface errors are the actual cause of difficulty.

4.3.2.4 IDEATION

Ideation is the step that most think of when discussing the creative process. Through ideation, the team envisions possibilities, or *speculations*, about how to achieve the goal state that was described during problem definition. No idea arrives full-blown. Instead, speculations are often fragments about what might work. A number of conditions need to exist for speculations about how to solve problems to flourish. Those conditions are based on problem solving behaviors (Section 3.4) and impediments (Section 3.5). Ideation relies on:

- *Being open to any possibility one's imagination suggests*—The imagination suggests options that are related in some way to the nature of what is being considered. Why it is related may not be readily apparent.
- *Generating large quantities of speculations*—One idea can spawn another. In a team setting, the interaction can spark multiple insights. A large number of speculations makes it easier to combine them in various ways to create possible solutions and makes idea selection a more successful step.
- *Freedom from habit*—Breaking mind sets which hold on to what has worked in the past allows more original insights to be considered.
- *Recording the notion*—Making a verbal and visual record makes it possible for the designer and others to grasp it.
- *Suspending judgment*—Whether an idea is good or not will not be known until the next stage, idea selection. Until then, it is better to assume that the speculation has merit in order to allow the flow of thinking to continue.
- *A willingness to play with ideas*—Staying with an idea and enjoying the process of exploring it can continue to generate additional insights.
- *Adequate time*—Starting early will make it possible to reflect, allow one's imagination to work and accomplish the preceding steps.

4.3.2.5 IDEA SELECTION

Idea selection is the process that is used to decide which of the speculations that were generated during the ideation step will be of greatest help to achieve the goal state that was described during problem definition. Doing a good job of idea selection relies on clearly defined goals and criteria. A well-defined problem enables the team to sift among all of the solution speculations to determine which is the 'best' by comparing each speculation to criteria and seeing how it satisfies their requirements.

Selection can also help to improve problem definition because comparing solution speculations can lead to a better understanding of the goal. After going on ideation excursions to envision the goal, the team can usually define the problem more accurately. In contrast, vague goal states and criteria that cannot be measured leave no basis for comparison. Too few speculations limit the comparisons and combinations that make idea selection robust.

4.3.2.6 IMPLEMENTATION

Implementation is the process that gives physical form to the chosen solution. Like the writer's text, visual media are the currency for creative thinking. Functional and appearance prototypes are used to embody all or a portion of the concept. Figure 4.1 portrays the range of media that can be used to represent, explore and explain ideas. Moving from an idea to the actual thing, each medium provides increasingly greater information. That increase in information comes at the expense of flexibility, time and cost. Simulation (Section 8.1) provides further guidance on how to go about visualization.

4.3.2.7 EVALUATION

Evaluation is the assessment of concept that was produced during visualization. Traits (e.g. significant aspects of performance) are compared to the objectives that were developed during problem definition. Conclusions about how well the solution concept compares to project objectives can be used to guide the team to produce further versions. Methods in Chapter 8 can be used to perform evaluation using observation. Methods in Chapter 10 can be used to perform evaluation as descriptive studies. Chapter 5 describes both approaches and how to choose the right study to use.

> How can I tell what I think 'til I see what I say?
> (E.M. Forster)

4.4 Rapid prototyping

The synthesis-analysis theme continues in the practice of rapid prototyping that draws together each of the seven steps just described. In rapid design prototyping, team members quickly make rough study models to approximate possible solutions. This is the repetition of Figure 1.4's upper loop of building a prototype and testing until the participants are satisfied that it is ready to implement.

Information does not need to be complete in order to perform rapid prototyping because the models may be based on incomplete and/or unverified data. Rather than refinement, the study models are made to generate insights into what the solution might be.

Each study model is an attempt to produce a closer approximation of some portion or all of the solution. Frequent iterations make this learning-by-doing method a rich, cost-effective development approach.

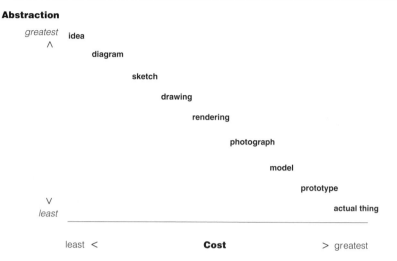

Figure 4.1 Visualization media—a leader in American design thinking, Jay Doblin, described how highly flexible, symbolic and inexpensive media such as diagrams and sketches are suited to the early stages in solving problem. The evaluation of concept configuration and function relies on more specific media such as prototypes. More comprehensive visualization requires greater time and expense. The challenge is to choose the most appropriate medium and use it well to capture or convey ideas.

Source: Faculty of the Institute of Design. Reproduced by permission of Illinois Institute of Technology

4.4.1 Building solution consensus

Prototyping can be used to create multiple iterations of an entire product or system concept in the form of tangible artifacts. 'Rapid' refers both to the short time between conceiving an initial notion and modeling it in physical form as well as the brevity between successive iterations. Rapid prototyping is used to develop a consensus among all who hold a stake in a product's outcome, including management, all corporate departments and users. Users can include expert and everyday end users, specifiers, operators and maintainers.

The physical model serves as a common focal point for communication among all concerned. Viewing a model triggers a response and compels them to action. Modifications to either existing products or new concept prototypes can be developed to explore new approaches and raise new paths of inquiry. Attention is focused on a central solution model, bringing human factors, design, engineering and other related knowledge to bear. Users are enlisted as collaborators to critique tangible prototypes as works in progress. Simple paper or foam mock-ups can effectively demonstrate initial form and control/display relationships. More refined appearance models can show finish and detail. Functional prototypes can demonstrate operating features. The result is a shared solution in which all participants are invested.

Rapid prototyping is an objective-driven, open process of collaborative solution construction that includes all stakeholders. Including all points of view makes it more likely that traditional assumptions about problems and expected solutions will be challenged and possibly overturned.

The initial design brief must be rigorous enough to provide focus for the team, capturing the essence of the desired outcome, yet flexible enough to encourage new thinking. Successful application of the rapid prototyping method has far-reaching results in user need satisfaction, the probability of consensus and the efficient use of resources (staff, supplies).

The satisfaction of user needs comes largely from extensive user involvement. Rather than extracting user models from marketing data or assumptions, co-construction involves actual and potential users in the development process. Such efforts are likely to recognize new opportunities in the marketplace. As users co-construct innovations with the other participants, they articulate previously unspoken desires and needs. These unarticulated needs can provide the foundation for additional new products, even entirely new markets.

The probability of consensus evolves from rapid prototyping's reliance on team synergy. Each corporate department better understands its counterparts as a result of the knowledge transfer during co-construction. Problems are more thoroughly explored from a broader variety of perspectives, leading to higher quality solutions that are readily championed by all involved. As participants work cooperatively toward the same goal, staff time and budget resources are used more effectively. Pertinent knowledge can be identified and used when it is needed. Rapid prototyping can often compress development lead-time, making faster entry into the market possible. The method introduces new challenges for all development participants.

4.4.2 Team success

Good teamwork is essential to research and development and rapid prototyping can be an effective tool for development teams. Team member success in rapid prototyping relies on:

- Low ego investment, a trust relationship among participants and good negotiating skills;
- Risk tolerance, or a willingness to commit to a solution direction based on impartial knowledge;
- Ability to interpret a solution without coaching, as models are necessarily incomplete compared with actual products;
- Willingness to accept a low level of finish in the models under development;
- Accountability, in which all participants are willing to accept and support the result;
- Ability to capture a variety of options during deliberations and account for the basis on which choices were made.

New tools will make the rapid prototyping process even more effective. More intuitive computer-aided design systems, finite element analysis and desktop manufacturing systems will all increase the ease with which concepts can be modeled in physical form. Growing comfort with interactive systems will make on-line simulation a more accepted means of demonstrating concepts for participants outside of technical disciplines such as design and engineering. Remote conferencing over image-based interactive networks will support off-site team members. Remote modeling, using flexible manufacturing systems to produce physical models at remote sites, would overcome reliance on physical proximity. With each iteration, the systems could act as a powerful recording secretary. That could allow for prototype alteration during the process while keeping a record of both the changes and the responses of all participants at each change.

Combined with corporate cultures and management strategies which support and implement their results, rapid prototyping shows promise as one of the new concept development methods which will build new business opportunities through insightful concepts aimed at serving new and untapped markets (Nemeth 1995).

4.5 Personal conduct

Fogler and LeBlanc (1995:11–24) identify four factors that make the difference between an ineffective and effective problem solver: solution procedures, attitude, aggressiveness and concern for accuracy. Effective problem solvers tend to share these traits in common.

- *Solution procedures*—Breaks the problem into parts, starts with one of the parts that can be understood, uses a few fundamental concepts as building blocks and uses heuristic methods as well as quantitative formulae and descriptions. Perseveres and keeps track of progress.
- *Attitude*—Believes the problem can be solved.
- *Aggressiveness*—Actively participates in the problem-solving process. Rereads and redescribes the problem. Asks questions of themselves and others, draws figures and sketches and writes equations.
- *Concern for accuracy*—Takes great care to understand facts and relationships and will check work for accuracy.

Traits that will further help individuals to function more effectively as development team members include: objectivity, awareness, risk tolerance, vision, sensitivity, interpersonal communication skills, time management and collaboration.

- *Objective*—Aware of personal values and convictions and how they influence problem perception. Biases can skew a research project through selective consideration of information.
- *Aware*—Alert and appreciative for the scope of the project as well as its working details. Working on a particular area of interest to the exclusion of others can compromise a project's ultimate success.
- *Risk tolerant*—Willing to accept whatever is encountered in learning despite the threat it may pose to current conditions. The status quo is sufficient only insofar as it remains a match for needs as they exist. Given the dynamic character of the world, conditions will change to the point that a new solution is a better fit—even one that has no relation to a firm's current offerings. Management author Rosabeth Moss Kantor (1983:237–9) considers a tolerance for uncertainty more important than a tolerance for risk. She values those with 'a longer time horizon, conviction in an idea, no need for immediate results or measure and a willingness to convey a vision of something that might come out a little different when finished.'
- *Visionary*—Sets a scale of approach that is suited to future, not just current, needs. Be willing to move beyond the popular assumptions about a situation in order to see things as they need to be. The team member who looks beyond the initial project brief can lead attention toward solutions that address broader issues and have longer-lasting impact.
- *Sensitive*—Has a keen attention to what is observed can make a significant difference in team performance. Carefully considers what is seen, heard, smelled and felt as evidence that can lead to deeper problem knowledge or insight. Sensitive team members share perceptions and integrate other team members' sensations into their own.
- *Skilled in interpersonal communication*—Able to listen to team members and those who provide information. Interpersonal communication skills improve project efficiency and accuracy. Putting what was heard into context, restating it and then asking for confirmation or correction verifies that what was heard was correctly understood.
- *Skilled in time management*—Sets a time line and focuses efforts on delivering results when they are needed. The practical nature of development eventually requires closure in order to put the product or service into production and deployment. Development projects rarely feel finished, as there is always something more to learn.
- *Collaborative*—Cultivates effective working relationships with those in and outside research and development. Development always occurs within an organization. Kantor finds entrepreneurs in an organization 'produce innovative achievements by working in

collaborative/participative fashion' by persuading, team building, seeking input, show-ing political sensitivity and being willing to share rewards and recognition. Chapters 13 and 14 provide information on the nature of relationships, how to manage change and how to communicate in an organization.

4.6 Summary

The ability to use synthesis (intuitive leaps to envision desirable ends) and analysis (thor-ough efforts to validate synthesis) provides those who work in development with the best means to solve problems.

Behavior influences how research is performed. Situation, experience and algorithmic and heuristic strategies affect problem solving behavior. Individuals will choose among skill, rule- and knowledge-based behavior to engage problems. Thinking incorporates plan-ning, storage and execution. Errors in planning (mistakes) are due to failure of expertise and lack of expertise. Even when necessary knowledge is known, problem solvers can still fail due to fixation, set and motivation, confirmation bias, availability heuristic and con-junction.

Companies launch development projects to generate revenue from sales, to develop a competitive advantage by meeting needs of those with an interest in the result and to gener-ate adequate margin. Firms will choose among defensive and proactive strategies to realize goals depending on corporate culture and the competitive market situation.

While the problems of development projects are typically described as 'ill-structured' and 'poorly defined,' development teams typically launch into a vigorous effort to engage a prob-lem through a process of continuous alternation between generation and evaluation.

Influences on problem solving include the roles of management, design, manufacturing, marketing, engineering and operational culture and resources. The situation and events of the real world causes the definition and understanding of a problem to change through time and to be affected by many interrelated parts. Real world influences also make data uncer-tain, incomplete and inaccurate and pose the potential for significant loss.

The process of development research is iterative, cumulative and utilitarian. Steps that investigators typically follow include: acceptance (resource review, qualification, clarifica-tion), definition, selection of approach(es), conduct of research, assessment and the articulation and modeling of results. Rapid prototyping provides a vigorous means to synthe-size and then analyze solution concepts. The method's use builds confidence among all of the stakeholders in a project that the concept will succeed.

Personal traits that contribute to success in problem solving include objectivity, risk tol-erance, awareness, vision, sensitivity and the ability to use interpersonal communications skills well and manage one's time. Each can be employed to pursue the process of human fac-tors research that is described in Chapter 5.

5   Human factors in research and development

What questions does human factors research answer?
Applied human factors research (5.1)

How do I get at the nature of problems?
Observation (5.2)

How do I determine causes and effects with some measure of confidence?
Inferential studies (5.3)

Which methods do I use? Why? When?
How to choose methods (5.4)

Many guidelines in handbooks are too general or vague for development purposes. With each project, an investigator or designer needs to create a unique problem and solution description. Even though guidelines and checklists are available, research and development team members still have to collect their own information. The best means to do that are research methods, which are procedures that enable the development team to create a successful solution.

Why use methods? Consider the story of the bar patron looking for his keys under the light of a street lamp late at night. When asked by a passer-by if he had lost his keys there under the street lamp he replied 'Well, no, but I'll never find them over there in the dark.' Just like the light under the street lamp, familiar territory is always more comfortable. However, the opportunities for discovery often lie far away, in the dark and unexplored unknown. Research methods lead the investigator away from the comfort of what is known into unexplored areas that are less comfortable but more likely to result in new discoveries.

As Chapters 2 and 3 explained, rational inquiry relies on evidence of one's own senses and logically deduced conclusions. However, rational conclusions depend solely on the strength of an individual's intellect. Rational conclusions also lack the benefit of evaluation to determine whether or not they are true. Scientific explanations, on the other hand, are the result of careful observation and rigorous testing against alternative explanations. Scientific explanations are designed to provide the broadest possible explanation over a variety of situations and 'tend to be more valid and general than those provided by common sense' (Bordens and Abbott 1998:15–24). The scientific method has evolved as the accepted practice to develop scientific explanations. The method is used to observe a phenomenon, form tentative explanations or statements of cause and effect, further observe and/or experiment to rule out alternative explanations and refine and retest the explanations.

Design research author Nigel Cross (1994:3) defines design methods as any procedures, techniques, aids or tools for designing that 'attempt to bring rational procedures into the design process.' However, methods are not laws. He cautions that these techniques can be treated too formally or followed too systematically. The same caution holds true for human factors methods in research and development. Methods are intended to lead the investigator into areas of discovery. It is up to the investigator to understand the strengths and limitations of methods and to be opportunistic while using methods in order to discover insights into problems and solutions.

Knowing which method to choose as well as why and how to use each method provides the structure that is necessary for a well-considered research effort. This chapter includes suggestions for identifying where to begin research and development work, how to go about research and how to select methods.

Science never appears to be so beautiful as when applied to the uses of human life ...
(Thomas Jefferson)

5.1   Applied human factors research

'Research is a systematic investigation (including development, testing and evaluation) designed to discover or contribute to a body of general knowledge' (National Institutes of Health 2002). Through inquiry the researcher seeks to learn about reality, which does not yield answers easily (Blumer 1969:22–3). Research can be basic (a search for general principles), applied (adapting general findings to classes of problems) or clinical (related to specific cases) (Friedman 2000:18).

Research is the principle method that is used to acquire knowledge about, and to uncover the causes of, human behavior. Section 1.6 described the nature of human factors practice which includes both basic research and applied research. Basic human factors research delves into the abilities and limits of human performance such as the influence of darkness on visual perception. Most of the basic research on human limits and abilities is conducted in universities. Applied human factors research makes use of basic research findings. Basic research (e.g. the limits and abilities of human vision) is applied in order to create products that are well considered with respect to human performance (e.g. highway markings, symbols and signage that can be sensed and perceived in all environmental conditions). Departments in firms and consulting firms that are involved in research and development typically perform clinical human factors research. Friedman (2000:18) contends this is so because time and budget allow for little else.

5.1.1   Human factors and design in research and development

As Chapter 3 described, both analysis (research) and synthesis (design) occur in parallel through research and development. Investigation, design and evaluation occur during three stages: exploration, definition and verification. Figure 5.1 portrays the general pattern of activity that occurs during research and development.

Both research (the collection of new knowledge) and design (the conception and modeling of speculations about how to solve problems) are performed concurrently. Applications vary from products to systems to structures, yet the process is similar. Project assignment provides resources, a schedule and a goal. Knowledge of the problem gained during analysis yields a problem definition and initial notions of a solution that can be evaluated in explorative testing. Information on human limits/abilities and technology is used to refine one or more concepts for comparison and evaluation by all who have a stake in the outcome. Information on production processes is used to refine the design concept into a pre-production prototype. That prototype can be used to collect validation information that will help to quantify expected performance benefits. Steps can be (and often are) repeated to obtain new information, to adjust the concept or to determine how well the solution performs.

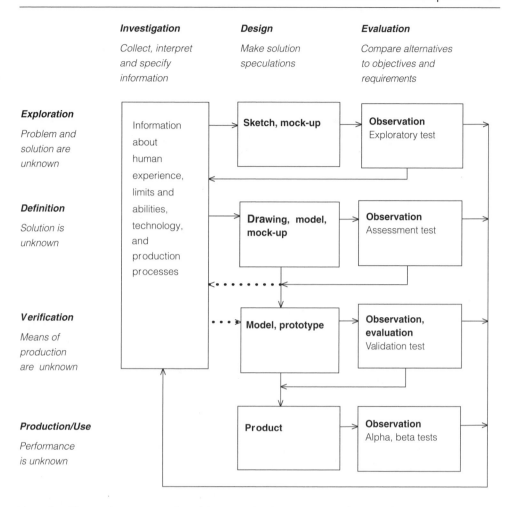

Figure 5.1 Human factors research and design in development—As Chapter 3 described, both analysis and synthesis occur in parallel through research and development. Human factors knowledge is built through investigation and applied through evaluation. Observation and evaluation methods are used through exploration, definition and verification. Usability tests can also assist evaluation from exploration through production and use. Evaluation knowledge is collected for future reference during investigation.

Section 1.4 described the development process, showing how research and development bridges the possible and the practical. The business problem such as how to become a leader in personal consumer healthcare electronics sets events in motion. The business plan is a response to the need that the business problem poses. Objectives (including budget) describe how the organization will pursue each function including marketing, sales, research and development and more. The research plan is a response to the research problem's questions about what needs to be learned. An outline of issues and questions can be used to define particulars such as the investigation structure, level of detail, specific questions and sample segments. Business problem, research problem and issues and questions determine what methods to employ and the logistics necessary to accomplish research (White 2001:2).

5.1.2 Human subject considerations

Performing research that involves humans differs from other kinds of research in two significant ways: reactivity and data type.

5.1.2.1 REACTIVITY

Human adaptive ability and motivation can significantly influence observations and data collection. Simply put: humans change their behavior when they know they are being observed.

The investigator and surroundings broadcast demand characteristics. For example, the investigator's clothing, the room's location and decor and nearby activity are all demand characteristics. The participant will observe and react to demand characteristics by using them to speculate about the nature of the research. Cues that the participant detects may cause a change in the participant's attitude about the research from being cooperative (willing to please) to being defensive or apprehensive and negative. Also, participants may be influenced by events outside of the project. For instance, participants may be more conservative in their safety preferences soon after a major airline accident (Bordens and Abbott 1998:95–101, Meister 1985:297–351).

5.1.2.2 DATA TYPE

No individual or group performs the same task in precisely the same way. The results of task performance will always vary in some way and to some degree. Because of that variability, the data that are generated on human performance can only be represented as probabilities. Simple skill-based repetitive tasks such as those represented in human error probability (HEP) data can be predicted with a fairly high level of confidence. That information finds use in some of the observational methods in this work including function allocation (Section 7.4) and the technique for human error rate prediction (Section 7.8).

More complex tasks involve rule-based and knowledge-based behaviors that resist reduction to probability. Instead, research and development creates conditions that allow for wide variation in behaviors and sets limits that are intended to protect humans from the effects of counterproductive behaviors (Meister 1985:297–351).

Methods in Chapters 6 through 8 that are useful for this purpose include critical incident study (Section 6.6), error analysis (Section 6.8), fault tree analysis (Section 8.2) and failure modes and effects analysis (Section 8.3).

5.1.3 Types of applied human factors research

Four types of research studies that are summarized in Table 5.1 have been brought into use in the field of human factors: observational, evaluative, descriptive and inferential.

Observational and evaluation studies differ from descriptive and inferential studies in that they do not control treatment condition. Treatment condition, the order in which subjects are assigned to perform certain tasks on certain pieces of equipment, is essential to both descriptive and inferential studies.

5.1.3.1 OBSERVATIONAL STUDY

Observation is the practice of studying oneself, others, things, events and phenomena as they are encountered. Rather than having the investigator influence conditions, effort is made to avoid investigator influence. The discussion of fieldwork in this chapter provides guidance on how to go about observation.

Table 5.1 Types of applied human factors studies—human factors research is a collection of study techniques from various fields that include operational research, engineering and the social sciences. The table below shows the relationship of human factors research methods to more widely known methods including observational, evaluative, descriptive and inferential studies.

Types of studies	Chapter	Methods
Observational		
Conditions and events in the	6	Analysis methods
operational environment are observed	7	Design guidance methods
as they are encountered	9	Interviews and questionnaires
Evaluative		
The effects of variables on human	8	Evaluation methods
performance are observed. Unlike	9	Interviews and questionnaires
inferential study, the effect of all variables	10	Usability: exploratory test
is observed simultaneously	12	Cost-effectiveness analysis
Descriptive		
Characteristics of subjects whose	10	Usability: Assessment and
performance is being evaluated as well		validation tests
as significant aspects of their performance		
are accounted for using descriptive		
statistics		
Inferential		
Variables and conditions are	11	Controlled studies
deliberately managed in order to		
isolate the effect of one or two		
variables on human performance		

Both Meister (1985:301) and Chapanis (1985) encourage greater use of direct observation in human factors studies. This is because a great deal of human performance research is conducted in laboratory environments. As a result, much is assumed but little is known about how people *actually* perform when at work in the real world.

The methods in Chapters 6, 7 and 9 are observational methods.

5.1.3.2 EVALUATIVE STUDY

Evaluative studies observe the collective effect of variables on human performance. For example, the evaluation of a new hand tool seeks to determine whether the tool is useful. Comparative evaluation seeks to determine which of many hand tool concepts is most useful. By contrast, experimental study of a new tool would isolate one or two aspects of performance to examine. Varying handle width or the grip position on the handle would be examined with regard to the amount of torque that a human subject could exert while using the tool.

Methods in Chapters 8, 9 and 12 and the exploratory assessment approach to usability studies that is described in Chapter 10 are evaluative methods.

5.1.3.3 DESCRIPTIVE STUDY

Descriptive studies are used to account for two kinds of information: the characteristics of individuals who are subjects of research and the results of their performance.

Sample subjects—The use of descriptive statistics makes it possible to design a sample of test subjects according to their characteristics. Descriptive study also makes it possible to discover patterns of characteristics that might not be evident during casual review.

Performance—The results of subject performance can be assessed according to overall traits of the group as well as each individual's performance relative to the group.

The assessment and validation usability studies described in Chapter 10 are descriptive studies. Chapter 10 covers descriptive statistics in the context of usability studies.

5.1.3.4 INFERENTIAL STUDY

Inferential studies examine the effect of one or two independent variables on human performance by establishing controlled conditions and following specific procedures. The conditions and procedures are necessary to establish a certain level of confidence regarding the independent variables' effect.

Inferential studies are used to examine the relationship of one or two variables on human performance. These studies are used to evaluate the significance of the differences that are found in human factors research. Statistical methods are primarily of value in experiments and experiments are rarely performed during system development. Even so, inferential studies can play a valuable role in some portions of research and development work. Chapter 11 provides a basic approach to controlled study and the inferential statistics that are used to analyze them.

Chapanis (1985) cautions that inferential study is often not appropriate for many kinds of development work. It can be time-consuming and expensive to conduct. It does not get at determining problems and it requires definite hypotheses about variables of interest. Changes brought on by the lab environment also affect problem study. Variables that are brought into the lab often change from those in the real world. The investigator may overlook or fail to include important variables. Even after the investment of time and effort, experimentation may yield statistically significant but practically insignificant results.

Observational and inferential studies in this book are now explained in greater detail.

> *Innovations are not safe, bounded, or easy.*
> *(J. Kotter)*

5.2 Observation

Observation is not only looking, but seeing. Not only listening, but hearing. Not only touching, but feeling. Observation is actively seeking information.

Observation methods in human factors are a loose collection of heuristic procedures that have been developed *ad hoc* by professionals in the discipline. The various methods can provide serendipitous discovery opportunities. That is, the methods will lead the investigator to explore portions of the problem domain where information that is useful may be encountered even though it was not anticipated.

Through methodical analysis, observation methods can be used to identify mismatches between users and equipment, software, procedures and environment. Observation methods provide data about human performance for design, alternative evaluations, trade-off decisions and allocation decisions. They can be used to predict operator–machine system performance and to evaluate how well it meets design criteria. The use of observation methods produces a base of knowledge about a concept as it develops. As work continues, knowledge about the problem becomes more specific and detailed. The collective knowledge can be used to brief new team members and to develop inputs for specifications and reviews (Chapanis 1985). The performance of observation methods can be tailored according to their purpose and the observer's point of view.

Methods that are used to observe, analyze, develop and evaluate user-centered products are also referred to as test and evaluation methods. Test and evaluation methods can be used to serve a variety of research and development needs. They can be used to establish the role of the human in the system. Their use can assist the design of interfaces to fit human capabilities and limitations and to determine how well the design fits. The methods can also be used to create training for humans in order to assure reliable, effective performance (Woodson 1981:893–9). Chapanis and Van Cott (1972:701–28) note that test and evaluation methods can be used to improve the quality of design decisions, to integrate hardware and personnel and to correct design deficiencies.

5.2.1 Process

As applied research, the methods are used to collect and apply knowledge in order to produce an optimal solution. That knowledge can also help to negotiate trade-off decisions that may compromise system objectives if those trade-offs are not executed well.

While methods vary, the general pattern for test and evaluation methods is to:

- Choose a purpose (e.g. find out what's wrong)
- Choose an appropriate methodology
- Specify a target population and choose a representative sample of individuals from it
- Choose appropriate situations to observe
- Choose appropriate data to collect
- Set-up test facilities or choose test sites
- Establish pilot test procedures
- Conduct the evaluation
- Analyze the results.

Each of the above steps builds a cumulative base of knowledge by following a methodical process. Information is collected to serve as inputs. Procedures are conducted to review inputs. Results are produced from the procedures that can serve as inputs for other follow-on methods.

5.2.2 Purpose

Most of the observation methods that are covered in Chapters 6 through 10 can be used for either prediction or assessment.

Prediction is the collection of information before a product is produced. The collected data are used to predict what will occur or will be needed as the concept evolves. For example, time line analysis (Section 7.2) can be used to collect information about task performance, to predict how long tasks will take and to identify tasks that might conflict if performed simultaneously.

Assessment is the examination of an existing product. Data are collected to determine how human performance or the performance of the system as a whole are affected by the product's use. For example, failure modes and effects analysis (Section 8.3) is used to determine the effect that failure of a component might have on a product's performance.

Many of the methods in Chapters 6 through 8 can be used for both prediction and assessment. For example, workload assessment (Section 8.5) can be used to predict the amount of work that an individual might encounter while using a new design. Subjects can view a demonstration of the new solution and be asked to predict the degree to which their workload would be affected. Workload assessment can also be used as an assessment tool. Subjects can use an actual product or prototype of a new solution and be asked to evaluate the actual workload that was experienced.

5.2.3 Traits

Regardless of phase, observational research methods are typically iterative, cumulative and utilitarian.

- *Iterative*—Steps may be performed more than once in order to complete or refine knowledge. Changes to the problem or problem understanding can lead to new paths the investigator must take;
- *Cumulative*—Steps build on each other. The accumulation of knowledge leads to wisdom regarding the problem that is greater than the individual portions of knowledge that are gleaned by each step;
- *Utilitarian*—The focus of team efforts is on learning the knowledge that is necessary to create a solution that is suited to the opportunity or problem at hand.

5.2.4 Issues

Observational methods require attention to such issues as observer reliability, agreement and correlation among observations.

5.2.4.1 POINT OF VIEW

Meister (1991:475–8) splits observational methods into two kinds based on the observer point of view: participant and non-participant.

As a participating observer, the investigator takes an active role in the collection of information. For example, interviewing is a participant observation method in which the investigator probes areas of interest to the project by asking direct questions of subjects.

During non-participant observation, the investigator has no contact with subjects. Non-participant observation can be a powerful tool to uncover insights about what people really do when they are not aware of being observed. Making digital still or video camera recordings of the staff working at an operational control center will show how individuals do (or do not) pay attention to security displays over the course of a full shift.

5.2.4.2 OBSERVER RELIABILITY

The reliability of a person to perform well as a data collector varies according to each individual (within observer) and among all of those who perform the role (between observer).

Within observer—Sources of observer bias include errors of omission (failure to score an observed response) and errors of commission (miscoding a response).

Observers can influence the behavior of those who are being observed by errors of expectancy (observer somehow communicates an expected outcome). The investigator may indicate attitudes about participants (e.g. that excellent performance is expected) or induce attitudes among participants (e.g. suggest that the group will get a reward for performance). The investigator may also create such effects by treating various participant groups differently (Bordens and Abbott 1998:95–100).

Individuals can also bias observations by being conspicuous, by personality and gender and by signaling the purpose for the observation. Observer drift is the decline in accuracy over time that can be corrected by supervisory attention (Meister 1985:304–8).

Between observer—The observer role can be viewed as the choice of a human to perform in the role of data collector. Retaining two or more observers requires attention to their accuracy and the stability of the measures they are using. Post-observation review can improve data collection reliability by using a consensus coding method to review film or tape of the observations. In that method, two or more observers view and review segments of the record. Review continues until they agree that a particular response has occurred and that the response fits a specific taxonomic category. Meister (1985:305) asserts that 'it is possible to achieve perfect reliability' using such an approach.

5.2.4.3 USE OF CORRELATIONAL STATISTICS

Because observational and evaluative studies take the operational environment as it is encountered, treatment conditions cannot be controlled. Statistics typically require control over treatment conditions. However, this does not preclude the use of statistics with regard to observational study. Meister (1985:308–10) proposes a number of opportunities to use significance statistics (such as analysis of variance) to determine whether:

- Contrasting conditions are inherent in system operation (e.g. day vs. night operations)
- A change of equipment affects performance
- Operation over an extended period of time affects performance
- Actual performance differs from performance that is directed by a standard
- One alternative performs more successfully than a number of alternatives.

5.2.5 Fieldwork

Much of an observer's work is spent in the field, outside of the office or laboratory. In the social sciences, fieldwork is the study of humans in social interaction. Human factors has translated social science methods into the study of how humans act and interact in the operational environment. Wolcott (1995:19) characterizes fieldwork as 'personal involvement to achieve some level of understanding that will be shared by others.' As both observer and interpreter, the investigator needs a clear idea of what to do.

Insight is the key to success in observation. Insight is the ability to grasp the meaning of what one sees and relate it to what one knows. Fortunately, there are many instances in which people make changes that signal what makes sense in a situation. It serves an observer well to look for such signals. Jones (1992:216) notes 'it is always worth paying a lot of attention to the crude adaptation that users make to their equipment and important to discover the reasons for them.' People adapt workstations as well as equipment. Design and human factors instructor Len Singer has likened the layer of notes and personal effects that an individual creates to personalize a workstation to a mold as it would spread to cover the surfaces of the original furniture.

5.2.5.1 OBSTACLES

There are four obstacles to insight while performing fieldwork: insufficient time, arrogance and bias, oversight and over-study.

Insufficient time—Allowing enough time to develop an authentic understanding of the problem area will make it possible to detect cues that are not obvious.

Arrogance and bias—Keeping an open mind will enable the investigator to remain ready to accept new experiences, particularly those that are contrary to what is already accepted knowledge.

Oversight—Taking in enough information about observed conditions will improve the investigator's understanding of the problem.

Over-study—As the opposite of oversight, over-study is attention that is paid to so many aspects of what is observed that the investigator loses the sense of what matters and does not matter. Experience will eventually lead an investigator toward the collection of data that are likely to ultimately prove most useful in the project.

5.2.5.2 REMEDIES

Wolcott (1995:91–5) recommends four guidance social behaviors for observers to avoid the impediments listed above: gaining entree, reciprocity, tolerance for ambiguity and personal determination.

Gaining entree—Obtain permission to observe human subjects, if necessary.

Reciprocity—Social scientists may take up residence in a community over the long term. Human factors work is normally more tightly defined in the nature and length of the relationship between observer and subject. Even so, it would be an error to assume that subjects have no reaction to being observed. Be courteous and use good judgment in dealing with those who provide the information that is being sought. This exchange of professional courtesy respects the goodwill that the subject has extended by both agreeing to participate and divulging information for the benefit of the study.

Tolerance for ambiguity—Accept that what is observed may not make sense and may likely conflict with what is known. Information that is learned at one time may make more sense when it is reviewed with other information. Be patient and allow items that appear to conflict to exist without immediately being reconciled. Keep track of documentation such as notes, tapes and photos. Later review may stimulate insights that were not evident previously.

Personal determination coupled with faith in oneself—New discoveries can undermine confidence in what one thought he or she understood. An investigator who breaks new ground is likely to be uncomfortable without the reassurance that normally comes from cues that others have covered similar material.

5.2.6 Case study: observation method selection and use

Human factors professionals have a key opportunity to protect human safety in high stakes environments. Invasive surgical procedures performed in a hospital catheterization laboratory are such an opportunity, particularly when the potential exists for the procedure to inflict extraordinary patient trauma. Nemeth (1996a) describes the methods that were used during a 1995 project to detect potential sources of cath lab system dysfunction.

5.2.6.1 INTRODUCTION: ANGIOGRAPHY

Cardiac and cerebral angiography is a diagnostic invasive procedure performed on patients who demonstrate symptoms of arterial disease and plaque build-up that could lead to a heart attack or stroke.

The cardiac catheterization lab where the procedure is performed can be considered as a complex workstation. While facilities vary in size and layout, the scenario is essentially the same at each location. The skin on the patient's upper thigh is shaved to remove hair, then treated with povidone iodine in order to create a sterile field. A large bore needle is inserted into the femoral artery. A long catheter is inserted through the needle via the thorax into the heart or carotid artery. Contrast dye is injected to show blood flow through arteries that feed the heart or brain. Strictures to blood flow, shown by X-ray images projected on a video monitor, indicate clogged arteries that may need angioplasty or by-pass surgery. The patient, whose help is needed in breath holding to stop movement during irradiation, receives local anesthesia and is awake during the procedure.

During the process of patient catheterization, the potential exists for one or more air bubbles to be introduced into the patient bloodstream via the contrast dye. Free to travel through the patient's body, the bubble may lodge in the nervous system or brain, causing paralysis or worse. Reports to the manufacturer that retained the investigator indicated that this had occurred in a number of instances. The cause was unknown, implicating all of the equipment in use at the time, including the product made by the manufacturer who sponsored this research project.

Consider the following questions, develop tentative answers, then read the remainder of the study to see how the investigator proceeded.

> *What elements are involved in providing this service?*
> *How could a diagnostic procedure impose a larger trauma than the patient's affliction?*
> *Who was responsible for ensuring the safety of the catheterization lab?*
> *What was the actual cause of trauma?*
> *What remedies would ensure that such trauma is not inflicted in the future?*

5.2.6.2 METHODOLOGY

The investigator's role was to assess the potential for any of the pieces of equipment used in the angiography to introduce air into the contrast medium. Project scope was also expanded to include recommendations for a new design approach. Four field visits to cath labs in a large metropolitan area provided the opportunity to observe procedures for hours on end, although the need to protect patient confidentiality precluded video or still photo documentation.

The investigator's first step was to perform an audit of the lab as a small-scale system, relying on the standard system model to account for personnel, hardware/software and procedures performed within a total environment. (For more information on system models, see Section 1.1.)

A critical incident analysis provided information on the nature and extent of incidents (i.e. introduction of air into the bloodstream), trauma, equipment in use and the date and time the incident occurred. A functional flow analysis made it possible to trace the flow of activity through preparation, operation and retrograde stages. Through that analysis, it was also possible to account for the tools and supplies that were handled, as well as the state of equipment in various stages during the procedure. Typically, items are removed from storage, prepared, staged for use, used and discarded or set aside for cleaning and storage.

Mapping the tasks among each of the staff members provided the basis for a simple workload assessment. The cardiologist and cath lab technician develop a closely harmonized routine. The lab systems technician, who operates imaging equipment in an adjacent control room, is linked through an intercom. In larger facilities, a circulating nurse assists the team by keeping a procedure log and by operating ancillary equipment. This step showed that it is not unusual for the circulating nurse to enter and leave the lab and even change shifts during the procedure.

Awareness of the rate of workflow also led to a simple operational sequence analysis. While some patients can be easily diagnosed in 45 minutes, others can take a great deal more time. Cardiologists also vary in technique and demeanor. The sequence analysis demonstrated that once the arterial diagnosis is complete, further diagnosis of heart muscle function might be performed dramatically faster. The rapid change in pace can quickly shift attention from one task to another, leaving little or no time to check equipment status.

An equipment review showed certain pieces of equipment were individually well engineered, but did not show the benefit of attention having been paid to issues such as user orientation or use in combination with other items. For example, controls on one item required for gross and fine adjustment were located on the device. When the device is used, it is kept outside the sterile field. At the same time, the operator must lean toward the sterile field to see clearly while operating the device. As a result, the operator must reach back and adjust controls entirely by feel. In addition, the device was mounted on a fixed pedestal with casters, which prevented height adjustment, making it more difficult for nurses of short stature to see displays and operate controls.

Finally, a failure modes and effects analysis made it possible to sort each of the above considerations into categories according to personnel, procedures, hardware, software and the total environment. Each section was organized into tabular format by five columns devoted to:

- Item of interest (e.g. 'Lab team' in the Personnel section)
- Observation (e.g. 'Some staff may be relieved by others during the procedure')
- Effect (e.g. State the reason why the staff member being relieved is a problem)
- Criticality (Estimate the degree to which the effect impedes mission accomplishment)
- Remedy (Specific changes which, if made to the product, promise to minimize or eliminate the failure).

5.2.6.3 ANALYSIS

It became apparent that when circumstances that were related to four elements coincided, the potential existed for an incident to occur:

Personnel—The circulating nurse could be directed to operate ancillary equipment with little advance notice;

Equipment—With no self-monitoring function, it was possible for a device to be unprepared for use without the staff knowing about it;

Procedure—The rate of procedure could accelerate abruptly, posing opportunity for oversight. In addition, no in-service training was provided for ancillary equipment operation after its initial installation, which resulted in instructions being passed to new team members by word of mouth;

Total environment—The variety and number of collateral tasks distracted some staff members.

The report linked the source of the potential problem to the effect it produced. It then connected each of the effects to solution recommendations that were intended to prevent the problem. The manufacturer found the logic compelling and adopted the recommendations.

5.2.6.4 RECOMMENDATIONS

Given some of the hospitals in which this equipment was used, the potential did exist for the circumstances that were just described to occur. The recommendations that were made to the manufacturer focused on changes to the product. Those changes would compensate for the potential ambiguity that may occur as a result of practices that vary among the many hospitals in which this procedure is performed. As a result, the manufacturer made plans to develop a new generation product with a design based on additional comprehensive human factors studies.

5.2.7 Workplace studies

Workplace studies is a body of research that consists largely of naturalistic studies (ethnographies) and is 'concerned with the ways in which tools and technologies feature in work and interaction in organizational environments' (Heath and Luff 2000:4–5). The research has evolved in recent years in response to the limitations of cognitive psychology and engineering in understanding how people work together.

Those limitations became evident in the failure of large information systems such as the United Kingdom National Health Service's value-added medical products (VAMP) that were supposed to have been well designed. Analysts who looked into such systems identified a number of reasons for the failures. First, designers have disregard for those whom they studied, treating 'the general practitioner, the user, as a judgmental, or cultural, dope.' Second, designers pay only 'cursory attention to circumstances.' Human-computer interface (HCI) designers 'misconceive the user' by oversimplifying what actually occurs in the daily work environment (Heath and Luff 2000:10–11, 59). Third, the mental models and methods they use are not suited to understand the complexity and sophistication of collaborative work. Sociologist Lucy Suchman (1987:178–89, 2001a: unpaged) has challenged the traditional notion that people act in predetermined goal-directed ways. Human behavior is far more complex, sophisticated, nuanced and fluid than designers understand. Workplace studies and conversation analysis have evolved to improve understanding in these areas.

Workplace studies and system design remain separate activities. Heath and Luff (2000:246) consider that relationship between ethnographic studies and design 'remains problematic' due to significant differences between the agendas, philosophies, thought processes and methods of naturalistic researchers and of designers. Workplace studies strive to produce rich ('thick') descriptions of group interaction as it is observed while designers focus on what needs to be done to create a future condition. Bannon (2000:237–9) finds that 'the relationship between ethnographic studies and system design is still problematic …' At the same time, he also contends that such studies may enable designers to question presumptions that are embedded in design's conventional problem–solution framework. While workplace studies research does not yet have a place in Part II as a human factors research method, its value in understanding complexities of people at work does have a role in human factors research.

The effective human factors researcher combines instincts, aware observation and analysis methods to detect possible sources of, or contributors to, a problem. Methods such as those described above can be used to qualify and quantify the investigator's initial observation and provide a substantive basis to improve product reliability, safety and ease of use.

5.3 Inferential studies

The refinement and precision of a controlled study enables the investigator to establish causation. Establishing controlled conditions and following specific procedures makes it possible to study the effect of one or two independent variables on human performance. The method's precision also enables the results that are obtained with a sample (e.g. 1,200 subjects) to predict the likelihood of occurrence in a broader population (e.g. U.S. adults) with a certain degree of confidence.

5.3.1 Process

A seven-step model, shown in Figure 5.2, reflects current behavioral research practice. The following steps provide guidance on how to perform human factors research.

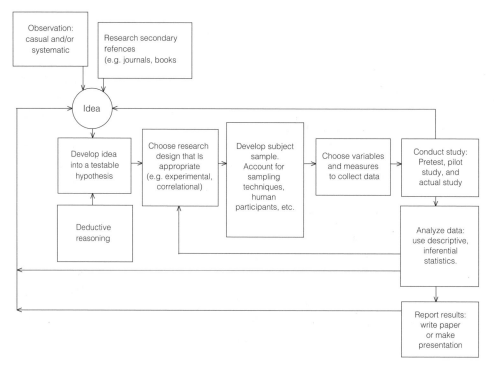

Figure 5.2 Inferential study research process—Bordens and Abbott (1998) provide a model of the research process that scientific study uses in the quest for knowledge.

Source: Adapted from *Research Design and Methods*. Bordens, K. and Abbott, B. (1998).

5.3.1.1 DEVELOP A RESEARCH IDEA AND HYPOTHESIS

The investigator chooses an issue to study. Deductive reasoning (see Section 3.1.1) is used to develop a statement (a hypothesis that links two variables) that is related to the issue which others can also study. An example of a human factors issue would be to study driver performance on behalf of a safety organization, a government highway bureau, an organization of insurance companies, auto manufacturers, or a regulatory agency. An investigator with the hunch that very young and very old drivers are more likely to have accidents might develop a hypothesis stating that changes in one variable (age) cause an increase in the incidence of accidents.

5.3.1.2 CHOOSE A RESEARCH DESIGN

The investigator chooses between a correlational or an experimental study. A correlational study would compare two or more variables and look for relationships among them. Correlations are useful when independent variables such as age and gender cannot be manipulated by the experimenter. In an experimental study, the investigator would manipulate a variable and look for concurrent changes in a second. A correlational study of state accident records might reveal that the above hypothesis is true, but an experimental study would be called for in order to learn why.

After drafting a plan for the study and organizing resources, running a pilot study would verify that the plan is valid. In the example of elder drivers, the hypothesis might specify reaction time or visual acuity as the cause for accidents.

5.3.1.3 CHOOSE SUBJECTS

Determining the traits of the sample subjects (e.g. age, driving experience, number) that are appropriate to the study, choosing a recruiting method and recruiting participants will make it possible to develop a sample of individuals who are available and qualified to participate in the study.

5.3.1.4 DECIDE ON WHAT TO OBSERVE AND APPROPRIATE MEASURES

A correlational study with the hypothesis that age causes an increase in the incidence of accidents might involve subjects using a driving simulator. The simulator would present driving scenarios under different conditions (e.g. daytime, night-time, good and bad weather, superhighway and narrow country roads) and with different kinds of hazards (e.g. the driver encounters debris in the way, construction equipment and barricades near the traffic lane, or an animal or a child runs into the roadway). Data could be collected on measures such as reaction time and errors.

5.3.1.5 CONDUCT THE STUDY

Subjects operate the apparatus during the scenarios and data is collected on their performance.

5.3.1.6 ANALYZE THE RESULTS

Data can be collected and analyzed using either of two types of analysis: description and inference.

Descriptive statistics can be used to describe the data that were collected in terms of average (e.g. median) and variability. See Section 10.3.9 for information on descriptive statistics.

Inferential statistics can be used to determine how reliable the collected data are. See Chapter 11 for information on inferential statistics.

5.3.1.7 REPORT THE RESULTS

A report of findings may be written and submitted for review by project colleagues, company management or clients. Results may also be submitted to a professional journal to be considered for publication.

The research process continues. Additions or refinements can be made to the original study, or further areas of inquiry can build on the original study's findings. Even with the benefit of scientific explanation behind it, the process is not fail-safe. Research can run afoul of three failures: faulty inference, pseudo explanation and circular explanation.

Faulty inference—Faulty inference is the incorrect identification of underlying mechanisms that control behavior. For example, Robert Yerkes' World War I intelligence testing of U.S. Army recruits yielded scores that indicated the average Army recruit had a mental age of 13 years – only one year above the upper limit for moronity. Yerkes made the faulty inference that many Army recruits were actually of low intelligence. In actuality, Yerkes' test failed to consider the effects of education among recruits. Many were from the rural Southern U.S. and did not have the same educational experience as other areas of the country.

Pseudo explanation—Pseudo explanation is the use of an alternative label for a behavioral event. Holding the opinion that males are talented at endurance tasks because they are physically stronger is a pseudo explanation. Only scientific study can say for certain how gender affects endurance.

Circular explanation—Circular explanation, which is also referred to as tautology, is the use of the behavior as its own explanation. For example: 'Certain individuals make errors because they are error-prone. Why are they error-prone? Because they are observed to make many errors.' Scientific study can identify the frequency and cause of errors (Bordens and Abbott 1998:11–3).

5.3.2 Preparation

There are many potential sources of error in research. Without correct planning, the results of research activity can easily be misleading. The following will help to plan an experiment that produces reliable, accurate results. Preparation also includes becoming familiar with professional organization codes of ethics. The Human Factors and Ergonomics Society includes its code of ethics in its annual directory and yearbook.

5.3.2.1 ENSURE CONDITIONS ARE REPRESENTATIVE

Conditions for a study need to be as close as possible to conditions in reality. This is particularly true of four study elements: the purpose for the work subjects will be asked to perform, study subjects and their motivation, tasks that subjects are asked to perform and environmental conditions.

Purpose—The purpose for the work to be accomplished needs to correspond with what is done in reality.

Subjects and their motivation—Subjects need to correspond to those who normally perform the task that is being studied. Their reason for and interest in participation should center on an interest in learning more about the area of study. Subjects can often be motivated by providing them with fast, accurate results of their performance.

Tasks—The procedures that a subject is asked to perform need to match what would normally be done in an operational setting.

Environmental conditions—The study physical surroundings (e.g. lighting, sound, setting) need to match the conditions that are typical of operational environmental conditions.

Each of these conditions will be explored below.

5.3.2.2 SELECT SUBJECTS

Individuals who participate as experiment subjects need to be chosen carefully according to a number of considerations, including experience, objectivity, skill, quantity and mix.

5.3.2.2.1 Experience – The investigator should use subjects who have an experience profile that matches that of expected users. Highly experienced users are not necessarily the best subjects. Experienced subjects who are more comfortable with versions of the equipment or material that they are accustomed to may not prefer significantly new alternatives. They may also have well-developed habits that make it more difficult to use a new version well.

5.3.2.2.2 Objectivity – In order to avoid the influence of ego involvement, the investigator should choose subjects who have no stake in the outcome of the test.

5.3.2.2.3 Skill – The investigator should select subjects according to their ability to perform required tasks. Performance ability should span the range from poor through good based on matters of record such as test scores.

5.3.2.2.4 Quantity – The investigator should choose the size of the sample based on the test's purpose.

In descriptive studies such as usability tests, Virzi (1992:457–68) finds four to five subjects provide 80 percent of the information to be learned. This smaller number of participants is allowable because the purpose in a usability test is to understand a problem's severity, not its frequency. Rubin (1994:29) explains that information about the size and character of a population is the basis for usability test sample development. However, because that information is often unavailable he recommends testing ten to twelve participants per condition and forty or more to ensure statistically significant results.

In a controlled study, frequency is an issue. To develop frequency data, a number of individuals are asked to perform a specific set of tasks using a specific piece of equipment under a specific set of circumstances. This unique set of tasks, conditions, equipment and observation method is referred to as a cell. A group of study participants are assigned to a particular cell (exposed to a unique set of conditions) and asked to perform one or more tasks. At least five subjects should be assigned to each cell. Morgan *et al.* (1963:38–50) recommend using thirty to forty subjects from a pool of prospects for results that will be analyzed statistically.

5.3.2.2.5 Mix – The investigator should assign members in varying combinations when using groups because individuals will perform better in some groups than others.

5.3.2.3 TRAIN SUBJECTS

Individuals who are to perform in an experiment need guidance to know what is expected of them. The investigator should ensure that groups who are operating different alternatives have been given the same preparation and instruct subjects in the task that is to be performed. The research team should measure performance objectively (e.g. score their performance) and continue training until further improvement is negligible.

5.3.2.4 DEVELOP TASKS

The procedures that subjects are asked to perform need to be comparable to actual operating conditions. The investigator must be aware of what actual conditions truly are. This includes work rates, the amount of work, work difficulty and operation (even though it may be a simulation that requires role-playing). It also includes issues related to medication such as drivers who are taking over-the-counter medications, pulling double shifts and using vehicles that are in disrepair.

5.3.2.5 CREATE ENVIRONMENT

Which conditions should be realistically portrayed? The investigator should pay particular attention to the conditions that will affect performance. The environment and the system mock-ups/prototypes will need the same tasks, stresses, motivation and results knowledge as those that are used under normal operating conditions.

5.3.2.6 SELECT MEASUREMENT CRITERIA

The investigator should choose experiment criteria that relate directly to the manner in which the actual system is expected to perform. While there are varying opinions on data collection, the traditional view is that two types can be collected: quantitative and qualitative.

5.3.2.6.1 Quantitative – Performance is recorded in terms that can be represented numerically, such as time and errors. Objective measurement criteria are often preferred over subjective criteria, because objective criteria can be used to report on actual performance. Quantitative data cannot, however, get at crucial aspects of human performance such as intention, motivation and preference.

Measurement scales can help to clarify information that is represented by numbers that result from measurement of a variable. Four kinds of scales can be used to account for objective data that are collected. They are ordered below according to scales that can provide increasingly larger amounts of information about the variables.

Nominal—Nominal measures are variables whose values differ by category. For example, sex is a variable that varies according to male and female categories. While the difference between the categories has to do with the quality of being male or female, data are collected according to the number of males and females.

Ordinal—Ordinal measures are variables that can be scored according to groups of values such as low, moderate and high. It is possible to rank the order of values. While their order is known the distance between the groups is unknown.

In the next two scales, spacing between values along the scale is known. This makes it possible to know whether (and to what degree) one unit of value is larger or smaller than another. These scales can be used to identify the amount by which various data points differ.

Interval—Interval measures include a zero point that serves as a convenient reference point. If preferences are collected from a sample, highly positive to highly negative responses might be shown along a 10-point interval scale. Zero would indicate the mid-point on the distribution (i.e. ambivalent, neither positive nor negative) rather then a lack of any preference.

Ratio—By contrast with interval measures, the zero point in ratio measures indicates the absence of any value of the variable. For example, if a ratio scale accounts for the number of responses, the zero point would indicate no responses (Bordens and Abbott 1998:84–5).

Choice of scales depends on three considerations: information that is yielded, statistical tests and ecological validity.

Information—When possible, the investigator should adopt the scale that provides the most information.

Statistical tests—The investigator should use more precise interval and ratio scales if they are to be used in more precise analyses such as statistical tests.

Ecological validity—The investigator should reflect what people do in real life situations. If a condition is binary (efforts to log-in to an on-line database either succeeded or failed) then a nominal scale would be appropriate even though it is less powerful.

Composite scales can be created to account for more than one measure and offset the weaker power of nominal and ordinal scales. For example, each category among ordinal values (low-medium-high) can be graded on an interval scale (1 to 10).

5.3.2.6.2 Qualitative – Information that cannot be quantified may be collected through reports from subjects of their own experience or opinion. There are some drawbacks to subjective measurement. Subjects may be unduly influenced by their own habits, experiences and preferences. Subjects may prefer the familiar over the novel. They may give answers that they believe are expected of them by the sponsor, investigator or fellow subjects.

McCracken (1991:9–29) distinguishes between the comparative value of qualitative and quantitative research methods.

> The quantitative goal is to isolate and define categories as precisely as possible before the study is undertaken and then to determine, with great precision, the relationship

between them. The qualitative goal, on the other hand, is often to isolate and define categories during the process of research.

Qualitative methods are means that are used to view a broad area using less precision in the interest of capturing its complexity. Quantitative methods view a narrow area with greater precision.

5.3.2.7 SELECT DATA COLLECTION MEASURES

Activity (e.g. control manipulation, display scanning) that the observer or experimenter will pay attention to is chosen. Aspects of performance to quantify are selected (e.g. time to actuate). Any measure in an experiment must meet certain requirements to be suitable for use. Two essential requirements are reliability and validity.

5.3.2.7.1 Reliability – A reliable measure is consistent and repeatable. Pairs of measures are developed and compared to their results in order to determine whether or not a measure is reliable. The measure that produces a more consistently accurate result is more reliable. Determining reliability needs to be completed prior to establishing validity.

5.3.2.7.2 Validity – A valid measure reflects the ultimate performance of a system in actual use. As this is the reason for any evaluation, measurement validity is crucial.

To ensure the measure is valid, it should reflect aspects of the problem that will be crucial and relevant to system performance when it is in actual use. Avoid compromising factors such as previous subject training and bias due to experience or ego involvement.

5.3.2.8 COUNTERBALANCE EXPERIMENT DESIGN

Human performance varies even when individuals try to perform tasks identically. This variation can introduce bias into a study. Counterbalancing is a deliberate arrangement of study conditions to avoid bias. Counterbalancing evaluation conditions minimizes bias by ensuring that tasks of equal difficulty are used on each solution alternative that is under evaluation.

Transfer of training is also an issue in experiment design. A subject who uses a product in a session develops some familiarity with it. Transfer of this familiarity (and possibly skills) can make the second alternative appear to be better. Counterbalancing also minimizes transfer of training effect.

A wayfinder product uses global positioning satellite data to enable an individual to determine location and the route to desired locations. The comparison of subject performance using two different designs for a personal wayfinder product (referred to as A and B) will illustrate how to counterbalance the study design. Six qualified groups of five subjects each (referred to as 1 through 6) will serve as the sample. Groups might be sorted according to experience using such products (i.e. Groups 1 and 4 inexperienced, Groups 2 and 5 moderate experience and Groups 3 and 6 experienced).

- Compare both systems at exactly the same time so that Groups 1 through 3 use Wayfinder A and Groups 4 through 6 use B. If simultaneous comparison cannot be done, have the groups use the wayfinder alternatives in counterbalanced order (e.g. have Group 1 use Wayfinder A, then B. Have group 2 use B, then A).
- Use the same subjects and systems to eliminate irrelevant differences among different subjects and equipment. Ensure each group uses the same version of Wayfinder A and B.

- Make sure that skill difficulty and task difficulty are kept separate. If Wayfinders A and B require different skills, have subjects perform the same task. Habit interference is the condition in which learned skills that are developed using the first alternative may interfere with the use of the second system. Having subjects perform the same task separates the variable task from the variable skills. If Wayfinders A and B require the same set of skills ensure that subjects perform tasks that are equal in difficulty.
- Decide in advance what (if any) statistical methods will be used. Choosing in advance will make it possible to identify variables of interest to the study and collect accurate, unbiased data for those variables. Chapter 10 provides further guidance on usability assessment and the use of descriptive statistics. Chapter 11 provides guidance on controlled study using correlation and inferential statistics.
- Evaluate using actual operational versions of the equipment, if possible. The difference in subject performance while using an initial prototype and the release version of the products can be significant (Rubin 1994:29).

> *To a man with a hammer, every problem looks like a nail.*
> *(Anonymous)*

5.4 How to choose methods

The chapters in Part II each describe chapters that can be used to perform human factors research. Each method in Part II has its own unique value. But which is the right tool to use? When is the time to use it? Table 5.2 lists the major human factors issues questions and the methods that are used to answer them.

The commitment of capital and operating costs to a new product or service requires confidence and support of the organization (stakeholders). To manage that process, research and development is broken into phases. Each phase accounts for a segment of effort that begins with a brief to launch the effort and concludes with a review to determine whether key information has been gathered. These so-called reality checks question whether the problem is understood and whether the solution concept has a high fidelity with reality. If reviewers are satisfied, confidence in the project grows. If they are not, either the team can be directed to perform further work or the project can be terminated. Table 5.3 portrays the development process, showing how both research and communications activity maps across each phase of activity. Methods are used to collect pertinent information in each phase. That information also serves as inputs for further definition and specification as the concept evolves in subsequent phases.

Well-considered research selects multiple methods to get at answers from more than one angle. This process of triangulation uses a number of methods together, as no single method can provide all of the knowledge that needs to be known. For example, the use of activity analysis might show that equipment operators perform certain actions such as fumbling for the correct control that could lead to accidents. The investigator might use critical incident study to determine whether this practice actually has been associated with near-accidents or accidents. A survey could be employed to find out how operators view the problem and what might be done to correct it. Link analysis could be used to demonstrate patterns of use that might be made easier or more efficient by arranging components differently. Simulation and usability assessment could be used to assess the flow and timing of control use to determine whether the new arrangement actually improves operator performance.

Table 5.2 Choosing methods—questions that a development team will have can be answered by one or more method. Both analysis and evaluation methods can be used to examine existing problems. Design guidance methods are used to configure a new concept.

Need to know	Method/section in this book
	Analysis methods (observational study)
What part of the problem deserves attention first?	Pareto analysis (6.1)
What purpose and needs will the solution serve?	Operational analysis (6.2)
How have others dealt with this?	Analysis of similar systems (6.3)
What do people actually do? How often?	Activity analysis (6.4)
How do people understand, use the product?	Verbal protocol analysis (6.5)
What is causing/may cause difficulties? Losses?	Critical incident study (6.6)
What threats to safety exist?	Management oversight and risk tree analysis (6.7)
What errors do humans make?	Error analysis (6.8)
What is the source of problems that cause failure?	Root cause analysis (6.9)
	Design guidance methods (observational study)
What functions must occur? In what order?	Flow analysis (7.1)
What order and duration of events can be expected?	Time line analysis (7.2)
What elements/arrangement comprise the product?	Link analysis (7.3)
What element can best perform each function?	Function allocation (7.4)
What will people do?	Task description (7.5)
What tasks/tools will be needed?	Task analysis (7.6)
How do people solve problems and make decisions?	Cognitive task analysis (7.7)
Is what we expect people to do reasonable?	Technique for human error rate prediction (7.8)
How will tasks be assembled into individuals' work?	Work design (7.9)
How can we choose among concept options?	Decision analysis (7.10)
	Evaluation methods (evaluative study)
How will an intended concept appear or perform?	Simulation (8.1)
How do people use this product?	Usability assessment (Ch.10)
What conditions could contribute to an accident?	Fault tree analysis (8.2)
What could fail, and how could it be prevented?	Failure modes and effects analysis/hazard analysis (8.3)
Do events happen in the order they need to occur?	Operational sequence analysis (8.4)
Is the work design feasible, safe?	Workload assessment (8.5)
What are individual traits, responses?	Interview (9.2), self-administered questionnaire (9.3)
What is value in this situation? How is it used?	Cost-effectiveness analysis (Ch.12)
	Descriptive study
How well will users perform with this product?	Usability assessment (Ch.10)
	Controlled study
What causes specific effects? With what certainty?	Controlled experimentation (Ch.11)

5.4.1 Methods support development activities

Chapter 1 described the research and development process. Each phase of the process is intended to achieve a certain result. The nature of these results may be more general (such as defining a goal) or more specific (such as exploring a solution concept). Throughout the process, development team members seek to create a solution that meets or exceeds expectations. While notions about a possible solution are more general in earlier stages, continued activity yields greater insight into problem and solution as work progresses.

As the tools that each member of the development team uses to contribute to team activities, methods are changed in each phase as problem and solution knowledge become more refined and specific. Table 5.3 portrayed how human factors methods are employed through the

Table 5.3 Human factors methods in the development process—through each phase, human factors research methods are used to collect increasingly defined and more specific information about both the problem and the solution. Prototypes are created in each phase to make ever-closer approximations of a desirable solution. Each phase begins with a brief, and ends with a report and review to verify that all necessary questions have been addressed. Confidence in the problem understanding and in the solution is low at the outset and grows as work on the project continues. Different methods are used according to what needs to be learned about the problem or solution.

	Concept exploration and definition		Concept demonstration and validation	
	What is the problem/opportunity?		What is the solution?	
need/opportunity identified	problem defined	requirements established	solution concept options explored	concept selected
need/opportunity statement	research brief	research report	design brief	design report
Confidence high ∧				

		Flow analysis (7.1)
		What functions need to occur, and in what order?
	Pareto analysis (6.1)	Time line analysis (7.2)
	What part of the problem deserves attention first?	How do the new procedures evolve through time?
	Operational analysis (6.2)	Link analysis (7.3)
	What should the solution include?	How are elements comprising the product related?
	Analysis of similar systems (6.3)	Functional allocation (7.4)
	How do current products/services compare?	Which element is best to perform each function?
	Verbal protocol analysis (6.5)	Task description (7.5)
	How do people perceive existing product?	What will people do in this solution concept?
	Activity analysis (6.4)	Task analysis (7.6)
	What do people actually do, and how frequently?	How will each person's job be designed?
	Critical incident study (6.6)	Simulation (8.1)
	What is causing/may cause safety problems?	How does the new idea look? How does it work?
	Management oversight risk tree analysis (6.7)	Technique for human reliability
	What might go wrong?	prediction (7.8)
	Error analysis (6.8)	Is what we expect people to do reasonable?
	What errors do humans make?	Work design (7.9)
	Root cause analysis (6.9)	How will tasks be assembled into work?
	What is the source of problems causing failure?	Decision analysis (7.10)
	Usability assessment (10)	How can we choose among concept options?
low	Is the product useful?	

early < Time

development process. Table 5.4 selects 'Phase I: Concept exploration and definition' from Table 5.3 to portray how human factors research activities are used to support the goals of a phase.

5.4.2 Introduction to Part II methods

Part II provides a toolbox of methods for human factors research in development that is drawn from a variety of sources, including work by Alphonse Chapanis (1959, 1996), Robert Bailey (1989), David Meister (1985, 1989, 1991), Wesley Woodson (1981) and Gabriel Salvendy (1997). Each method finds use in one or more areas of activity, including requirements determination, concept exploration, concept demonstration and validation, full-scale development, production and operation and support.

Engineering manufacturing and development		Production and deployment
Is the solution feasible?		Is the solution successful?
solution validated, optimized	*product/service prepared for production/release*	*product/service in use*
development brief	development report	customer service reports

		Cost-effectiveness analysis (12) *What is value in this situation? How is it used?* Verbal protocol analysis (6.5) *How do people understand, use the product?* Usability assessment (10) *Is the product useful?*
Verbal protocol analysis (6.5) *How do people understand, use the product?* Fault tree analysis (8.2) *What conditions could contribute to an accident?* Failure modes and effects analysis (8.3) *What could fail, and how could it be prevented?* Simulation (8.1) *How does the new idea look? How does it work?* Operational sequence analysis (8.4) *How do all elements work together through time?* Workload assessment (8.5) *Is the job design feasible, safe?* Usability validation (10) *What is the result of using product?* Controlled studies (11) *What causes specific effects? With what certainty?*		

Time > late

Table 5.4   Methods support activities—each development phase is intended to achieve a certain result (e.g. define a concept). Activities are repeated in each phase, yielding greater insight (e.g. 'problem as given' becomes 'problem as understood' in later phases). Methods are the tools that each specialty uses to contribute to team activities. Methods are changed in later phases, as problem and solution knowledge become more defined and specific. In the example below, both human factors and design team members have knowledge and methods that are unique to their practice. Both collaborate to perform the activities shown in order to explore and define concepts. Section numbers indicate portions of the book that describe activities and methods.

Phase	*Concept exploration and definition*	*Section*	*Design*		
Activities	Acceptance	4.3.2.1			
	Analysis	4.3.2.2			
	Problem definition	4.3.2.3			
	Ideation	4.3.2.4			
	Idea selection	4.3.2.5			
	Implementation	4.3.2.6			
	Evaluation	4.3.2.7			
Methods	Operational analysis	6.2	Visualization		
(human factors)	Analysis of similar systems	6.3	•	Diagram	
	Activity analysis	6.4	•	Sketch	
	Verbal protocol analysis	6.5	•	Rough model	
	Critical incident study	6.6	•	Prototype	
	Mgt. oversight risk tree analy.	6.7			
	Error analysis	6.8			
	Root cause analysis	6.9			
	Simulation	8.1			
Knowledge	Physiology	2.4.2	Materials		
	Behavior	2.1	Manufacturing processes		

5.4.1.1 OBSERVATION METHODS

Chapters 6, 7, 8 and 9 provide a selection of methods that can be used to analyze, design and evaluate development work.

Chapter 6: Analysis methods are used to develop a logical structure for new solution concepts. They can also be used to revise aspects of an existing product or service or to discover possible sources of problems.

Chapter 7: Design guidance methods are used to establish what the new concept can and must be, what functions will be assigned to what elements and what tasks humans will perform. They are used to analyze existing systems or products and to model design alternatives.

Chapter 8: Evaluation methods are used to verify whether one solution concept among many alternatives will be preferable. They can also be used to investigate an existing product or service in order to discover improvement opportunities.

Chapter 9: Interviews and questionnaires elicit information from individuals and groups that may not be available through observation either in the field or in controlled settings.

Chapter 10: Usability assessment is used to evaluate the fit between task and product, as well as to discover opportunities for product improvements.

5.4.1.2 CONTROLLED STUDIES

Chapter 11: Controlled studies describes the particular needs and circumstances to perform evaluation in a setting that is free from outside influences.

5.5 *Summary*

Even though guidelines and checklists are available, research development team members have to collect their own information. The greatest resources to do that are research methods. Knowing which method to choose and why, and how to use the method provides the structure that is necessary for a well-considered research effort.

Scientific explanations are the result of careful observation, rigorous testing and are designed to provide the broadest possible explanation over a variety of situations. As a result, scientific explanations tend to be more valid and general than explanations that are provided by common sense. The scientific method has evolved as the accepted practice to develop scientific explanations. The method is used to observe a phenomenon, form tentative explanations or statements of cause and effect, further observe and/or experiment to rule out alternative explanations and refine and retest the explanations.

Four kinds of studies comprise human factors research: observational, evaluative, descriptive and inferential. Observation is the practice of studying oneself, others, things, events and phenomena as they are encountered. Methods in Chapter 6, 7 and 9 are observational methods. Evaluative studies observe the collective effect of variables on human performance. Chapters 8 and 9 include evaluation methods. The usability assessment exploratory test in Chapter 10 and cost-effectiveness analysis in Chapter 12 are also evaluation methods. Descriptive studies account for the characteristics of those who are subjects of research as well as the results of their performance. The assessment and validation usability studies described in Chapter 10 are descriptive studies. Chapter 10 covers descriptive statistics in the context of usability studies. Chapter 11 provides a basic approach to controlled study and the inferential statistics that are used to analyze them.

Each method finds use in one or more areas of activity, including requirements determination, concept exploration, concept demonstration and validation, full-scale development, production and operation and support.

Part II

Human factors methods

6 Analysis methods

What part of the problem deserves attention first?
Pareto analysis (6.1)

What is expected to be accomplished?
Operational analysis (6.2)

How have others dealt with this problem?
Analysis of similar systems (6.3)

What do people actually do and how often?
Activity analysis (6.4)

How does a user perceive and react to the use of a product?
Verbal protocol analysis ('thinking aloud,' 'directed dialog') (6.5)

What is causing difficulties? Losses? Why?
Critical incident study (6.6)

What threats to safety exist?
Management oversight and risk tree analysis (6.7)

What errors do humans make and how might they be rectified?
Error analysis (6.8)

What is the source of problems that cause error and failure?
Root cause analysis (6.9)

This chapter provides a set of methods for use in getting at the nature of a problem. Each builds the knowledge necessary to make decisions about what the product should be.

By their nature, systems are not easy to perceive. Parts of systems are often in many locations. Some parts, such as software, have no physical counterpart. As a result, it is necessary to embody the system in some form that can represent its structure and character. The methods in this chapter make that possible.

Each section in this chapter follows the same format so that information is easy to find and understand: introduction, preparation, materials and equipment, process, results and example.

- *Introduction*—Explains how the method was developed and different approaches that have been taken to perform it.
- *Preparation*—Describes what needs to be done before performing the method.
- *Materials and equipment*—Accounts for physical items that are necessary including materials and equipment.
- *Process*—Enumerates the steps that are involved in performing the method.
- *Results*—Describes the outcome that performing the method produces.
- *Example*—Leads the reader through a brief discussion in order to illustrate the method's use.

The methods that are included in Chapter 6 are used to create context. Context includes the information that is inherent in an existing product and how it is used. Context can also include the situation and requirements that bring a new product into existence. In either case, the information that is collected through the use of analysis methods can be used to develop a solution model by employing the methods that are provided in Chapter 7.

6.1 Pareto analysis

Determine which aspects of a problem deserve attention first.

Nearly all situations involve multiple aspects that will need attention. For example, a safety program for rail transportation stations can encompass shelters, walkways, auto traffic and parking areas, platforms, guard devices, lighting, surveillance systems, notification systems and much more. Where does a team start? It is not unusual for research and development teams to engage the parts of a problem that are most familiar, the easiest to recognize, or the closest to the team's skills. Yet, parts that are most familiar or comfortable are not necessarily the most important. Pareto analysis can be used to determine which aspect of a problem will most benefit the solution if it is improved.

Based on the work of Italian sociologist and economist Vilfredo Pareto (1848–1923), the Pareto principle contends that '80% of the trouble comes from 20% of the problems.' Fogler and LeBlanc (1995:91–104) point out that Pareto analysis 'shows the *relative* importance of each individual problem to the other problems in the situation' (emphasis in original). Because of this, Pareto analysis can be used to make research and development more efficient. Addressing the issues that matter most will help to make the development effort effective. Because Pareto analysis identifies those portions of problems that are most significant, the method is best used at the beginning of the project. This ensures that efforts will be directed at what is most necessary to support project objectives.

6.1.1 Preparation

Step 1—Collect information on each major portion of the problem. Each individual and department who has an interest in the outcome is a potential source of information about it, although they will likely seek to influence the outcome. Stakeholders are a rich (although not necessarily objective) resource. Through Pareto analysis, the team will be able to consider all points of view and compare each aspect of the problem objectively.

If the team is working on a revision for an existing product, break the solution apart and classify it into constituent parts (e.g. assemblies, sub-assemblies) by using the method of system description in Chapter 1.

Step 2—Identify which aspect of the assignment poses difficulty or may be a source of dysfunction. If that knowledge is not available, performing a critical incident study (Section 6.6) may be a way to obtain it.

Step 3—If the system does not exist yet, performing an operational analysis (Section 6.2) may enable the team to identify requirements for the product that can be prioritized.

6.1.2 Materials, equipment and environment

No particular materials or equipment are required for this method.

6.1.3 Procedure

The procedure for Pareto analysis is summarized in Table 6.1.

Step 1—Identify what aspects of the problem cause loss. Problem aspects can include operating cost, number of accidents, number of fatalities, lost work days, cost to repair, user complaints, cost to produce, number of repairs needed, cost of rework, inspection rejects or fines. 'Time to complete' can be used to identify which problem aspect needs immediate attention because it will take the longest amount of time to finish. Costs or revenue are often one of the most significant considerations. 'Lost revenue' is frequently used.

Step 2—Create a bar graph as shown in Figure 6.1 that will be used to compare each aspect of the problem. Quantify each aspect in terms of dollar cost to provide a common basis for comparison. Company records can provide such cost-related information. Cost information is also available for losses other than operating costs. For example, 'number of lost work days' can be quantified in terms of lost productivity as well as the cost of health care treatment insurance claims. The cost of a fatality has been established by case law and includes costs such as loss of future earnings and loss to family.

Step 3—Plot the cost of each problem aspect that causes loss on the bar graph.

Step 4—Compare each problem aspect. Identify the aspect that causes the greatest amount of loss. That aspect presents the greatest obstacle to achieving product or system objectives and deserves immediate attention to correct it.

6.1.4 Result

The method identifies which aspect of a problem will require the organization's immediate attention. It enables the team to set priorities according to the organization's objectives.

6.1.5 Example

The customer service department of any company is a ready source of information on problems with products that are in use. Customer service representatives are trained to thoroughly collect accurate information on customer problems with products and to maintain records of it. Those records will

Table 6.1 Pareto analysis—determine which aspects of a problem deserve attention first. Figure 6.1 provides an illustration of Pareto analysis.

Identify problem elements
Select problem elements that impose a loss on the organization (e.g. 'lost revenue' for a software producer)

Create a bar graph
Identify criteria that are of greatest import to problem
Plot the value of problem elements according to the loss that each causes

Select problem element
Determine which of the elements imposes the greatest burden on the organization
Select most significant problem to address

6.1 Pareto analysis

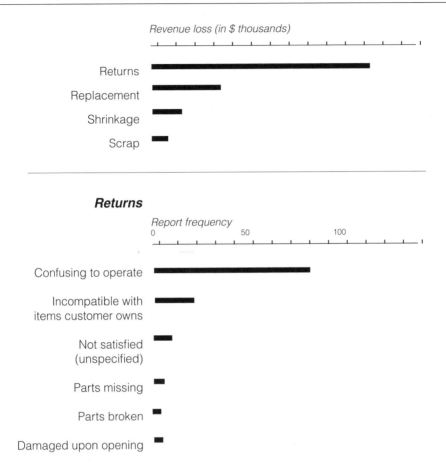

Figure 6.1 Pareto analysis comparison—Pareto analysis can be used to identify problem elements according to the loss that each causes. Attention can then focus on significant problem areas. In the example below, the upper bar graph shows that customer returns are a significant factor in lost revenues. A review of customer records reveals six categories of reasons for returns which are plotted on the lower bar graph. The graph shows that customer confusion over installation and operation is a key cause of the returns problem. Development team attention can be directed to learning more about what confuses customers as well as how the product and its documentation can be improved in order to minimize customer confusion.

reveal difficulties such as manufacturing and material problems. They may also reveal difficulties that customers have with product use and performance.

For example, a research and development team that is assigned to create a new generation product may intend to implement a new process or technology. The team might check customer service records while performing initial research into possible problems with the existing product. Review of customer service records might reveal that product returns by customers are causing the greatest loss of revenue. Figure 6.1 shows the relative loses that each aspect of the problem incurs. The costs represented by the 'returns' category can be broken down according to the reasons for returns. That comparison reveals that the foremost cause is 'confusing to operate.' User confusion is a problem that research and development can address by looking in the nature of the complaint, (i.e.

why was it confusing?), its frequency (i.e. number of times the complaint occurred), its composition (who was having the problem?) and its final outcome (e.g. 'refund issued').

The comparison in the bar graph leads the investigator to ask further questions. Why was customer dissatisfaction so high? Was it related to issues of user behavior or physical ability? If it was, what could be changed in order to get a different outcome for the user? Answers can be pursued using other methods in this text such as interviews and questionnaires.

6.2 Operational analysis ('mission analysis')

Discover what is currently, or is expected to be, accomplished.

Operational analysis (OA), also known as mission analysis or systems requirement analysis, is used to create profiles and scenarios to obtain information on operational situations and events that users, operators and maintainers will experience. Meister (1985:16–9, 163–71) considers operational analysis 'a prerequisite to determine system functions, particularly in systems like aircraft which have very distinct sequential phases in their operations such as take-off and landing.' OA also discovers 'important behavioral implications of requirements, such as environmental conditions.'

Operational analysis can be used to explore the most abstract aspects of products. For that reason, it is often used early in the development process. The use of OA answers the question 'why should this product exist?' The statement of purpose that results defines the product's scope and purpose. Operational analysis can be used in two ways: to either create a new product that does not exist or build a rationale for a product that already exists but does not have an expressed purpose.

Why perform operational analysis? Its use places design attention on the fit between need and solution and among solution elements. Woodson (1981:910–22) points out that

> Although an emphasis on primary users may be important during the conceptual phase of system development, the ultimate success of the system operation invariably depends on the effective performance of all subsystems and components, new or old, including all people who may be involved, directly or indirectly.

That also includes the cost of operation and maintenance that may be overlooked during development. Failure to use the method can result in products that are made without regard for the evolution of new needs, for the use of new technology, or for coordination with other products.

Beevis (1999:38–47) describes two types of OA: narrative and graphic. Narrative OA employs 'a written or point form of discussion' while graphic OA provides 'the mission information in graphic form.' The latter approach is better suited to top-down analyses as detailed diagrams that depict multiple events can become too complex to understand.

Use of OA lays the foundation for particular design guidance methods in Chapter 7 including flow analysis (Section 7.1), function allocation (Section 7.4), task description (Section 7.5) and task analysis (Section 7.6). Like Pareto analysis, operational analysis is most often used at the beginning of a project.

Operational analysis is one means to gauge the potential effect of paradigm shifts that were mentioned in Section 4.2.4. Much attention in research and development is focused on the product. By contrast OA focuses attention on the context; that is, considerations other than the product. By looking beyond the product to other concerns, the research and development team can better see how valid their assumptions are with regard to the needs and nature of technology application.

Beevis (1999:40–1) contends that 'the advice of experts with operational experience of similar missions and systems is essential' to OA, which can be too subjective if the amount of data that is brought to bear is too limited.

6.2.1 Preparation

Collect information on current users, equipment and procedures. Information of interest in operational analysis includes personnel skills, education, physique, training and experience as well as hardware size, power, controls, displays and procedural operational sequences and interactions.

Interview others who have experience in the same field or have experience with an existing or similar system. Ask about what is done, how and why. This audit of decision and reasoning behind it has been referred to as naturalistic decision making, or walkthroughs. The process enables the investigator to learn about latent hazards before loss occurs in the same way as critical incident study (Section 6.6) and management oversight and risk tree analysis (Section 6.7). Cognitive task analysis (Section 7.7) probes further into how people solve problems and make decisions.

6.2.2 Materials, equipment and environment

Account for any materials such as client memoranda, requests for proposals (RFPs), bid invitations and specifications and statements of requirements that spell out what the product will be expected to do.

6.2.3 Procedure

The procedure for operational analysis is summarized in Table 6.2.

Step 1—Review the collected information for descriptions of goals, functions, performance parameters, inputs, outputs, operating conditions and limitations.

Step 2—Describe what the product will do in a single statement that summarizes the purpose of the product. Why should the product exist? What is its purpose?

Table 6.2 Operational analysis—discover what is currently, or is expected to be, accomplished.

Audit requirements
Review the collected information for descriptions of goals, functions, performance parameters, inputs, outputs, operating conditions, and limitations.
Describe purpose
Describe what the product will do in a single statement that summarizes the purpose of the product. Why should the product exist?
Describe what the product will do in statements that describe each subordinate function that will be necessary to accomplish the product's purpose. Why should the component parts exist?
Assess potential problems
Anticipate and enumerate failures and emergencies that may occur. Estimate their severity, and the potential each has to affect how well the product will operate.
Determine requirements for persons
Describe those people who are expected to use the product and any general traits that are known about them.

Step 3—Describe what the product will do in statements that describe each subordinate function that will be necessary to accomplish the product's purpose. Why should the component parts exist? What is its purpose?

Step 4—Anticipate and enumerate failures and emergencies that may occur. This includes actions or failures to act by people and interactions among people. Estimate their severity and the potential each has to affect how well the product will operate.

Step 5—Describe those people who are expected to operate and maintain the product and any general traits that are known about them.

6.2.4 Results

The operational analysis method spells out:

- The context in which the product is expected to operate, including what is expected to occur and the influences that are expected from other products and the total environment;
- What the product will do, including specific performance parameters or specifications that have been set by the client;
- What the product and its constituent parts can and cannot do;
- Potential causes for degraded performance or failure and what might be done about it.

It can be effectively used to create a diagram or an artist's concept rendering in order to depict the solution scenario that operational analysis creates. An operational sequence diagram can also help. The use of operational sequence analysis (Section 8.4) produces such a diagram.

6.2.5 Example

Even though health care is an essential need for all, certain types of health care are less available in remote, underdeveloped areas. A system of mobile health care units, operated by governmental health agencies, non-governmental organizations (NGO), the military and private health care firms could bring needed medical care to remote sites. Such units might also be concentrated in a specific site in the event of a natural disaster. Operational analysis would be the first step to develop such a mobile care system.

1 BACKGROUND

- Goals—The restoration to health of largest number of patients possible. The prevention of threats to public health.
- Functions—Respond to natural disasters, epidemics or general health needs. Augment locally available health care professionals. Provide specialty equipment beyond that which is typical for basic medical clinics.
- Performance parameters—Use procedures that are in keeping with the requirements of local medical authorities. Keep the cost of procedures within operating limits of the agency that funds the unit's deployment.
- Inputs—Medications, medical supplies.
- Outputs—Prescriptions, treatment regimens.
- Operating conditions—Environmental conditions will vary according to area of deployment. The potential range of environmental conditions can vary from tropical to arctic.

- Limitations—The need for mobility will make it difficult to use heavy, sensitive, bulky equipment. The solution will not be able to remain permanently on site to conduct ongoing care.

2 PURPOSE

Provide professional quality health care to under-served populations in remote areas, including disaster relief.

3 SUBORDINATE FUNCTIONS

Four subordinate functions will fulfill the system's purpose.

- Diagnosis/triage—Determine history, condition of care recipient. Determine and prescribe needed care, when and how care can be provided.
- Treatment—Provide care appropriate to current visit.
- Documentation—Develop, maintain and retrieve diagnosis and treatment records.
- Transition—Enable local health or social organizations to continue attention to needs after unit is relocated through follow-on care.

4 FAILURES AND EMERGENCIES

Five failures can affect the system. The first would render care unavailable, making the impact on the mission severe. For the remaining four, initial diagnosis and documentation can be provided but treatment may be inadequate. That would have a moderate impact on the mission.

- Unit unable to reach location
- Type of care that is required is unavailable
- Medication unavailable or corrupted
- Sterile procedure violated
- Information unavailable

5 PEOPLE

- Mobile unit care providers—Physician or physician's assistant, primary care nurse. Provide diagnoses, treatment. Provide instruction to local care provider in current procedures and follow-on care for those who have visited the unit for care.
- Local care provider—Manage preparation, documentation in advance of visit and follow-on care after visit.
- Care recipients (patients)—Those in need of care.
- Support—Central supply, vehicle maintenance and administrative staff assist to the care providers.

The development team has the opportunity to create a range of solutions to meet these initial requirements. Pedestrian-carried back packs, bicycle or tricycle units, motorcycles, three and four wheel all-terrain cycles, autos, all-terrain vehicles, vans, trailers, helicopters and delivery by airborne cargo parachute drop are all potential means of transport. Solutions may range from general care 'fly away' kits mounted on any vehicle to specialized vehicles that are used to provide treatment specific to installed equipment. Robust high bandwidth communications might be used for consultation with specialists as well as for transmission and retrieval of patient records and medical information.

6.3 Analysis of similar systems

Discover how others have addressed a related problem.

Analysis of similar systems (AoSS) is used to discover the most significant features of other products or systems that are similar to a current product or one that is being developed. All traits of interest among existing comparable systems are compared side-by-side. Which traits matter? They are the traits that have the most direct implications for meeting system objectives. For example, in a kitchen appliance such as a microwave oven, the controls and display have a significant effect on the product's use. Differences in controls and displays and their configuration (some of them quite subtle) can affect criteria such as 'ease of use.' Through AoSS, the investigator methodically explores each feature that has implications for his or her own area of interest.

An in-depth review of each product's traits of interest is shown in a comparison chart. This creates an analysis that is both broad (according to the number of products reviewed) and deep (according to the number of features compared). The use of AoSS provides insight into what works successfully and what does not. Thorough comparison can yield insights as to why one item may be better (e.g. easier to operate) than other items.

The method can be used to explore new areas of interest by analyzing products that are similar in *function*, as well as those that are similar in traditional categories. For example, a team may be at work on the development of a new remote control device for surgical procedures. Even though remote manipulators for use in handling hazardous materials are not medical products, such devices may be very informative when they are considered among other remote control devices.

AoSS can be used at any time during the development process, but is most often used early in development.

AoSS can be performed using a comprehensive set of traits or a single significant trait. For example, software product user interfaces can be surveyed with regard to their information structure and decision trees, control mechanisms and display characteristics. On the other hand, AoSS could study a single significant trait such as display type font, size, color, spacing and leading in the interest of examining legibility.

6.3.1 Preparation

Systems or products under consideration can be reviewed from the viewpoint of either performance or physical attributes. Chapanis (1996:86) recommends consulting information sources to learn how other products or systems do or do not perform.

Collect records of productivity, maintenance, training and accident or incident reports and investigations in order to look for information on product operability and maintainability. Records can also reveal information on the numbers of people, skills and training that are required to operate the product and on the problems that are encountered in design and in use. If records are not available, consult those who may have personal experience with other systems. These may include operators, maintainers, instructors, or recent students of similar systems.

To compare physical features, identify the products that are similar to the one under consideration. For example, if the project involves a new generation music entertainment device that uses disk storage, include portable CD-ROM players in the review sample. Examine aspects of the products that have implications for the new product under development. For example, there may be some concern over the method of disk retrieval, insertion, activation, retrieval and return to storage. How does the design of each unit accommodate the process? How do devices other than personal CD-ROM players that rely on disks accommodate the same activity?

6.3.2 Materials and equipment

It is important to have as much access as possible to the products to be compared. Direct contact makes it possible to experience characteristics such as method of operation, feel and more.

6.3.3 Procedure

The procedure for analysis of similar systems is summarized in Table 6.3.

Step 1—Review activity analysis (Section 6.4), critical incident studies (Section 6.6) and accident investigations in order to determine which aspects of the systems under consideration are significant.

Step 2—Collect information on each candidate product or system according to the aspects that have implications for the new product. Note how each product varies. Pay attention to differences that may exist among user groups for each product.

Step 3—Create a format that includes each category of interest. Sort pertinent information for each product or system that is considered into each category.

Step 4—Review results from the comparison.

Step 5—Develop conclusions with regard to desirable or undesirable attributes and their implications for the product under development.

When performing the AoSS, note how the features of each product affect its performance. Are terms understandable or ambiguous? Do some parts of a complex assembly work well while others do not? Account for both strengths and weaknesses. Note how each feature contributes to or might detract from the way a product performs. Ensure that aspects of products that are compared actually have an effect on performance. This will help to avoid mistaking the inconsequential for the consequential.

Table 6.3 Analysis of similar systems—discover how others have addressed a related problem.

Develop categories of interest
Determine which aspects of the product or system under consideration are significant.
Derive relevant information
Review activity analysis, critical incident studies, and accident investigations.
Collect information
Review each candidate product or system according to the aspects that have import for the new product.
Compare results
Review each aspect (e.g. control type) and compare how each product varies according to it.
Develop conclusions
Look for patterns of similar traits, and for distinctive or unique examples within the examples.
Reflect on the success or failure of the feature for that particular product and implications for the product under development.

6.3.4 Results

Chapanis outlines five results from the use of AoSS:

- Identifies operator and maintainer problems to be avoided
- Assesses desirability and consequences of reallocating system functions
- Identifies environmental traits that affect personnel
- Assesses work load and stress levels, present skills and the effect of training on design
- Estimates future manpower requirements.

A table can be helpful in order to compare each aspect among all of the products and systems that are considered. Such a side-by-side comparison can make it easier to detect patterns that lead to insights with regard to similarities and differences.

6.3.5 Example

AoSS can be used in the design of a new portable educational and entertainment device to download audio through the Internet and play it. A team assigned to such a development project would begin by searching for current personal electronic products that can be carried and connected to data sources. Items that fit that description might include compact disk players, cassette tape players, still image cameras and video cameras.

The team would then follow these steps. Observe and analyze patterns of how individuals use these products noting the effect that wear and user errors have on each product. Examine each product for information that is of particular interest to the user interface: the kind, amount, size and location of symbols and typography on the display and body; the kind, amount, shape and number of controls. Note the method of operation for each product.

Create a table to compare significant product features. In this instance, the features may be size, weight, steps to operate, display symbols, display labels, control buttons and switches. Analyze each product for information that fits the categories of interest. Review the information to extract patterns from your observations. What are the strengths and drawbacks of each product? What implications does the information have for the new product? What ideas does the team have that might overcome the other products' drawbacks (e.g. inadvertent control activation, confusing or ambiguous labels or displays, clumsy handling) or to add value (e.g. simplified control operation, compact form to make it better for carrying on the person). List features to include and to avoid to guide the design of the new product.

6.4 Activity analysis

Discover and quantify what those who are working actually do.

Activity analysis (AA) is used to account for anything that a person does while performing a task. The purpose of AA, which Chapanis and Meister term activity sampling, is to determine the activities that occur in any situation, as well as how often the activities are performed. AA is of particular help in situations in which operators perform a variety of tasks but the order of those tasks is not predetermined. AA is most useful during early phases of development.

The investigator who performs AA acts in the role of objective observer, free from judgment or interpretation. In a sense, he or she serves simply as a camera, noting what is seen at specified intervals. Activities must be identifiable by distinguishing cues. Such cues include flipping a switch, moving a lever, or pressing a button. Observations are made at timed intervals. For that reason, the investigator is required to rely on the use of a timer for consistency. Observations can be made at regular (periodic) or irregular (random) intervals.

Morgan *et al.* (1963:8) find that AA 'is particularly useful for judging the relative importance of various operator activities and the demands they make on the operator's time and effort.' AA also helps in 'revamping operational procedures, redesigning equipment, unburdening operators and redistributing duties and staff responsibilities.' As a result, operators are assigned the type and amount of tasks that are best suited to their abilities. Work is safer, more reliable and more efficient.

Two issues can affect the use of AA: simultaneity and task conflict.

Simultaneity—Recording a single discrete task at each observation can overlook the fact that other events may also occur at the same time.

Task conflict—The method's singular focus on one task that is being performed can cause the investigator to overlook other activities that may conflict with it.

Chapanis (1996:87) points out that the use of AA documents how tasks change with alternative designs, verifies manning requirements, clarifies how operators spend their time and assesses workloads and stress levels, as well as the desirability and consequences of system function reallocation. Chapanis' (1959:27–39) variation on activity analysis, 'Activity Sampling on a Unit Basis,' can be used to follow a specific item through a process (e.g. the production of camera ready advertising artwork).

Flow analysis (Section 7.1) and time line analysis (Section 7.2) are other methods in this book that are also used to explore the implications of procedures and time.

6.4.1 Preparation

6.4.1.1 Learn the role in order to understand the reason for activities that are expected to be observed.

6.4.1.2 Develop a form similar to Figure 6.2, listing the activities that are expected. Allow sufficient space to add hash marks that indicate how often the activity occurs. Chapanis recommends specifying about twenty-five or fewer different activities. Allow room on the form for the observer to note unexpected activities that occur.

6.4.1.3 Obtain the cooperation of those who have a say in the work situation. Get management, supervisor and union steward approvals in advance. Ensure the subject understands that the purpose of the study is to improve conditions or products and not to evaluate his or her personal performance.

6.4.1.4 If direct observation is not possible or appropriate, arrange in advance to obtain or make a photographic or video recording of the activity for later viewing and analysis.

6.4.2 Materials and equipment

Still photo or video 'stop action' recording can be a valuable aid to activity analysis. Use of a camera frees the investigator to pay more attention to other aspects of the method such as activity classification. While initially aware of the camera, workers eventually become accustomed to it and tend to

Activity:	TAXI DRIVER — COLUMBIA TAXI		
Driver:	B. PHILLIPS		
Observer:	P. RENAUD		
Date: AUG. 15, 2001	Time: 0710		
Shift: MORNING	Sampling interval: 15 SEC.		
Comments: HEAVY DOWNTOWN TRAFFIC			

Activity	Tally	Sum	Percent of Total
Driving	ЦНt ШН ШН ШН II	22	31
Calling dispatcher	ШН II	7	10
Start/chg/ending fare	ШН III	8	11
Payment transaction	IIII	4	6
Using map	I	1	1
Checking time	IIII	4	6
Waiting	ШН ШН I	11	16
Eating	II	2	3
Transition	ШН I	6	9
LOADING LUGGAGE	III	3	4
CONVERSATION	II	2	3
Total		70	100

Figure 6.2 Activity analysis record—activity analysis relies on the use of a form to list up to twenty-five activities that the investigator expects to observe. Space is allowed for the observer to add a hash mark to indicate how often each occurs. Allow room on the form for the observer to note unexpected activities that occur. The example shows how a form would be organized to sample the activities of a commercial driver in order to develop a transaction console. In this instance, a taxi driver's activities are observed over a standard work shift. Multiple observation sessions among a diverse sample of drivers, driving conditions, and types of driving (taxis, vans, busses, etc.) would provide information on the activities a transaction console might support.

disregard it. More than one camera can be employed to capture more information from different views. Cameras can be set to take images at precise intervals (e.g. every 5 seconds or every 15 seconds). Their images can be used after the fact to serve as an accurate record of the activity. This is particularly helpful when an expert is available to review the images. Interpreting the images afterward takes additional time that manual methods do not require.

The use of a timing device is essential for AA. A watch or clock with an elapsed time function that can be easily changed is helpful, particularly when doing random observation. When using video recording equipment to create a record of the observation for later reference, include a digital display in the field of view to show elapsed time.

To conduct random sampling, use a random number table to ensure that observer habit does not bias the frequency of observation.[1] Follow the series of numbers the table provides as the length of time to wait between samples.

6.4.3 Procedure

The procedure for activity analysis is summarized in Table 6.4.

Step 1—Observe the subject, usually an operator, as he or she performs normal work. Observe the work over a long enough time so that it can be considered to be representative of the job. Chapanis (1959:27–39) recommends that 'the sampling interval should not be longer than the shortest unit of activity' for a complete description of an individual's activity. To ensure data are reliable and valid, make roughly 1,000 or more observations.

Step 2—Start the timer and, when the time elapses, note what the subject is doing at that exact moment. Be careful not to disregard what is observed. Activities such as waiting and checking documentation or calling a colleague to figure out what to do are valuable observations even though they are not included in a job description. Make a hash mark on the line of the activity that is observed, or write in another activity if it was unexpected.

After the observation session:

Step 3—Add the total number of observations for each task and figure the percentage how frequently the activity was observed compared to all of the other observations. For example, in Figure 6.2, eleven observations of 'waiting' activity produced an estimate of 16 percent of the time that the driver spent in this activity.

Step 4—Create a frequency chart (e.g. bar graph) that graphically demonstrates the activities and their relative frequency of observance. Individual activities are aggregated into activity frequency tables and graphs that show the frequency of each activity and the percentage of time spent in them,

Table 6.4 Activity analysis—discover and quantify what those who are working do. Figures 6.2 and 6.3 provide an example of an activity analysis record and summary.

Have the subject (usually an operator) perform his or her normal work.

Observe the work

Observe over a long enough time that it can be considered to be representative of the operator's job.

Start the timer.

When the time elapses, note what the subject is doing at that exact moment.

Make a hash mark on the line of the activity that is observed, or write in another activity if it was unexpected.

Analyze the observations

After the observation session:

Add the total number of observations for each task.

Figure the percentage how frequently the activity was observed compared to all of the other observations.

Depict the analysis

Create a frequency chart (e.g. bar graph) that graphically demonstrates the activities and their relative frequency of observance.

Aggregate individual activities into activity frequency tables and graphs that show the frequency of each activity and the percentage of time spent in them.

Develop state diagrams to show the probabilities of activities following other activities.

state diagrams showing the probabilities of activities following other activities. Figure 6.3 shows the graphical representation of Figure 6.2 frequency percentages.

Step 5—Review and assess the results. Are there types of activity that occur which were not expected? Does the equipment and training provided support the observed activity? Is there an opportunity to adjust the role (e.g. job enrichment)?

6.4.4 Result

Three results can be obtained from AA. The frequency record provides an estimate of the percentage of a worker's time that is spent on each observed activity. Analysis and compilation of recorded observations can provide information on the average length of time spent on the activity as well as the sequence in which the worker performs the parts of his or her job. The method can also discover activities that occur even though they are not programmed (Chapanis 1959:27–39).

6.4.5 Example

Many work roles center around display and control consoles including bus and taxi drivers, airline flight crews, rail dispatchers and air traffic controllers. Changes to procedures often adjust the character of the job. Yet, unless attention is directed to update the physical environment, equipment, tools and surroundings, they can become an impediment to performance. AA can be used to determine changes that are needed to update procedures, equipment and facilities.

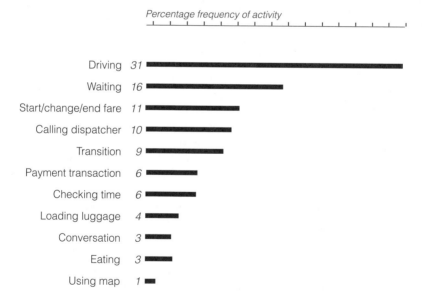

Figure 6.3 Activity analysis summary—Figure 6.2 shows a taxi driver's activities observed over a standard work shift. The bar graph above translates the percentages into a display that represents the proportionate amount of time the driver spent in each activity. The type and frequency of activities can be used to inform the development of a driver transaction console. The frequencies shown indicate map review may not be as important as the development team thought it was. Further review might reveal that this observation was conducted in downtown traffic and well-known destinations and that maps are consulted more frequently on most other routes.

Transit drivers of busses, vans and taxis perform a variety of similar tasks. Improving their performance can lead to safer, more efficient road transportation and can ultimately ease traffic congestion. To develop a console that assists drivers with the variety of activities they perform, an AA could be performed throughout a standard shift. Activities that would be anticipated include dispatch monitoring and response, route verification, suggestions on topics of interest to passengers (e.g. restaurants) and fare collection. Activities that might occur in addition may include informal conversation with other drivers, waiting and finding the way back after getting lost.

It was noted above that AA can miss tasks that are performed at the same time and tasks that conflict. For example, driving and calling a dispatcher or driving and using a map have significant safety implications. Recordings of activity performance may show situations that are hazardous. Rather than activity-based analysis methods, time-based methods are better suited to the analysis of simultaneous or conflicting tasks. Two such methods are time line analysis (Section 7.2) and operational sequence analysis (Section 8.4).

6.5 Verbal protocol analysis ('thinking aloud,' 'directed dialog')

Explore user mental processes regarding the product under evaluation.

Originally developed by Newell and Simon in 1972 for the purpose of problem solving, verbal protocol analysis (VPA) provides a window on the user's mental model of what is going on while performing an activity by saying any and all thoughts that occur. Also referred to as 'thinking aloud' and 'directed dialog,' VPA can be used in any instance that requires an understanding of how a person perceives, operates, or uses a product, system or space.

Moroney and Bittner (1995:361–438) note that VPA can be used to debrief subjects in order to learn about meaningful information related to workload assessment (e.g. strategy shifts that subjects used in order to maintain their performance during an event). VPA can be used in tandem with video (or other kinds of) playbacks up to weeks after an event in order to reveal strategies that subjects used. VPA can also be used as an approach for cognitive task analysis (Section 7.7) by using verbalization to discover how an individual thinks through a task.

Most recently, VPA has been employed in usability studies on graphical computer software interfaces, instructional materials and training materials. Lewis (1982) found that the method has both advantages and some limitations. The advantages of the method include:

- *Focus*—User hesitation and confusion can be used to identify what must be fixed.
- *Rationale*—Users can reflect on what caused difficulty at a number of levels.

The method's limitations include:

- *Accuracy*—User may not be aware of all mental processes, or find it easy to describe.
- *Timing*—Talking aloud may cause task performance to take longer than normal.
- *Labor-intensive*—The 'detailed record of complicated behavior' that the method produces can be difficult to summarize. Working with one participant at a time can be time-consuming.

Praetorius and Duncan (1988:293–314) echo Lewis' 'accuracy' comment by cautioning that verbal reports are a limited source of information, as 'only some of the knowledge we possess about our performance of a task may be readily accessible in verbal form.' Simply put, not everyone has a full grasp of why they do what they do and can clearly explain it. Subjects can be trained to be more informed about their thought processes. That training, though, imposes a time and cost burden on the research process.

Meister (1991:435–539) contends that the method has verbalization and free expression in common with interviews (see Section 9.2), yet is freer 'because verbal protocol is under no control other than that of the speaker himself.'

6.5.1 Preparation

Select item(s) that are to be used (e.g. product, manual, software program) during the procedure. Decide on what tasks the user is to perform using the item(s). Tasks should be significant enough to offer a meaningful role for the user to perform, yet fit within the amount of time available. Tasks that are trivial may cause the user to become disinterested. For more complicated activities, it may be necessary to prepare a sheet describing the tasks that are to be performed.

6.5.2 Materials, equipment and environment

Verbal protocol analysis can be used in either a controlled environment (e.g. office or laboratory) or in the field. Pencil and paper notation can be used to record observer notes. The observer can use a clock (either routine digital display or elapsed time display) to keep track of when events occur and how long the subject takes to perform activities.

Videotape recording is the preferred documentation method because it can be used later as a reference. The taped record also saves the observer from having to take voluminous notes while watching and asking questions. When setting up sessions that involve the use of a display (e.g. computer screen), place the video camera on a tripod and aim it over the user's shoulder at the screen. In that location, the camera can capture user comments, screen state, cursor movements and elapsed session time in one frame. Because of the need for data reduction, the observer should still make pencil and paper notation. Finding tape segments of interest can be made easier by correlating observer notes to session elapsed time on the clock.

Both Rubin (1994) and Lewis provide guidance to help observers to create an environment that elicits good results. They recommend: retain real users (not simply experts), remind the subject that it is the product and not the user that is being evaluated, allow user frustration (up to the point that the subject may quit) and remain neutral (in order to avoid coaching the subject).

6.5.3 Procedure

The procedure to perform verbal protocol analysis is summarized in Table 6.5. Verbal protocol analysis can be performed using either of two approaches: single subject, or co-discovery by two subjects.

6.5.3.1 SINGLE SUBJECT

An observer is on hand to guide the user and can perform in one of two roles: active intervention or neutral observer. In active intervention, which is helpful in early explorations, the observer actively questions subject on nature of, or reasons for, steps the user took during task performance. As a neutral observer, the investigator watches the subject perform tasks and intervenes only when the user cannot proceed without assistance. This second approach is most appropriate later in the development process when it is necessary to accurately measure task performance. In either case, the investigator needs to listen to the subjects with unconditional positive regard. Comments, utterances or gestures that imply anything other than interest and positive regard for what the subject says can inadvertently influence the observation.

Table 6.5 Verbal protocol analysis ('thinking aloud,' 'directed dialog')—explore user mental processes regarding the product under evaluation. Rubin (1994:217–19) provides two approaches to the evaluation of a product using verbal protocol analysis. The investigator can arrange for one individual to act as an observer, or arrange for a pair of subjects to work together.

As a sole observer

Choose method
　　Select either active intervention or neutral observer approach.

Ask the user to perform a task
　　Ask the user to speak aloud while using the product, materials, or system that is under evaluation. If using active intervention, query user regarding actions and thoughts.

Observe, record user performance and remarks
　　The constant stream of comments provides insight into the individual's reflections on his or her thought processes.

Using two subjects for co-discovery

Select subjects
　　Pair two test subjects to work together to perform tasks.

Ask the user to perform a task
　　Ask the subjects to speak aloud while using the product, materials, or system that is under evaluation.

Observe, record user performance and remarks
　　The constant stream of comments provides insight into the users' reflections on thought processes.

Step 1—Choose method. Select either active intervention or neutral observer approach

Step 2—Ask the user to perform a task and to speak aloud regarding what crosses his or her mind while using the product, materials, or system that is under evaluation. The task can be as simple as 'please show me how to operate this video camera' to more complex work. If using active intervention, ask the user about his or her actions and thoughts.

Step 3—Observe, record user performance and remarks. The constant stream of comments provides insight into the individual's reflections on his or her thought processes.

6.5.3.2 TWO SUBJECTS

Two test subjects can be paired to work together to perform tasks, in what is referred to as 'co-discovery.' The approach is best suited to early stage exploratory studies. It can also serve to moderate the discomfort that some subjects experience when asked to think aloud alone. As both subjects will interact with each other, the observer normally assumes a 'neutral observer' approach.

Step 1—Select subjects. Pair two test subjects to work together to perform tasks.

Step 2—Ask the users to perform a task and to speak aloud to each other regarding what occurs to them while using the product, materials, or system that is under evaluation.

Step 3—Observe, record user performance and remarks. As in the single subject approach, the constant stream of comments provides insight into the users' reflections on thought processes.

6.5.4 Result

The use of VPA demonstrates the approach and assumptions that an individual brings to a situation. It can reveal the process that a person follows while attempting to pursue an objective and can detect obstacles that users encounter in the use of a product. Study results can provide a revealing insight into how those who are intended to use a product actually perceive and use it.

6.5.5 Example

Most large transportation facilities present a challenge to any visitor who needs to find a particular location. The problem is compounded when the buildings at the site have been constructed at various times through the years. Architectural styles, traffic patterns and cues are mixed. Variations in cultures among visitors and employees of various organizations can make it even more of a challenge. Decisions must be made (sometimes quickly) regarding fare transactions, means of transportation and route selections. Signage and symbols that are meaningful for the institution but not meaningful for the visitor can add to the confusion.

VPA can be used to determine the needs and questions a visitor has when arriving at and traveling through the facility. Questions, observations, errors, discoveries, guesses and more become evident as a visitor talks aloud in a stream of consciousness commentary. Where is the visitor confident or reticent? Where is a decision made? On what information is the decision based? Is the decision correct?

The results of such an investigation can be used to develop an information system that might include signage, landmarks, printed material and information displays that would serve the needs of those who are unfamiliar with a complex facility. This is especially important in the public environment that is populated by a wide variety of users.

6.6 Critical incident study

Discover sources of dysfunction in existing systems.

Critical incident study (CIS) is used to look for the cause of human–product problems in order to minimize loss to person or property. CIS is useful when problems occur but their cause and severity are not known. These problems can be manifest as errors (lapse in the application of rules), mistakes (lapse in the application of knowledge), or accidents (an event that occurs in loss that may or may not be recoverable).

Loss is what makes events 'critical.' A gas build-up that causes an explosion and fire that destroys a warehouse containing costly inventory incurs significant property loss. A patient who went into surgery for the amputation of a diseased leg only to have the healthy leg removed suffers a loss to personal health and well being. Loss can cost money, livelihood and lives. CIS can be used to explore what might go wrong in order to avoid loss before it occurs. It can also be used to figure out what went wrong if a loss has already occurred. In this second instance, the method may be conducted when legal proceedings are contemplated or are underway.

In order to perform a CIS, the investigator looks for information on the performance of activities (e.g. tasks in the workplace) and associated implements (e.g. controls and displays). Operators and records can provide such information. The investigator may ask operators to recount their own experience or their recollection of others' experience while using a product. Records can also provide documentation of what has occurred. That can be of particular help when those who have first hand experience are not available.

Chapter 2 provided an overview of human capabilities and limitations. Both physiology (sensation, neuromotor ability) and psychology (perception and cognition) play significant roles in the assessment of critical incidents. CIS is used to focus on possible causes of human error due to *commission* or *omission*. Commission is an intentional act that a person or persons perform. Causes for commission can include misinterpretation of input, faulty reasoning (incorrect response) and an attempt to perform a required task but was not able to due to lack of skill. Omission is an act that was not performed due to oversight. Causes for omission can include missed input, failure to connect input with needed output, high workload or an inability to perform a task even though it was known output was needed.

Error can be thought of according to the model of human information processing that includes three stages described in Chapter 2: input (attention, sensation), mediation (perception, cognition, memory, thinking and learning) and output (response). Spurious causes can induce either commission or omission. For example, an accidental bump can cause an operator to commit the error of pressing the wrong control. A distraction can divert the operator's attention from the performance of a required task, causing its omission.

The occurrence of an error does not automatically result in loss. It does not necessarily result in an accident. What actually happens upon exposure to a hazardous situation depends on a series of five steps that Ramsey set forth in his 1977 accident sequence model:

- Perception of the hazard (based on sensory perception)
- Cognition of the hazard (based on information processing)
- A decision whether or not to avoid the hazard (based on attitude/behavior)
- The ability to avoid (based on biomechanics or motor skills)
- Luck (which can intervene regardless of the foregoing steps) (Christensen 1985b).

An individual may fail to sense, recognize, decide to or be able to avoid, a hazard. Even so, luck may still intervene and prevent an accident. Avoidance of loss results in near errors, near mistakes and near accidents. Chapanis found that all three are as valuable as actual errors, mistakes and accidents. That is because each is a set of circumstances that allows for the possibility of an error, mistake, or accident to occur. When many errors (or near errors) occur, the point of view the method imposes is that fault lies with the design of the product, not with human carelessness. The investigator conducts a CIS in order to learn what is wrong with the product that may cause it, or has caused it, to induce problems. Near accidents occur much more frequently than accidents and are as informative as accident information. They are a richer source of information than accidents, as people are more willing to talk about close calls than situations in which loss has occurred. Chapanis also notes that the method has some limitations. For example, it cannot be assumed that events occur as often in reality as they are reported. They may occur more frequently or less frequently. In addition, some kinds of incidents tend to be easily forgotten. and recollection of event frequency can fade with time. Furthermore, subjects can selectively report incidents (Chapanis 1959:76–95).

Subjects tend to shy away from identifying themselves or others with a negatively perceived event such as an accident. Ensuring anonymity makes it possible for interview subjects to describe how a product operates or what another person did without being divulged as the source of information. This is a particular issue in communities that share a strong bond among their members such as pilots or police officers.

Fault tree analysis (Section 8.2) is similar to CIS and is also a method that is useful in the prevention and investigation of accidents.

6.6.1 Preparation

Arrange interviews with individuals who may have knowledge of hazards, unsafe practices, or events that could have, or actually did, result in injury and loss. It can also be beneficial to have operators or maintainers write down first-hand accounts of what they observed. Maintaining reports in a database makes it easier to search by keyword in order to discover patterns.

6.6.2 Materials, equipment and environment

Obtain records such as investigation accounts and accident reports from departments or agencies that have responsibilities for safety and accident reporting. Sources for such records can include company safety departments, insurance companies, or regulatory agencies.

6.6.3 Procedure

Table 6.6 summarizes the steps that are used to perform critical incident study.

Step 1—Interview individuals who have experienced problems or who have observed others who have had problems. Ask questions from the viewpoint of 'Tell me what you know about what happened.'

Step 2—Alternately (or in addition) review reports of situations in which problems have occurred. Use a database of reports that may reveal clusters of difficulties that are related to a certain aspect (e.g. set of conditions or piece of equipment). Look for events that appear with some frequency. Note how often and under what conditions events occur.

Step 3—Create categories of these frequent events, noting the kind of problem and the item that is associated with it (e.g. a piece of equipment such as a display).

Step 4—Develop possible explanations for the source of the difficulty (e.g. display showed measurements in meters, but user expresses distance in terms of yards).

Table 6.6 Critical incident study—discover sources of dysfunction in existing systems.

Interview individuals/review reports
Speak with those who have experienced problems, or who have observed others who have had problems.
Alternately, or in addition:
Review accounts of situations in which problems have occurred.

Analyze findings
Look for events that appear with some frequency.
Note how often and under what conditions the events occur.
Create categories of these frequent events.
Note the kind of problem and the item (e.g. piece of equipment such as a display) that is associated with the problem.
Sort events into categories.
Review categories, looking for patterns of occurrences that are related to system elements (i.e. hardware, software, procedure, persons).
Develop explanations for the possible source(s) of the difficulty.

Propose remedies
If appropriate, recommend actions that will minimize or eliminate cause of loss.

Step 5—Develop remedies to change the product design or system operation in order to eliminate the problem (e.g. convert information displayed to meters or train users to rely on distance measurements in meters).

6.6.4 Result

CIS identifies possible sources of serious user–product difficulties. The recommendations for improvements promise to eliminate the potential for the situation to result in loss. Only actual use demonstrating that the product no longer induces the problem ensures that it is now safe.

6.6.5 Example

The field of health care offers sobering examples of problems between humans and products that can result in morbidity (illness and injury) and mortality (death). The following brief review of the cardiac catheterization lab case study in Section 5.2.6.1 demonstrates how CIS can be particularly effective.

Recall that invasive surgical procedures performed in a hospital catheterization laboratory present the potential to inflict extraordinary patient trauma. Cardiac and cerebral angiography is a diagnostic invasive procedure performed on patients with symptoms of arterial disease and plaque build-up that could lead to a heart attack or stroke. A large bore needle inserted into the femoral artery, a long catheter inserted through the needle through the thorax into the heart or carotid artery and contrast dye is injected to show blood flow through arteries which feed the heart or brain.

Reports on patient morbidity and mortality are required at health care facilities. Executed by professional staff, the reports provide a thorough and informative record of conditions, what was observed, what equipment was in use and what result occurred. An analysis of reports on file with the manufacturer reviewed information on the nature and extent of incidents, the trauma that occurred, equipment that was in use at the time the trauma occurred and the date and time the trauma occurred.

The reports consistently reported the presence of air in the patient bloodstream. How could it get there? Direct observation and interviews with cath lab staff revealed that during the process of patient catheterization the potential exists for one or more air bubbles to be introduced into the patient bloodstream via the contrast dye. Free to travel through the patient's body, the bubble could lodge in the nervous system or brain, causing paralysis or worse. Reports to the manufacturer that retained the investigator indicated that this had occurred in a number of instances. The cause was unknown, implicating all of the equipment in use at the time, including their product.

CIS record review revealed that the air embolism had occurred when a certain model of one device was used. CIS played a crucial role in the discovery of when those circumstances occurred. Review of records showed that even though staff and location varied, a certain type and model of equipment was consistently involved in incidents. Directed questioning of health care professionals at various sites also revealed compensating behaviors that staff members had developed in order to get the unit to function without introducing air into the contrast dye stream. Flow analysis (Section 7.1) and failure modes and effects analysis (Section 8.3) were also employed to analyze workflow in cardiac catheterization labs. It became apparent that when specific circumstances coincided the potential existed for an incident to occur involving a number of elements in the cath lab. Recommendations focused on changes to the new model that would prevent the unit from being used if it was not in a ready to use state.

6.7 Management oversight and risk tree analysis (MORT)

Discover threats to safety and reliability.

Management oversight and risk tree analysis (MORT) incorporates a logic tree approach to investigate accidents and to evaluate safety programs. The value for research and development is that the method can be used to model the requirements that are necessary to create a safe, reliable solution concept.

Gertman and Blackman (1994:242–69) describe MORT as a method that can be used to visualize the need for procedures and hardware in order to match the abilities of personnel at all levels in an organization. MORT relies on the development of logic trees (similar to fault tree analysis diagrams) to portray lines of responsibility, barriers to unwanted energy, events, priority gates and assumed risks as well as controls (such as codes, standards, regulations and plans) on human workplace performance. MORT is used to investigate problems that involve people as well as multifaceted problems (such as programmatic problems) that have long causal factor chains. Questions can be used to probe specific control and management factors. As new information is discovered, it can be used to update the logic trees. Logic trees can also be used to plan hazard analyses and to review life cycle safety requirements for processes and facilities.

Jones (1999:699) considers MORT a useful method to make explicit the organizational effects on error. Freeman (1991:140) considers MORT to be a product safety analysis method and he favors it for two reasons. Unlike fault tree analysis (Section 8.2), MORT can be used in a positive sense. By using the top event in the hierarchy to describe a successful condition, an organization can build a set of circumstances that models safe conditions. In addition, unlike root cause analysis (Section 6.9), MORT can be used to anticipate adverse outcomes and then act to identify, analyze and control them before loss is sustained.

Because it focuses on one system, MORT may not account for hazards that may be inherent in other related systems that can have an effect on the system that is under study. Information that is provided as input for MORT may be limited by the methods, such as root cause analysis, that are used to obtain it.

6.7.1 Preparation

MORT relies on knowledge about the elements and hazards in a system that can be gathered using methods such as operational analysis (Section 6.2), root cause analysis (Section 6.9), and flow analysis (Section 7.1). Evaluation methods such as fault tree analysis (Section 8.2) and failure modes and effects analysis (Section 8.3) can also be used to provide input.

6.7.2 Materials, equipment and environment

Information on product elements and their specifications is important to the construction of a logic tree.

Inexpensive tracing tissue can be used to quickly develop and revise fault tree diagrams. Illustration or desktop editor software programs can be used to create smooth versions of tree diagrams, particularly for those that include multiple and extensive branches.

6.7.3 Procedure

Gertman and Blackman's (1994:242–69) account of Nertney's original MORT method places it in the realm of process control accident analysis. Because Freeman's (1991:140–52) approach lends itself to new concept development, it is described here. Steps to perform it are summarized in Table 6.7.

Table 6.7 Management oversight and risk tree analysis—discover potential threats to property and life. Freeman (1991:140–6) describes a version of MORT that can be used to anticipate hazards and to create programs including equipment, staff and procedures in order to protect potential threats to person and property.

Describe top event

Describe the desirable condition such as 'injury free operation of a manufacturing facility.'

Identify assumed risks

Account for each of the hazards of operation to which people and property will be exposed.

Develop management systems

Develop the roster of management systems in order to eliminate or to diminish the consequences of assumed risks.

- Safety engineering elements are inherent in system design, such as physical barriers and job design.
- Safety management elements such as training and inspection are the means that are used to operate a system so that it will remain as safe as possible.

Identify rewards

Account for the positive benefits that result from the program such as higher productivity or lower costs of injury and workers' compensation.

Expand to task level definition

Complete the program to the task level so that each aspect can be implemented in a safety engineering or safety management program.

Step 1—Describe the desirable condition or 'top event,' such as 'injury free operation of a manufacturing facility.'

Step 2—Account for each of the assumed risks, which are hazards of operation to which people and property will be exposed.

In a manufacturing facility, this may include kinematic hazards. Kinematic hazards are the locations where machine components come together while moving that result in the possibility of pinching, cutting, crushing operations occurring to any object that is caught between them. Such hazards can include sets of gears, belts running onto pulleys, matching rollers, shearing operations and stamping operations where forming dies close on each other.

Step 3—Develop the roster of management systems in order to eliminate or to diminish the consequences of assumed risks. Management systems can include safety engineering and safety management programs to protect person and property from exposure to hazard. Safety engineering elements are inherent in system design, such as physical barriers and job design. Safety management elements such as training and inspection are the means that are used to operate a system so that it will remain as safe as possible.

Step 4—Account for the rewards, which are positive benefits that result from the program. Benefits might include higher productivity or lower costs of injury and worker's compensation.

Step 5—Expand on the initial MORT analysis in order to flesh out each element. Complete the program to the task level so that each aspect can be implemented in a safety engineering or safety management program.

6.7.4 Result

As Figure 6.4 shows, MORT results in a logic tree that can be used to demonstrate a safety program that makes the best use of hardware/software, personnel and facilities. Gertman and Blackman (1994:269) report that typical MORT applications include industrial safety, fire protection, industrial hygiene, vehicle and traffic safety, emergency preparation, environmental protection and nuclear power generation (radiation protection, reactor safety).

6.7.5 Example

Among the examples in which MORT is employed, Freeman (1991:140–52) describes the method's use in a retail facility in order to minimize the potential for patrons to slip or trip and fall. This example has implications for architecture, interior design and product design of fixtures and displays. Retailers invite customers into their facility to consider and purchase merchandise. Causes for falls can include floor material with a low coefficient of friction, floors that are wet or oily, floor coverings that are loose or torn, disorientation due to lighting or mirrors and obstructions such as floor display fixtures. Results of falls can be significant, particularly for more elderly customers.

The top event in this case would be a minimum number of fall accidents and injuries for store customers.

Assumed risks that management knows it will have to accept include inclement weather (resulting in slippery surfaces near doors), floors that are wet from maintenance or product spills, falls from one level to another, poor walking surface design (floor openings such as drains), floor obstructions (including merchandise being added or removed from displays) and elderly customers.

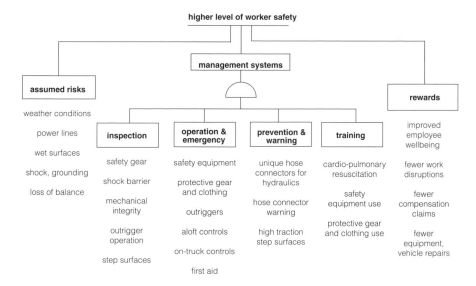

Figure 6.4 Management oversight and risk tree—Freeman (1991:6,140–6) describes a use of management oversight and risk tree analysis as a means of safety analysis. The example shows how a development team would acknowledge the assumed risks that are part of a new concept for a lift that is used in fieldwork such as trimming trees. How many systems and which kinds of systems to include depends on the nature of the risks that are inherent in its operation. In this instance, four kinds of management systems would be necessary to reduce the consequences of assumed risks. Each of the management systems includes staff procedures as well as safety engineering measures such as physical guards. The AND logic gate indicates that all four systems must be included to be effective.

Management systems that the facility can employ include inspections, warnings and emergency, training and equipment and operations.

Inspections—Review the condition of flooring by conducting slip tests, examine coverings, check protective equipment including railings and gratings and pay particular attention to higher risk areas such as rest rooms and food service areas.

Warnings and emergency—Communications among store employees would make it possible to identify and correct hazards quickly. An escort service could be made available to assist elderly customers, possibly in a pedal- or battery-powered cart.

Training—Management, sales and maintenance staff can receive training in slip and trip hazard identification and correction.

Equipment and operations—Equipment could include barricades and warning signs to prevent customers from walking into hazardous areas. Programs might include exterior walkway snow removal and special response teams that can be called to clean up a slippery floor.

Rewards that could result might include reduced damage claims and lawsuits, more efficient operations without the need to attend to accidents and improved customer goodwill.

> To err is human …
> (Alexander Pope)

6.8 Error analysis

Discover the nature and frequency of human error.

Error analysis (EA) is the examination of a product to determine all of the errors that individuals might make while using it.

Recall that Chapter 2 defined error as the failure of a person to satisfy requirements. Park (1987:191–258) refines that description to 'the failure to perform a task within the designated time and under specified conditions.' EA is used to discover the kinds of errors that humans make, how often those errors occur and to develop remedies. Errors can be caused by the product or by the human. Situation- or system-caused error can be the result of a design inadequacy. Human-caused error can result from an individual trait such as lack of skill. EA takes the point of view that human-caused error is actually situation- or system-caused error. For example, if the human caused error was due to lack of skill, the individual should have either received sufficient training or possibly not have been hired to perform the job (Meister 1985:16–79, 163–71).

EA relies on neutral observation in order to collect data on user performance. To be neutral, the observer provides little or no intervention in order to assist the user. Either discrete tasks such as procedures or continuous tasks such as tracking tasks can be observed. EA tends to be more successful in the analysis of discrete tasks. That is because it is easier to identify a specific task and to determine whether the task goal has or has not been satisfied.

On the assumption that the time allowed to perform a task is sufficient, Park identifies four EA analyses that may be conducted: error effects analysis, probability of error and error effects, criticality ratings and elimination of error sources.

- *Error effects analysis*—The probable effects of human error are classified according to four categories: safe (error will not result in major degradation), marginal (error will degrade system to some extent), critical (error will degrade the system) and catastrophic (error will produce severe degradation that will result in loss of the system).

- *Probability of error and error effects*—Review sources of information on human performance (including other similar systems) to estimate the extent of system degradation that results from human errors.
- *Criticality ratings*—Indicate the potential for specified errors to degrade the system.
- *Elimination of error sources*—Tasks that receive high probability and criticality ratings are examined to determine likely sources of errors.

Park provides formulae to compute probability and criticality ratings. However, the lack of precise task information makes it difficult to specify such circumstances with a certainty. Instead, EA finds use as a descriptive procedure. After errors are identified and insignificant errors are eliminated from consideration, it can be helpful to query subjects whether they have observed the errors, how often and what the consequences were (Meister 1985:16–79, 163–71).

EA is often used in assessment-level usability studies (see Chapter 10). The method is of particular benefit in the improvement of products that have been developed to an intermediate stage. Quantifying the type and frequency of errors can be used to substantiate recommendations for product improvement. Rank ordering errors that are most detrimental makes it easier to assign scarce resources more efficiently.

It would help at this point to understand how EA fits among the host of methods that are used to discover and explore human–system problems. Verbal protocol analysis (Section 6.5) is used to understand how a person perceives, operates, or uses a product, system or space. However, VPA can create some difficulties in accurate data collection. For example, speaking may slow subject performance. Also, subjects may edit their actions while performing tasks for the investigator. Three other methods are of benefit in the discovery of dysfunction (which can include human error). Critical incident study (Section 6.6) is used to discover the cause of human–product problems. Management oversight and risk tree analysis (Section 6.7) is used to detect hazards and develop remedial action. Fault tree analysis (Section 8.2) will explain a technique to discover what conditions may combine to create a hazard and possibly cause loss. Only EA focuses specifically on human error.

6.8.1 Preparation

Review preliminary design information and task requirements. Identify tasks that a subject is to perform using either an existing or a proposed product. The tasks may include all of the work a user is expected to perform or may be confined to tasks that are believed to be error prone. Other sources of information on human error can help to identify tasks with a higher probability of error.

Identify the behavior that the operator will perform such as rule application, use of memory, hypothesis development or data evaluation. Estimate the types of errors that might occur such as omission, commission and out-of-sequence errors (see Chapter 2 for additional classes of errors). Estimate the likelihood that the errors will occur in order to eliminate highly unlikely errors. Identify where the error might occur when operating or maintaining the product. Spell out the consequences that would result if the error is made.

Prepare a setting that minimizes interruptions. Recruit enough subjects in order to develop an adequate amount of data to compile. Chapter 10 provides further guidance on sample recruiting. Dumas and Redish (1994) and Rubin (1994) provide guidance on preparations for such assessments.

6.8.2 Materials, equipment and environment

Provide subjects with three items: a list of tasks, documentation (e.g. manual, instructions) and product(s) or prototype(s) sufficiently complete that very little interpretation is needed in order to use it to perform tasks. Automatic data capture devices and video recording can assist error compilation and analysis.

6.8.3 Procedure

Follow the series of steps that is drawn in part from Scerbo (1995:72–111) and summarized in Table 6.8 to perform EA.

Step 1—Orient the subject. Provide the subject with the product and documentation. Ask the subject to perform the series of tasks that is listed on the prepared sheet.

Step 2—Observe subject performance. Note the tasks the user performs as well as errors and the steps and the length of time that the subject takes to correct them.

Step 3—Document and compile error data. Review session documentation. Develop a table of errors according to categories and the number of times the errors are observed. Note tasks that are performed in error and the time that is needed to recover the error.

Step 4—Assess error data. While the team agrees that an error occurred, discuss why it occurred. The cause is not always obvious. Identify potential causes by collecting the comments of others who observed the session and interviewing the subject after the session is concluded. Rank errors and error sources according to those that are most detrimental to performance.

Step 5—Develop remedial actions. Consider one or more of the remedies: error control, error identification and recovery or minimizing error consequences.

Table 6.8 Error analysis—discover the nature and frequency of human error.

Orient the subject

Provide the subject with the product and documentation.

Ask the subject to perform the series of tasks that is listed on the prepared sheet.

Observe subject performance

Note the tasks the user performs as well as errors and the steps and the length of time that the subject takes to correct them.

Document and compile error data

Review session documentation.

Develop a table of errors according to categories and the number of times the errors are observed.

Note the tasks that are performed in error and the time that is needed to recover the error.

Assess error data

Identify potential causes by collecting the comments of others who observed the session. Interview the subject after the session is concluded.

Rank errors and error sources according to those that are most detrimental to performance.

Develop remedial actions

   Error control
   Error identification and recovery
   Minimize error consequences

Error control—Reduce the likelihood of error by reducing the number of opportunities to commit errors. This approach is most often interpreted as reducing the degrees of freedom by making a more limited set of actions available to the individual, requiring less skill and less training. For example, most people cannot make a straight cut while using a saw. Using a mitre box removes the freedom for the saw to wiggle and wobble, performance improves and errors are reduced.

Error identification and recovery—Make errors detectable and correctable (e.g. through the use of augmented feedback) before an unrecoverable loss is incurred. For example, large ships experience a long delay between a helm rudder order and actual ship movement. Waiting to see where the ship will go is hazardous, particularly in close quarters and in aggressive currents. Augmented feedback such as a predictive display shows where the position of the vessel will be based on steering orders. Timely corrections to steering orders can be made by observing the predictive display.

Minimize error consequences—If errors are made, make the product more forgiving by making it possible to recover from the errors. Shoulders, median strips, barriers and contoured landscaping are all employed to lessen the severity of consequences if an auto drives out of its lane on a highway.

Park (1987:191–258) offers further alternatives to task elements and tasks that can be employed to eliminate errors. He recommends: reallocate functions to humans or machines (see Section 7.4), make design changes to subsystems or equipment (with emphasis on limits upon functional links between humans and machines), change procedures or elements of the environment, rearrange operational equipment, employ incentives or job aids or warning or alerting devices and revise personnel training or supervisory methods.

6.8.4 Result

Error analysis produces remedies that seek to lessen consequences or to eliminate errors. The method can be used to document errors that occur in operational products, in competitive products and proposed concept prototypes. Tables of error types and the time needed to correct them can be used to guide new generation product development or to improve concepts before they are produced as operational products.

6.8.5 Example

Early personal computers opened a world of opportunity for those who were involved in the performance of knowledge work. Unlike large centralized mainframe computers, PCs made it possible for users to work on a computer by themselves. Users could create, manipulate and transfer data files by entering strings of alphanumeric characters. They could also delete data files, which was frequently necessary given the limited memory capacity of early machines. Unfortunately, it was not unusual for a user to delete an important file. With no way to retrieve it, the important file had vanished. Such unrecoverable errors were a serious impediment.

By using user-oriented design improvements, firms such as Apple Computer eventually developed features that protected users from such loss due to errors. Within years, development of the graphical user interface (GUI) made it possible to recognize and manage files through the use of symbols rather than by entering code strings that were vulnerable to typing errors. In addition, dialog box caution messages ('Are you *sure* you want to delete this file?') and retrievable 'Trash' and 'Recycling Bin' files made it possible to recover files that were deleted in error. These second and third steps made the systems more forgiving of errors that users were certain to make sooner or later.

6.9 *Root cause analysis*

Determine the source of problems that cause error and failure.

Any event that results in loss to person or property is inevitably followed by questions such as 'How could that have happened?' and 'What can we do to prevent it in the future?' Root cause analysis (RCA) is used to discover the source (root cause) for a sentinel or adverse event—popular terms for an event that signals a hazardous condition exists. For example, the collapse of a construction worker in an excavated pit is a sentinel event that indicates that there was insufficient oxygen to support life within that confined space. A series of conditions had to come together in order for the final result to occur. RCA seeks to account for those conditions, then develop remedies to ensure they do not occur again. Remedial actions to prevent another downed construction worker might include better crew training in compliance with safety regulations regarding work in confined spaces.

RCA was made popular in Japan and has found use in the United States as a total quality management tool to discover the source of product defects (Melsa 1999:269–302). RCA has more recently become popular in the United States as a means to discover health care procedural flaws and to rectify medical error.

Other human factors methods share a common approach with RCA. Management oversight and risk tree analysis (Section 6.7) is used to anticipate potential threats to safety and remedy them before they are encountered. Cognitive task analysis (Section 7.7) methods are used to explore how decisions are made. Critical incident study (Section 6.6) is used to discover the possible causes of loss or near-loss based on previous occurrences. The technique for human error rate prediction (THERP) (Section 7.8) and fault tree analysis (Section 8.2) are used to determine the relationship and likelihood of events that may contribute to dysfunction.

Gertman and Blackman (1994:230–99) account for six methods that are most often applied to root cause analysis:

*   *Causal factor (walkthrough) analysis*—During causal factor analysis, an observer watches as individuals re-enact the series of steps that led up to an adverse incident. To obtain preliminary information, the investigator determines what the person was doing when the problem occurred and decides on a task of interest. Information is collected on relevant procedures and materials. Interviews are conducted with others who have performed the task (but who will not be observed) in order to understand how the task will be performed. The investigator produces a guide that outlines how the task will be performed, indicating task steps, key controls and displays in order to know what to look for and to more easily record observations. Subjects are chosen who normally perform the task of interest. The investigator observes subjects as they walk through the task, records the subject's actions and (if appropriate) use of displays and controls. After completing observations, the investigator summarizes findings, consolidates noted problem areas and identifies probable contributors. Table 6.9 summarizes the steps that are involved in causal factor analysis.

Figure 6.5 shows the cause and effect diagram that can be used to analyze elements, or drivers, that may influence a situation or outcome. Also known as a fishbone or Ishakawa diagram, it accounts for each system element. The diagram's structure is used to enumerate possible causes that contributed to the adverse outcome.

Table 6.9 Root cause analysis: causal factor analysis—Gertman and Blackman (1994:230–99) describe causal factor analysis as one of the six root cause analysis methods that can be used to discover the source of problems that induce error and failure.

Obtain preliminary information

Determine what the person was doing when the problem occurred

Decide on a task of interest

Obtain information on relevant procedures and materials

Interview others who have performed the task (but who will not be observed) to understand how the task will be performed

Produce a guide

Outline how the task will be performed

Indicate task steps, key controls and displays to know what to look for and to more easily record observations

Become thoroughly familiar with the task and the guide

Conduct observation

Choose subjects who normally perform the task of interest

Observe subjects as they walk through the task

Record actions and, if appropriate, use of displays and controls

Summarize

Summarize and consolidate noted problem areas

Identify probable contributors

- *Change analysis*—Change analysis is a six-step process that is used to investigate obscure causes and organizational behavior breakdown and multifaceted problems with long causal factor chains (such as administrative or procedural problems).
- *Barrier analysis*—Barrier analysis is used to investigate issues that involve complex barriers and controls such as equipment failures and procedural or administrative problems.

Three of the methods require some amount of training in advance of use.

- *Management oversight and risk tree analysis (MORT)*—As Section 6.7 describes, MORT is used to investigate people problems and multifaceted problems that have long causal factor chains (such as programmatic problems). Questions probe specific control and management factors. Dynamically updated logic trees portray lines of responsibility, barriers to unwanted energy, events, priority gates and assumed risks as well as controls (such as codes, standards, regulations and plans) on human workplace performance.
- *Human performance evaluations (HPEs)*—HPEs are used to look into problems in which people have been identified as the cause.
- *Kepner-Tregoe problem solving and decision making (K-T analysis)*—Kepner-Tregoe analysis is used to conduct a thorough analysis of causes and corrective actions.

Latino (2000:155–64) recommends the use of automated root cause analysis software programs that have been developed for industrial applications in response to regulations that are enforced by the U.S. Occupational Safety and Health Administration (OSHA) and Environmental Protection Agency (EPA). Such software provides for 'data collection and organization, team member selection, logic sequence representation, validation of hypotheses, identification of causes and recommendations.'

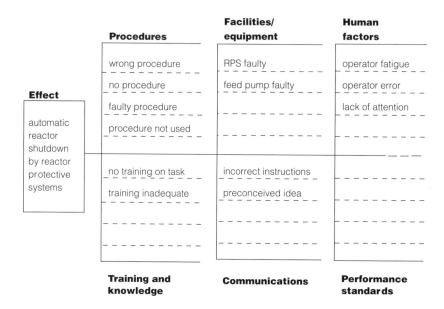

Figure 6.5 Cause and effect diagram—a cause and effect (also known as a fishbone or Ishakawa) diagram is used to account for the system elements and root causes that produce an effect. In this case, a nuclear reactor was automatically removed from service and shut down. Contributing aspects of a system (drivers) include procedures, facilities and equipment, human factors, training and knowledge, communications, and performance standards. Note the similarity to Figure 1.1, which relates element selection, performance, and interaction to overall performance and result.

Source: *Human Reliability and Safety Analysis Handbook*. Gertman, D. and Blackman, H. Copyright © 1994. John Wiley and Sons. This material is used by permission of John Wiley & Sons, Inc.

RCA can be used to probe beyond symptoms in order to look for the underlying causes for problems. There are some issues about the method that investigators should be aware of.

- There may be many causes for a failure, not one. In fact, as Section 2.8 described, Reason (1997) and Shappell and Wiegmann (2000) contend that it is most likely that there are multiple causes for any adverse event.
- Temporary or latent causes may elude the investigator's attention. Events may happen at one point in time and manifest symptoms much later.
- Remedies may address one cause but fail to prevent other related causes.
- RCA assumes a direct relationship between cause and effect. However, it is often a collection of seemingly unrelated circumstances that convene to cause an event.
- Because a sentinel event is used to initiate the method, RCA is retrospective. When the event involves the loss of life, it is a high cost that other risk management methods such as management oversight and risk tree analysis (Section 6.7), fault tree analysis (Section 8.2) or failure modes and effects analysis (Section 8.3) might have been used to anticipate and possibly prevent.
- Conditions continue to evolve. It is only when the results of RCA are linked to a remedy that results can be fed forward in order to try to prevent further loss.

6.9.1 Preparation

Obtain information on the sentinel events that may have previously occurred and the event that has most recently occurred. Sources for information can include records, participants, witnesses, users and devices that may have collected data before, during and after the event.

6.9.2 Materials, equipment and environment

No particular materials or equipment are required for this method.

6.9.3 Procedure

Table 6.9 summarizes the following steps from Gertman and Blackman (1994:230–99) that are used to perform causal factor analysis.

Step 1—Obtain preliminary information on what the person was doing when the problem occurred.

Step 2—Decide on a task of interest.

Step 3—Obtain information on relevant procedures and materials (e.g. system drawings, block diagrams, piping and instrumentation diagrams). Interview others who have performed the task (but who will not be observed) to understand how the task will be performed.

Step 4—Produce a guide that outlines how the task will be performed. Indicate task steps, key controls and displays to know what to look for and to more easily record observations.

Step 5—Become thoroughly familiar with the task and the guide.

Step 6—Choose subjects who normally perform the task of interest.

Step 7—Observe subjects as they walk through the task. Record actions and, if appropriate, use of displays and controls.

Step 8—Summarize and consolidate noted problem areas. Identify probable contributors.

6.9.4 Result

RCA discovers all evident causes for a sentinel event and is used to propose remedies to prevent the event from occurring in the future. A table that lists problem effect, the cause of each effect and a solution can be used to demonstrate a problem and solution structure. A cause and effect diagram (Figure 6.4) portrays elements (drivers) in a system and accounts for the root causes within each that may have led to the incident.

6.9.5 Example

The Joint Commission on Accreditation of Healthcare Organizations (JCAHO) (1999:150–1) offers a tabular approach to discover the causes for sentinel events. Inquiries focus on levels of analysis covering what happened to why it happened (proximate causes, processes involved, underlying systems). Each level is expanded to detail possible reasons, questions that those possibilities evoke and findings in answer to the questions. Strategies to reduce risks and to measure the outcome of those strategies provide a course of action that can be implemented and monitored.

The case of an infant abduction from a nursery provides an example. Exploration of why the event happened includes processes that were involved. Possibilities include patient care processes. During the investigation, discussion of proximate causes identified a process deficiency (weak or missing step) of identification tags being attached to the bassinet but not the infant. Examination of processes led to the question 'what is currently being done to prevent failure at this step?' The finding that nothing was being done led to a risk reduction strategy to redesign purchasing procedures for the new tags and a measurement strategy to monitor the inventory of tags on units to ensure that a sufficient supply is available.

6.10 Summary

Analysis methods are useful for getting at the nature of a problem. Each builds the knowledge necessary to make decisions about what the product should be.

Pareto analysis determines which aspects of a problem deserve attention first.

Operational analysis, also called 'mission analysis,' discovers what is currently accomplished in a situation or is expected to be accomplished by a proposed product.

Analysis of similar systems discovers how others have addressed a related problem.
Activity analysis discovers and quantifies what those who are working actually do.

Verbal protocol analysis, also called the thinking aloud method or directed dialog, explores user mental processes regarding the product under evaluation.

Critical incident study discovers sources of dysfunction in existing systems.

Management oversight and risk tree analysis discovers threats to safety and reliability.

Error analysis discovers the nature and frequency of human error.

Root cause analysis methods pursue the source of problems that cause error and failure.

Analytic methods are used to build an informed context. Context can include the information that is inherent in an existing product and the way it is used. Context can also include the situation and requirements that bring a new product into existence. In either case, the information that is collected through the use of methods for analysis can be used to develop a solution model by employing the methods that are provided in Chapter 7.

7 Design guidance methods

What functions need to occur and in what order?
Flow analysis (7.1)

What order and duration of events can be expected?
Time line analysis (7.2)

What elements comprise the product and how are they related?
Link analysis (7.3)

Is a person, or hardware or software best suited to perform each proposed function?
Function allocation (7.4)

What will people do in this proposed solution concept?
Task description (7.5)

What tasks, training, equipment and information will be needed to perform a job?
Task analysis (7.6)

How do people solve problems and make decisions?
Cognitive task analysis (7.7)

Is what we expect people to do in this concept reasonable?
Technique for human error rate prediction (7.8)

How will tasks be assembled into work for individuals to perform?
Work design (7.9)

How can we choose among concept options?
Decision analysis (7.10)

The analysis methods in Chapter 6 are used to assist research and development team efforts to learn about a new area of interest, to determine what deserves attention, to find out what actually occurs, to discover what may be going wrong or could go wrong and to learn what others in similar situations have done. The evaluation methods in Chapter 8 make it possible to evaluate how well the decisions that were made during the design process actually turn out.

As in Chapter 6, each section in this chapter follows the same format so that information is easy to find and understand: introduction, preparation, materials and equipment, process, results and example.

- *Introduction*—Explains the background of how the method was developed and different approaches that have been taken to perform it.
- *Preparation*—Describes what needs to be done before performing the method.
- *Materials and equipment*—Accounts for physical items that are necessary including materials and equipment.
- *Process*—Enumerates the steps that are involved in performing the method.

- *Results*—Describes the outcome that performing the method produces.
- *Example*—Leads the reader through a brief discussion to illustrate the method's use.

This chapter explains methods that are often considered the core of human factors involvement in research and development: what humans are expected to do. Laughery and Laughery (1997:329–54) refer to the kinds of methods that are discussed in this chapter as 'analytic techniques' while David Meister (1985:7–41) refers to them as 'design aiding techniques.' They are called design guidance methods here because each uses knowledge about the capabilities and limits of humans in order to guide the design of products, systems and services.

Functional analysis

It will help at this point to give a brief overview of one of the most valuable approaches to human factors research: functional analysis. Functional analysis is used to account for the series of events that is necessary for a product to fulfill its purpose. Functions are expressed in concise statements of what is (or needs to be) done in the product or system. The statements that are used to describe functions should make no mention of what element performs the function or how the function will be accomplished. Functional analysis opens the way for any discipline such as mechanical engineers or chemical engineers to consider each function and recommend the best way to accomplish it. The approach also makes it possible to see patterns of functions. Seeing this larger perspective is likely to result in more substantial and comprehensive solutions rather than incremental or piecemeal solutions (Bailey 1989:188–194). Who or what performs each function is determined by function allocation (Section 7.4). Simulation (Section 8.1) is used to explore how those function assignments can be realized in a physical form.

Many authors have created variations of functional analysis. Each method is typically referred to by its popular name that is based on the criteria that is used to perform the analysis. For example, time line analysis is used to arrange functions according to time. Three approaches to functional analysis are presented in this chapter under their commonly used names: flow analysis (Section 7.1), time line analysis (Section 7.2) and link analysis (Section 7.3). Table 7.1 shows a selection of the more popularly used functional analysis methods and diagrams.

Functional analysis pattern—Buede (1999:997–8) describes four elements that are common to all functional analysis methods:

- Functions are represented in a hierarchy using either a top-down decomposition or bottom-up composition approach. Most often, a hybrid of the two approaches will produce the best results. For example, a top-down decomposition of traditional retail sales could be represented by the functional modes of operation, customer service, and maintenance. Customer service would be further decomposed into reception, presentation, review/selection, transaction, preparation and departure. Each activity such as review and selection would be further divided down to the level of each task that a sales person would need to perform (e.g. describe features, compare features, elicit customer needs, match product features to needs, invite questions, recommend product, encourage selection, confirm confidence in selection). A bottom-up composition follows the reverse approach. Each task that needs to be performed is accounted for, then aggregated into clusters. Through successive layers of hierarchy, the functions are tied together into a complementary whole.

Table 7.1 Functional analysis methods—A variety of methods have evolved in order to perform functional analysis. Diagrams have also been developed in order to depict functions and serve as a medium for review and analysis. The methods below are organized according to Laughery and Laughery's classification scheme. Three functional analysis methods are described in this chapter: flow analysis (Section 7.1), time line analysis (Section 7.2) and link analysis (Section 7.3).

Type	*Method*	*Diagram*
Flow	Functional requirements definition	'Block' diagram, functional flow diagram
	Process analysis	Flow process chart
	Flow analysis	Flow diagram
	Action-information analysis	Action-information analysis diagram
	Decision-action analysis	Decision-action analysis diagram
Time line	Time line analysis	Time line, Gantt chart
Network	Link analysis	Interaction matrix, link diagram
	Program evaluation and review Technique (PERT), critical path method (CPM)	PERT diagram

- Diagrams are used to represent the flow of data or items through process modeling (structured analysis, data flow diagrams, N2 charts), behavior modeling (control flow diagrams, function flow block diagrams, behavior diagrams, finite state machines, state charts), object-oriented modeling (computer science and software engineering approach to represent a system as objects that interact). Graphical methods make the process explicit and portray patterns that are otherwise difficult for the observer to detect. Most of the figures in this chapter are examples of these diagrams.
- Processing instructions describe how to transform inputs to outputs. Step-by-step statements in each diagram in this chapter describe activity that occurs from the beginning to the end of the process.
- The control flow sequences the termination and activation of functions, making the process efficient.

Functional analysis methods in this section are most often applied to the development of products that people operate, maintain and use.

Types of functional analysis—Functional analysis is an umbrella term for a variety of methods. It was first termed process analysis during early studies of manufacturing process functions. Chapanis' (1959:51–62) earlier work in this area described process analysis as an all-inclusive method that could be used to record the steps that comprise a process (often within the context of manufacturing). Process analysis results were portrayed by any of four types of diagrams: process charts, flow diagrams, multiple process charts and link analysis diagrams. Laughery and Laughery (1997:329–54) use functional analysis to describe the method in order to reflect more formal systems analysis procedures that can be applied to a wide variety of processes. They sort functional analysis methods into three categories: flow analysis, time line analysis and network analysis.

- *Flow analysis*—flow analysis is used to examine the flow and order of materials, actions, or information. Process analysis, functional flow analysis, action information analysis and decision-action analysis methods can be considered flow analysis methods.

- *Time line analysis*—time line analysis is used to examine events with respect to time.
- *Network analysis*—network analysis is used to examine the structure or relationships among elements in a system. Network analysis can be used to learn about elements that are abstract (e.g. organizational departments) or concrete (e.g. equipment, people).

Link analysis is considered to be a network analysis method by some and a flow analysis method by others. It fits as a flow analysis method because results can be used to determine the effect on the flow of work activity. However, link analysis is also used to examine the relationships among elements, which is network analysis. Chapanis (1959:51–62) describes link analysis as a method to analyze the interaction of humans among each other and with controls, displays and equipment. Based on that interaction among elements, link analysis is described in this text as a network analysis method.

Functional analysis diagrams—To perform any of these functional analyses, the investigator defines a set of functions (i.e. what will be done) and depicts their relationships in the form of diagrams. This visual portrayal makes it possible to detect patterns that can be used to understand the flow of activity through the entire product. Pattern recognition also makes it possible to partition products into components such as subroutines and modules.

Whether this approach to diagramming will continue to serve investigators well is open to question, though. Systems have traditionally been represented by hierarchical diagrams that describe high to low order functions that are partitioned into orderly series (e.g. Figure 7.3 and 7.4). Recently, sophisticated products have begun to diverge from the neatly decomposed and ordered patterns into more complex and varied patterns. Such products will challenge research and development teams to find new ways to represent them.

> *Everyone designs who devises courses of action aimed at changing existing situations into preferred ones.*
>
> (Herbert Simon)

7.1 Flow analysis

Analyze the nature and sequence of events in a product or process.

Flow analysis is used to account for the functions that the product is to perform and to portray their order. According to David Meister (1985:7–21) the major value of flow analysis is to prepare the groundwork for function allocation (Section 7.4). Flow analysis is a 'detailed examination of the progressive travel of either personnel or material from place to place and/or from operation to operation' that is 'used for representing order information about serial events.' (Laughery and Laughery 1997:329–54)

The flow analysis methods that are described here follow essentially the same method: identify functions, determine their order and depict the ordered set of functions. In order to make this valuable method easier to understand and to use, this section will sort through the many approaches, terms and visualization methods that have been developed for flow analysis.

7.1.2 Flow analysis

As described earlier in this chapter, flow analysis is used to examine the flow and order of materials, action or information.

7.1.2.1 FLOW ANALYSIS METHODS

Four methods are considered to be flow analysis methods: functional requirements definition, action-information analysis, decision-action analysis and functional analysis systems technique (FAST).

Functional requirements definition (FRD)—Functional requirements definition is similar in some ways to action-information analysis. A zero order level diagram (the top tier of the functional hierarchy) in both methods looks quite similar. However, while FRD depicts required functions, action-information analysis portrays (as its name indicates) the actions and information that occur in a current product or prototype for a proposed solution. Woodson (1981:913–22) describes FRD as a way to refine the results of operational analysis (Section 6.2) through the use of functional block diagrams to depict 'basic system organization and function.' FRD block diagrams look similar to flow diagrams although FRD is not considered a flow analysis method. The block diagram analysis followed essentially the same approach as the flow diagram procedure that is described in this section. The direction in a FRD block diagram indicates a relationship between elements and not necessarily a flow between them. For example, a regulation that requires a safety restraint would be indicated by an arrow that connects the two items.

Action-information analysis—Chapanis' (1996:98–100) action-information analysis is a flow analysis method that can be used to identify 'the information that is needed for each action or decision to occur.' Rather than tracing the flow of material or persons, it accounts for the flow of information. As a result, it is a valuable method to use in the development of intelligent products and interfaces.

Decision-action analysis—Decision-action analysis is Chapanis' variant of action-information analysis that includes decisions that can be phrased as questions with binary (yes/no) choice alternatives. This author has found that imposing the requirement for answers to be clearly yes or no (whether a condition has or has not been satisfied) often results in much clearer thinking about processes. As a result, the use of decision-action analysis often produces results that are more thoroughly considered and (thereby) better.

Functional analysis systems technique (FAST)—Functional analysis systems technique is a variation of flow analysis that focuses on events and their order. The method relies on the development of verb-noun statements (such as 'emit light') to describe functions that are to be performed. The statements are written onto cards and arranged into a branching tree structure. A top-level basic function begins the tree. Questions as to 'how the product does what it does' produce function branches that are arranged to the right of the basic function. 'Why' questions can be used to track the rationale for each function from the end of the branches leading up to the basic function. Functions that do not prompt strong answers to the how or the why questions are candidates for deletion.

Operational sequence analysis (OSA) is also considered to be a flow analysis method. OSA is presented in Section 8.4 as an evaluation method that adds the aspect of time to flow analysis and depicts multiple sequences of information and decisions that must be performed through the creation of an operational sequence diagram.

7.1.2.2 FLOW ANALYSIS DIAGRAMS

It can be difficult to understand the difference among flow methods and diagrams as well as the purpose for each. This section will describe the types and uses of more popular flow analysis diagrams: flow diagram, flow process chart, information flow charts and operation-decision diagrams and decision-action diagrams.

7.1 Flow analysis

- *Flow diagram*—A flow diagram (Figure 7.1) is used to show the physical setting in which functions are carried out. The orthogonal top-down projection ('bird's-eye view') makes it possible to map relationships between objects in scale measurement. The pictorial drawing in the lower half of Figure 7.1 trades off the accuracy of a two-dimensional projection for the ease of understanding that comes from a three-dimensional perspective projection.
- *Flow rrocess chart*—A flow process chart (Figure 7.2) is used to show how a product or material, a person or equipment flows through a process.
- *Operational sequence diagram*—Operational sequence diagrams portray the combined flow of activity and information among all system elements. The operational sequence diagram is described in operational sequence analysis (Section 8.4).
- *Information flow charts, operation-decision diagrams and decision-action diagrams*—Information flow charts account for the flow of information in a product. A functional flow diagram, as Figure 7.3 shows, depicts the nature and direction of functions that are to be accomplished. Functional flow diagrams are also referred to as functional block diagrams.

The addition of decision points to a functional flow diagram can result in operation-decision diagrams and decision-action diagrams. As Figure 7.4 shows, decision-action diagrams represent the flow of actions through a process. Decisions are either satisfied or not satisfied and paths are created to indicate either outcome.

7.1.3 Decision-action diagram

Research and development involves a great deal of work in the area of intelligent products including user interfaces. The decision-action diagram is described here because of its value in intelligent product development. Some background on symbol and layout conventions will help to explain the procedure to develop this type of flow analysis diagram.

Flow analysis diagrams are simple directed graphs. Verbal labels that are placed within or next to conventional symbols are used to describe functions. Arrows that connect the symbols indicate the direction of information and decision flow. Each of three symbols lends structure to the flow analysis through a meaning that governs its use in the diagram: circles, rectangles and diamonds.

- *Circles*—A circle is used to represent activity starting points and end points. The choice of which activities to use as start and end points is what sets the product apart from other products and from the total environment.
- *Rectangles*—Actions are represented by oblong or square forms.
- *Diamonds*—Decisions, representing conditions that are either satisfied or not, are represented by diamonds. Unlike start/end points and actions, decisions have two subsequent paths. One is marked with a plus (+) and the other with a minus (−) sign. If the condition is met, the path marked by a plus is followed. If the condition is not satisfied, the path marked by a minus is followed. Sorting out where decision points should be in the diagram and how to phrase each action or decision are often the greatest challenges of flow analysis.

Typically, the first generation of a flow diagram will consist of higher order functions (e.g. 'prepare the auto' for a road trip). The second version of the flow diagram will enlarge the pattern of functions by starting to detail the functions more specifically to include second

Exhibit flow diagram

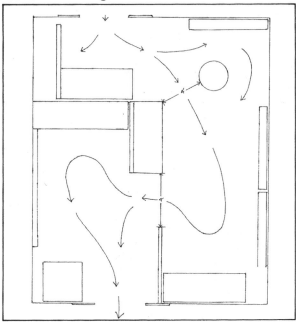

Pictorial drawing of visitor flow

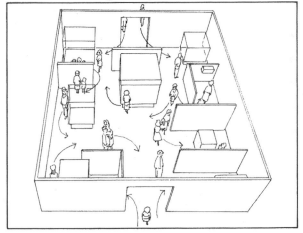

Figure 7.1 Flow diagram—a flow diagram is used to represent the flow of a process in the physical setting where it is carried out. The path describes the flow of an item or information in the
context of a physical setting. In this example, a controlled exhibit floor plan describes the
flow of visitors. The illustration below it is a pictorial representation of an exhibit with an
uncontrolled flow.

Source: *Good Show!* Wittenborg, L. Reproduced with the permission of the Smithsonian Institution. Illustrator:
S.D. Schindler.

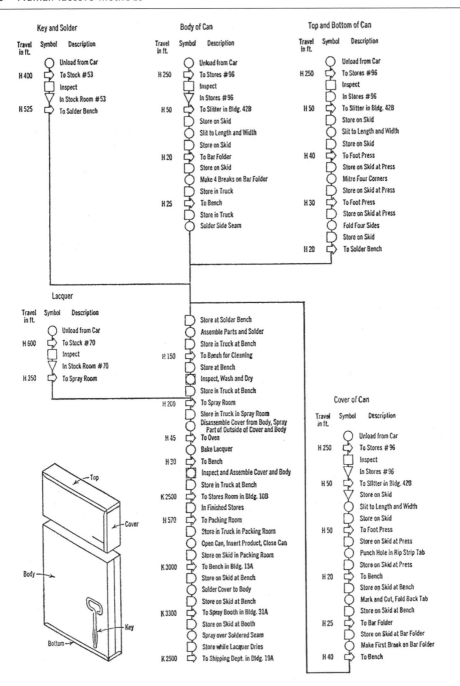

Figure 7.2 Flow process chart—a flow process chart is used to show how a product or material, person, or equipment flows through a process. This example depicts the production, filling and sealing of a rectangular can to ship instruments. Symbols indicate operations (circle), delay (half round), storage (triangle), inspections (square) and transportation (arrow).

Source: *Motion and Time Study.* Barnes, R. Copyright © 1968. John Wiley and Sons. This material is used by permission of John Wiley & Sons, Inc.

Figure 7.3 Functional flow diagram—a functional flow diagram depicts the nature and direction of functions that are to be accomplished. Assignment of numerical labels makes it possible to identify related functions as they are expanded into more elaborate lower order depictions. In this example, the flow of functions has to do with the preparation and the dispensing of real ale. Figure 7.4 shows how a decision-action diagram refines these general functions into tasks and adds decisions.

Preparing a cask for service

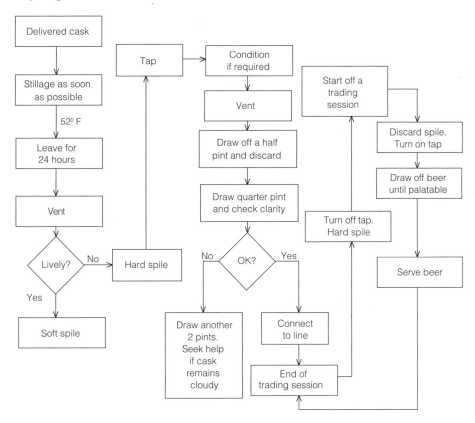

Figure 7.4 Decision-action analysis diagram—the United Kingdom's Campaign for Real Ale provides guidance to proprietors on the best process to assess and prepare cask conditioned beers for dispensing. Installing a soft spile (porous wood plug) in the keg allows the yeast in the beer to ferment gently, infusing the beer with a fine carbonation. Installing a hard spile (hard wood or plastic plug) stops the process. Failure to understand or follow the process can result in beer that is flat or clouded by suspended yeast.

Source: CAMRA *Guide to Cellarmanship*. Reproduced with the permission of the Campaign for Real Ale. London.

order functions. For example, more specific second order functions that would accomplish 'prepare the auto' would include 'fill the gas tank,' 'change oil and filter' and 'fill tires with air.' The process continues through subsequent diagrams, becoming more specific with each version, until all functions are accounted for. Each path in the diagram needs to lead to the 'end' function. It may also be necessary to add a decision function that asks 'Done?' in order to know the process is complete and can be terminated. Each order in the entire diagram should have the same level of task complexity (e.g. a high level order that includes general functions should not include a highly specific low level function). Diagrams will change many times from the first through the final version. Patterns usually need to be rearranged many times until they make sense.

As they evolve, flow charts can become quite complex and large. One way to manage a large size chart is to break it into parts. To do that, it is helpful to use a higher-order chart to show overall system structure. Portions of that chart can be shown to end in a lettered code, such as 'A' in a circle. Another chart that begins with a start point labeled 'A' in a circle depicts a lower order set of activities that is related to the higher order set. In this way, a number of separate charts can be shown to relate to each other.

7.1.4 Preparation

Perform and review the results of either operational analysis (Section 6.2), analysis of similar systems (Section 6.3), or activity analysis (Section 6.4). Each of these methods reveals functions that are either inherent in an existing product or are necessary to be performed in a concept under development.

7.1.5 Materials and equipment

Inexpensive tracing tissue can be used to quickly develop and revise functional block diagrams. Illustration or desktop editor software programs can be used to create smooth versions of block diagrams, particularly for those that include second and third order functions.

7.1.6 Procedure

Table 7.2 summarizes the steps to perform decision-action analysis.

Step1—Review the functions revealed by the use of either of the analysis methods: operational analysis (Section 6.2), analysis of similar systems (Section 6.3) or activity analysis (Section 6.4).

Step 2—Create a first order diagram.

- Select five to seven functions that represent the highest order aspects of what the product will perform.
- Arrange the functions in order of their precedence (e.g. performance over time). Unlike functional block diagrams, the information flow diagram will also need to include a 'start' and a 'finish' step, as well as decision nodes.
- Trace through the possible function paths from start to finish to ensure that their order and precedence are accurate.

Step 3—Create a second order diagram

Table 7.2 Decision-action analysis—determine the nature and flow of information and decisions in a current or proposed product.

Review the functions that were revealed by the use of either of the analysis methods: operational analysis or analysis of similar systems.

Create a first order diagram

Select five to seven functions that represent the highest order aspects of what the product will perform.

Arrange the functions in order of performance through time.

Trace through the possible function paths from start to finish to ensure that their order and precedence are appropriate.

Create a second order diagram

Add functions and decision nodes that are necessary to accomplish those that are shown in the first order diagram.

Trace the paths again to ensure the precedence and relationship choices are sound.

Continue with a third order diagram

Depict functions that will be necessary to accomplish the second order functions.

- Add functions and decision nodes that are necessary to accomplish the functions and decisions that are shown in the first order diagram. These 'lower order' functions begin to describe the product more completely.
- Trace the paths again to ensure the precedence and relationship choices are sound.

Step 4—Continue with a third order diagram

- Depict functions that will be necessary to accomplish the second order functions. The process can continue until the team is satisfied that functions have been accounted for. Functions should be refined to the level that they can serve as input for function allocation (Section 7.4).

7.1.7 Result

The final version of the information flow diagram depicts the flow of action and information throughout a current or proposed product. The symbols, their arrangement and the paths that connect them represent the type of activity, activity flow and the supporting activities that are necessary to accomplish higher order functions.

7.1.8 Example

The United Kingdom's Campaign for Real Ale (CAMRA) promotes the brewing and enjoyment of cask conditioned beers. Cask conditioning, which has seen a revival as a result of CAMRA's efforts, imparts a finer natural effervescence to ale than the use of carbon dioxide (CO_2). A development team that is assigned to develop a new ale container that improves on the process of handling real ale would first document the process in a decision-action diagram. The topic of fresh beer offers a refreshing break at this point in the book and provides the opportunity to discuss a few improvements to the example diagram.

A publican (bar proprietor) has to pay more attention to handling real ale than pasteurized beer that is kept under CO_2 pressure. That is because real ale's effervescence relies on the activity of live yeast present in each keg. The yeast fermentation rate must be managed correctly so that each keg has the right amount of effervescence. To assist publicans, CAMRA provides guidance on how to assess the condition of the ale and to prepare it for dispensing. Figure 7.4 diagrams the process that the Brooklyn Brewery in New York followed.

As a decision-action analysis diagram, it includes decision nodes that are either satisfied or not satisfied. For example, the 'Lively?' step determines which slender wood spile (peg) to select. If the ale is effervescent, the publican inserts a hard spile in the keg to arrest fermentation. If is it not effervescent, insertion of a soft spile in the keg allows fermentation to continue. The diagram should return to the 'Lively?' node following 'soft spile,' to indicate monitoring until the ale is ready for hard spiling.

The 'OK?' decision actually asks whether the pint drawn from the keg is clear of yeast sediment (cloudy beer is less attractive and can present 'off' flavors). If it is clear, the ale is ready to connect to serving lines and be drawn by hand pump to waiting customers. If it remains cloudy after drawing two more pints, the beer may be contaminated and need to be returned to the brewery (a tragic outcome, indeed).

7.2 Time line analysis

Analyze events with respect to time to determine their order and duration.

Time line analysis is a functional analysis method that combines functions or tasks with time-related information in order to predict workload levels for a solution that is under development. The subject can be any element (humans, hardware, software and procedures). The method is used to sort through tasks, to determine the order in which they are to be performed and to estimate the amount of time that will be taken to perform them. Laughery and Laughery (1987:329–54) describe the method's use in 'workload prediction, analysis and scheduling' to 'help anticipate or solve resource allocation problems.' Chapanis (1996:115–6) describes time line analysis as a 'relatively minor and easily performed extension of task analysis' that can be used to create plots showing the temporal relationships among tasks, their length and timing. It is used to discover possible conflicts and it provides input for workload analysis and early estimates of personnel numbers. The analysis can be used to assess and predict the task performance of an individual or a team. Diagrams can be organized by operator (with multiple work assignments nested by operator). A line chart that parallels the time line can also represent the percentage of workload.

Time line analysis can be used in two ways: as a flow analysis method and as a follow-on to task analysis (Section 7.6).

- *Flow analysis*—As a flow analysis tool, time line analysis can be used at a more general level to estimate the order and duration of functions. It can also be used in combination with activity analysis (Section 6.4) in order to detect conflicting or simultaneous events.
- *Following task analysis*—As a refinement for task analysis, time line analysis can be used to organize the tasks that individuals are expected to perform in terms of precedence and duration. It can also detect potential cognitive overloads, such as too many visual or aural tasks required at the same time.

Charlton (1996a:185) portrays time line analysis as a means 'to identify how long tasks and task components will take and if they can be accomplished with the allotted time' assuming that 25 percent of that available time will be left uncommitted. Arriving at a total task time that is no longer than 75 percent of the time available allows for a buffer (slack time) that lessens the potential for operator overload.

Time estimates can be based on four sources that Woodson (1960:965–6) describes:

- Similar tasks that are currently performed in operating products
- A sample of operators who perform (or have performed) the task, using a detailed description of what is to be performed
- Expert opinion (e.g. 'equipment engineers'), in the event no experienced operators are available.
- Investigator ('the human engineer') breaks the task into its simplest elements to use in the estimation of complexity and duration.

Time line analysis can also be used as a project management tool. For example, the Gantt chart is a widely used time analysis technique that can be used in project task planning and management. A 2-axis Gantt shows a graduated time line along a horizontal axis and each project function such as planning, production and delivery along the vertical axis. An arrow is used to show the current date. Comparison of planned activity to the actual date will reveal whether the project is on, ahead of, or behind schedule.

7.2.1 Preparation

Time line analysis is performed using the same procedure whether it serves as a flow analysis method or as a follow-on to task analysis. Start by obtaining information on functions or tasks, their sequence, start and stop time and duration.

7.2.1.1 FOR USE AS A FLOW ANALYSIS METHOD

Review the results of operational analysis (Section 6.2) and flow analysis (Section 7.1) for functions and activity analysis (Section 6.4) for time-related information.

7.2.1.2 AS A METHOD TO REFINE TASK ANALYSIS

Review the results of task description (Section 7.5) or task analysis (Section 7.6). Expert opinion can be sought regarding task time estimates, 'skill levels, difficulty ratings, repetition frequencies and perceptual-motor channels' (Chapanis 1996:79–141).

7.2.2 Materials and equipment

Inexpensive tracing tissue can be used to quickly develop and revise diagrams and tables. Illustration or desktop editor software programs can be used to create smooth versions of tables and diagrams, particularly for those that are more complex.

7.2.3 Procedure

Table 7.3 summarizes the steps that comprise time line analysis.

Step 1—Organize the required tasks according to a linear sequence. Assign each task a unique number.

Step 2—Create a diagram to represent the series of tasks. Time line analysis diagrams such as Figure 7.5 typically list functions or tasks along one axis and show bars that are laid off along a time line that indicates duration and timing.

Step 3—Identify a time interval of interest (e.g. 5 seconds, 5 minutes). The greater the function or task complexity, the smaller the time interval will need to be.

Table 7.3 Time line analysis—analyze events with respect to time to determine their order and dura-
        tion. Time line analysis diagrams such as Figure 7.5 typically list functions or tasks along one
        axis and show bars laid off along a time line that indicate duration and timing.

Organize the required tasks in a linear sequence

Assign each task a unique number.

Create a diagram to represent the series of tasks

Identify a time interval of interest

The greater the function or task complexity, the smaller the time interval will need to be (e.g. 5 seconds versus 5 minutes).

Estimate task duration

Take particular account of high-difficulty tasks, which can take longer to perform.

Sum task times to be performed during each interval

Review the diagram

Look for opportunities to improve timing, duration, or assignment.

Step 4—Estimate task duration, taking particular account of high-difficulty tasks.

Step 5—Sum task times for each interval.

Step 6—Review the diagram for instances in which there are significant numbers of high difficulty
tasks at 75 percent loading or task time amounts to more than 100 percent of the time allotted
(Charlton 1996:181–99). In such cases, review the diagram for opportunities to revise timing, duration or assignment.

7.2.4 Result

Time line analysis produces a diagram of the functions or tasks that are expected to be performed
in a product shown in sequential order by time. The method also makes it possible to review and
assess activities that are performed at the same time.

   Review of the time line analysis plot will result in assignments that fit within allotted time, do not
conflict and allow enough time for complexity or difficulty. The results can serve as an input for workload assessment (Section 8.5).

7.2.5 Example

Emergency response teams need to work collaboratively. The faster a victim is treated, the more likely
that the trauma's effect will be minimized. Depending on the condition of the victim and the severity of
injury, a trauma victim may only have a short time before delay affects recovery. This is especially true
in the case of trauma that affects the supply of oxygen to vital tissues such as the brain.

   Multiple resources are brought to bear when the need for emergency services occurs. Products
or procedures that can lessen the time or use it more effectively will benefit the patient. A research
and development team that is assigned to improve emergency response would include a time line
analysis in their work.

   Figure 7.5 portrays the general activities that occur during the response scenario. In most cases,
a citizen will observe an event (e.g. accident, seizure, cardiac arrest) and may act to report the event
to authorities, possibly by phone. An operator who is responsible for incoming phone calls would
determine the nature of the need and correspond with an emergency services dispatcher who

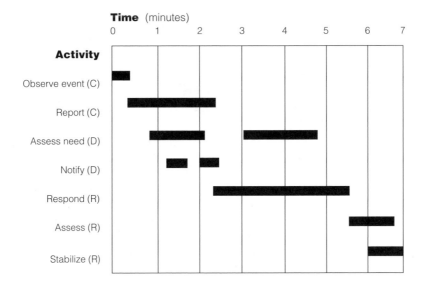

Figure 7.5 Time line diagram—time line diagram makes it possible to evaluate the activities of two or more persons who are involved in a collaborative activity. Time line diagrams can also be organized by operators (showing nested multiple activities). Percent of workload activity can also be represented in a line chart diagram that parallels the time line. In this case, individuals participate in an emergency rescue. Time intervals are set according to the time period under consideration and the level of precision that is necessary. Actions are taken by a (C) citizen, (D) dispatcher, (R) rescue crew. Time matters in rescue situations. In many cases, faster treatment lessens the degree of trauma that the victim incurs.

would identify the type of response that is required and assign a crew, vehicle and equipment. The responding crew would proceed to the site where the victim was reported to be, assess the victim's condition and take action to stabilize the victim. Prevention of any further deterioration in vital functions will prepare the victim for transportation to a facility that can provide more extensive care.

In this case, a time line would pose certain issues that could be used to improve the response process. Could the operator and dispatcher function be merged? Could units already on patrol (e.g. community service vehicles, police) have data displays that would show dispatcher activity and victim location? Is there a way to enable crews in the field to hear directly from a reporting citizen instead of going through a centralized operator and dispatcher?

7.3 Link analysis

Determine optimal arrangement of elements.

Link analysis (LA) is used to assess the relationships among a selected number of elements. Chapanis (1959:51–62) describes the method as 'a kind of flow diagram in which the linkages between various components are expressed in physical terms—relative frequencies between components.' Thomson (1972:422–3) considered it 'a technique which provides information needed to produce an acceptable arrangement of men and machines in a system' that 'can be found only by optimizing different types of links (such as communication and movement) that are important to the particular system being designed.'

7.3 Link analysis

While flow analysis (Section 7.1) is used to depict arrangements and dependencies among elements, LA takes flow analysis a step further. LA uses interaction frequencies to determine affinities among elements and to configure arrangements based on those relationships.

As a form of network analysis, the performance of LA produces an undirected graph that shows the connections among nodes. Network links (the 'link' in link analysis) can be used to represent physical contact, visual observation, spoken messages and more abstract relationships such as time. Nodes can be used to represent specific elements (e.g. people, equipment, displays, controls) or more abstract elements (e.g. cases in a study).

LA has traditionally been used to arrange equipment and individuals in a workplace and to arrange controls and displays on a console. New applications for LA are being developed beyond workspaces and consoles. Virtual and on-line displays and controls offer much more flexibility than hardware and equipment. This flexibility makes it easier and more compelling to rely on link analysis to offer optimal layouts. It also opens the way to refine configurations from generic 'one serves all' designs to arrangements that are better suited to specific groups and individuals. For example, Stuster and Chovil (1994:8–15) describe Harris' application of the method to criminal intelligence analysis. They also recount their own application to better understand outbreak epidemiology. Using the computer-supported link analysis software programs Criminal Network Analysis© and Enhanced Criminal Network Analysis©, they evaluated the cases that were involved in an outbreak of measles and derived relationships that suggested patterns of infection and where and how they occurred.

Other network analysis methods such as critical path method (CPM) and program evaluation and review technique (PERT) are similar to link analysis and used for 'planning, scheduling, distribution and layout or arrangement' (Laughery and Laughery 1987:329–54).

7.3.1 Preparation

Observe the operation of products that are currently in use. Observe the use of mock-ups or new concept simulations.

Perform activity analysis (Section 6.4) and task analysis (Section 7.6), or review the results of those methods. If necessary, review the results of operational analysis (Section 6.2) to determine link importance.

If creating a layout, obtain the physical measurements of the available area whether it is a room, a console or a display.

7.3.2 Materials and equipment

Inexpensive tracing tissue can be used to quickly develop and revise diagrams and tables. Illustration or desktop editor software programs can be used to create smooth versions of tables and diagrams, particularly for those that are more complex.

Displays can be simulated through the use of illustration or layout programs. Graphical controls embedded with hyperlinks and displays can be portrayed and operated by subjects. Simulation (Section 8.1) offers a number of other techniques.

Good relationship data collection is an important aspect of link analysis. Seek opportunities to use object methods to collect information on interaction. For example, eye tracking can be used to accurately capture data on eye scan across a display. Combined with keystroke and mouse operation, the data would provide reliable support to estimate links between the operator, control elements and display elements.

Computer supported software programs such as Enhanced Criminal Network Analysis© can be used to facilitate the link analysis process.

7.3.3 Procedure

Seven steps, based on Chapanis (1996:116–23), are used to perform LA and are shown in Table 7.4.

Step 1—List all items of equipment and people. Identify each using a unique number or letter (for use in the interaction matrix and the link diagram).

Step 2—Measure (or estimate) frequencies of linkages between operators and between operators and equipment. Linkages can be any instance of interaction (e.g. spoken message, physical control actuation, visual display status check, audio signal, physical relocation). Woodson (1960:965–6) recommends coding the link (e.g. control, visual, audio) to differentiate types of interaction in the diagram.

Step 3—Measure/estimate the importance of each link based on its value to the purpose of the product.

Step 4—Figure the frequency/importance value for each link by multiplying its frequency by its importance.

Step 5—Plot equipment/operator layout using highest links to establish their configuration. Successively add operators and equipment with lower link values.

Step 6—Fit the layout into the plan of available space.

Table 7.4 Link analysis—determine the elements that comprise the product and how they are related. Link analysis has traditionally been used to arrange equipment and individuals in a workplace, and to arrange controls and displays on a console. New applications for link analysis (e.g., custom configuration, criminal case analysis) are being developed beyond workspaces and consoles.

For facility layout

List all items of equipment and people

Measure (or estimate) frequencies of linkages between operators and between operators and equipment

Measure/estimate the importance of each link

Figure frequency/importance value for each link

Plot equipment/operator layout using highest links. Successively add operators and equipment with lower link values

Fit layout into plan of available space

Evaluate layout quantitatively

For control/display layout

List all items of controls, displays and people

Measure (or estimate) frequencies of linkages between people and equipment

Measure/estimate the importance of each link

Figure frequency/importance value for each link

Plot control/display layout using highest links. Successively add controls and displays with lower link values

Fit layout into available space on display

Evaluate layout quantitatively (e.g. error rate, time to complete tasks)

Step 7—Evaluate the layout quantitatively. Collect data on variables of interest such as time to perform task, traffic flow analysis or productive output.

7.3.4 Result

The performance of LA produces three items: an interaction matrix, a schematic link diagram and a layout recommendation.

7.3.4.1 INTERACTION MATRIX

The matrix or 'activity relationship diagram' (Figure 7.6) accounts for each element and the degree and kind of interaction it shares with every other activity or function. Letters in the matrix indicate how important being close is, using commonly accepted terms: A (absolutely necessary), E (especially important), I (important), O (ordinary closeness), U (unimportant) and X (not desirable). Importance in this case is a function of number of trips a day between stations and direction of the trip.

Activity/function	Supplies container loading	Delivery loading area	Restroom	Waste disposal	Cold storage	Service counter—3	Service counter—2	Service counter—1	Fountain	Grill	Preparation area	Ingredients storage	Unloading area	Container storage
Container storage	E	O	O	I	E	U	U	U	U	U	O	E	E	
Unloading area	A	A	U	O	I	X	X	X	X	X	I	A		
Ingredients storage	A	I	U	X	E	I	I	I	I	E	E			
Preparation area	O	O	X	O	E	I	I	I	I	A				
Grill	U	U	X	O	I	I	I	I	I					
Fountain	U	U	X	O	I	I	I	I						
Service counter—1	U	I	O	O	O	E	E							
Service counter—2	U	I	O	O	O	E								
Service counter—3	U	I	O	O	O									
Cold storage	O	U	U	O										
Waste disposal	E	O	O											
Restroom	O	O												
Delivery loading area	E													
Supplies container loading														

Key: Link importance

A   absolutely necessary
E   especially important
I   important
O   ordinary
U   unimportant
X   not desirable

Figure 7.6 Link analysis interaction matrix—an activity relationship diagram shows the frequency and type of interaction among elements in a dairy. Codes indicate the degree of interaction between elements from highest to lowest (A, E, I, O, U, X). Figure 7.7 shows the link (or string) diagram that can be created from the information in a matrix such as this.

7.3.4.2 SCHEMATIC LINK DIAGRAM

Relationships in the interaction matrix can be used to create multiple possible versions of a link or 'string' diagram as shown in Figure 7.7. The diagram represents each element, its relative location in a two-dimensional layout and the type and degree of its interaction. Shape codes show the types of functions that are present in the facility. It can be most helpful to start the layout using the element having the strongest (most frequent, most important) relationships, then locate elements with the strongest relationships to it nearby. Elements with successively less important relationships can then be added until all of the elements have been portrayed.

7.3.4.3 LAYOUT RECOMMENDATION

A two-dimensional illustration (often a plan view) represents the display, console, work station or work area configuration that will best meet system objectives.

The study of abstract elements (e.g. social or work relationships) can be represented in diagrams that show clusters or hierarchies of relationships.

7.3.5 Example

The configuration of space in a work environment has a significant influence on productivity. Products such as open landscape furniture systems, flexible and moveable wall systems and relocatable work-stations can be easily configured in any manner that an organization requires. Space planning has evolved as a practice in architectural programming and construction to manage such issues.

What exactly is the best arrangement? The correct answer to that question can mean the difference between a productive facility and one that isolates the workers who need to collaborate. Link analysis can be used to ensure that the configuration supports, rather than impedes, the human organization. To begin, list all items of equipment and people including full time, part time and flex

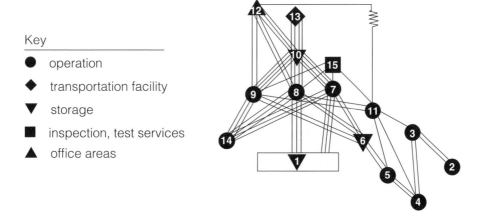

Figure 7.7 Link analysis diagram—the link (or string) diagram shown indicates the elements in a dairy facility. Their interaction is based on an interaction matrix, shown in Figure 7.6. Arrangement of elements is based on link intensity and importance. The strong relationship between carton storage (1) and filling machines (7, 8 and 9) sets the basic scheme. The truck loading area (13), cooler (10) and filling machine (8) also exert a significant influence on the arrangement. 'A' relationships are absolutely necessary and indicated by four lines. An 'X' (not desirable) relationship is shown by a zig-zag line.

Source: Adapted from *Introduction to Industrial and Systems Engineering*. Turner, W., Mize, J. and Case, K.

workers. One automated means to measure the frequencies of linkages between operators and between operators and equipment is to install a digital camera connected to a pressure sensitive mat (or a similar kind of motion detector) at key locations. The time- and date-stamped images will provide a record of the timing, frequency and nature of the interactions. Such records might also lead to discoveries beyond simple interaction frequencies. The data collected might show that support staff for the teams have a much higher frequency of interaction than expected. That could lead to an improved arrangement for work teams and those who are responsible for their support.

The importance of each link might be based on productivity. For example, work team interactions may figure heavily in an organization. Interaction between work team members might be rated higher than service interactions (e.g. maintenance staff).

Multiply the frequency by the importance to obtain the frequency/importance value for each link. Plot equipment/operator layout using highest links. Successively add operators and equipment with lower link values. At this point, clusters will develop based on stronger and weaker link values. In a space-planning problem, this can lead to an arrangement such as work team bays located around a central core of shared support staff.

Fit layout into plan of available space. Install a pilot layout and have a work team shift into it for a trial period. Then, evaluate the layout by collecting data on quantitative measures (e.g. traffic flow analysis) and qualitative measures (e.g. interviews or questionnaires as described in Chapter 9). A successful analysis and design can optimize the placement of individuals and teams, the flow of work and the use of resources. Poor or no analysis can impede a working group's productivity and (at worst) render it dysfunctional.

7.4 Function allocation

Assign functions to humans, hardware and software.

Function allocation (FA) is used to assign the performance of each function to the element (humans, hardware or software) that is best suited to perform it. It normally follows the performance of flow analysis (Section 7.1), time line analysis (Section 7.2) or link analysis (Section 7.3).

Management, requirements and designers influence FA decisions. Management can dictate certain functions that will be performed by humans, hardware, or software. Requirements can specify a function that is beyond the ability of humans to perform (e.g. lift 500 pounds ten feet in height). Designers can make allocation decisions based on human, hardware or software capabilities (Bailey 1989:188–94). Similar influences were described in influences on research and development (Section 4.2).

Bailey catalogs six strategies that have been used through time to make allocation decisions: comparison, leftover, economic, humanized task, flexible allocation and allocation by users.

- *Comparison*—Comparison allocation assigns functions according to a standing list of human and machine capabilities, what human factors pioneer Paul Fitts called 'men are better at, machines are better at' ('MABA-MABA') lists (Sheridan 1998a:20–5). Unlike the 1950s when Fitts did his work, machine capabilities are now changing at a rapid rate.
- *Leftover*—Leftover allocation assigns as many functions as possible to the computer, then allocates the remaining 'leftover' functions to the human.
- *Economic*—Economic allocation bases assignment on which element incurs the least cost to accomplish the function, ignoring life cycle cost.

- *Humanized*—Humanized task allocation is the obverse approach to leftover allocation: a worthwhile set of functions for an operator to perform is assembled and the remaining functions are allocated to the computer.
- *Flexible allocation*—Users assign functions according to values, needs and interests. Flexible allocation is used much more frequently as product intelligence improves.
- *Allocation by users*—Bailey's preferred approach to allocation is to create alternate ways to accomplish the same set of activities, so that the operator can choose the type and amount of allocation according to conditions.

Woodson (1960:965–6) also contributes considerations for how to make allocations, based on capabilities similar to those outlined in Chapter 2. They include: environmental constraints (humans are vulnerable to environmental stressors), sensory isolation (humans must receive inputs above sensory thresholds), speed and accuracy, physical strength, storage capacity, surveillance (what has been referred to earlier in this book as the monitor role), the interpretation of and response to unexpected events (humans are highly adaptive), fatigue (hardware is subject only to failure and not fatigue), learning and cost.

The basis for making allocation decisions is an ongoing source of discussion among human factors practitioners. Interactions between humans and machines are complex and are not well understood. Sheridan (1998:20–5) concludes that no clear sense has evolved among practitioners as to whether a rational basis for allocation can exist. Many variables are used to define goodness in a system after the fact: 'performance, efficiency of operation, safety, aesthetics, justice … As a result, such knowledge may remain too difficult to convert into an independent objective basis for allocation decision making.'

David Meister (1985:21–9) allows that allocation decisions are often made informally (e.g. the use of Fitts' list) yet strongly favors a more quantitative approach in light of the stringent performance requirements that must be satisfied. Meister is concerned that, without a comprehensive data bank of knowledge about human performance, our grasp of human abilities is insufficient. In addition, the difference in the way that humans and machines and their performance are described prevents an equitable comparison and assignment (Meister 1989:181–219).

The amount and rate of function reallocation has significant implications for human performance and, ultimately, product performance. This is particularly true in complex systems such as aircraft that are experiencing changing conditions. The operation of an aircraft is an example of a complex system that continually confronts changing conditions such as relationships with nearby aircraft, flight path assignment, weather, fuel and trim. Traditionally, the allocation of functions was considered to be static. Once allocated to human, hardware or software, the assignment could not be changed. However, the growth of intelligent products has made assignment flexibility possible in many circumstances. To take advantage of that flexibility, Sharit (1997:301–39) suggests that allocation should be thought of as a continuum between static allocation (i.e. each element performs a function assigned to it) and dynamic allocation (tasks are assigned and reassigned as ongoing events dictate).

Sheridan (1998:20–5) uses 'supervisory control' to describe cooperation along 'qualitative scales of degrees of human and computer control.' He suggests that 'insofar as it is feasible and practical, the human should be left to deal with the "big picture" while the computer copes with the details' (Sheridan 1999:591–628). To illustrate, let us return to the example of aircraft operation. Under a supervisory control arrangement, functions can be managed in three different ways:

7.4 Function allocation

- *Performed solely by one element*—The pilot flies the aircraft using manual controls.
- *Shared by two elements*—The pilot and an automated (fly-by-wire) system control the aircraft at the same time. Higher order directions by the pilot control lower order adjustment of control surfaces by the automated system.
- *Traded by handing control back and forth*—The automated flight control system is assigned control of the aircraft based on predetermined rules. Pilot control can be reassumed by choice or when rule parameters are exceeded (e.g. hazardous weather conditions).

The choice to employ dynamic allocation models should be approached with caution. Such allocation models must be developed and implemented with genuine appreciation for their implications. The human operator *must* be prepared to assume functions if and when they are reallocated, which can occur either by the operator's choice to intervene or by the system's reversion to a default condition.

Intervention—If automated systems are inadequate, the operator must assume control. If human performance jeopardizes the safety of the system, the system may need to be allowed to assume control. When should that happen? Can the operator make decisions adequately? Sheridan (1999:606–8) cautions that such conditions are not yet well defined as 'we continue to lack sufficient understanding of human decision making under stress, trust, situation awareness and mental workload.'

System default—A system can place the operator in immediate position to assume control, unannounced. For example, an autopilot is intended to allow pilots to allocate control over an aircraft to its avionics during routine conditions. If conditions change (e.g. ice build-up on the wings), the avionics may not be programmed to detect the change in aircraft stability. If the change exceeds the autopilot's ability to maintain stable flight, control may be instantaneously released to the pilot. In such circumstances, it is nearly certain that the pilots will not be able to successfully understand and correct the situation in time to rectify it.

7.4.1 Preparation

Sharit (1997:301–39) catalogs three system decomposition approaches that are used to prepare for FA:

Flow analysis methods—flow analysis methods represent the sequence of events using flow process charts, operational sequence diagrams, functional analysis systems technique (FAST).

Time line analysis—time line analysis represents the evolution of events using Gantt charts. Gantt charts plot time along a horizontal axis and functions along the vertical axis. Bars aligned with each function show start and stop dates and duration. Tracking the current date makes it possible to determine whether any function is on, ahead of, or behind schedule.

Network analysis—Network analysis analyzes relationships among events, entities and processes using computer-based simulation such as systems analysis of integrated network of tasks (SAINT).

Step 1—Collect information from or perform flow analysis (Section 7.1) or time line analysis (Section 7.2).

Step 2—Interview experts or those who can provide information on systems that are similar to the concept under development.

Step 3—Collect and review information on the performance capabilities for state-of-the-art hardware and software. Vendors, product registries and trade publications that conduct side-by-side comparisons are often good sources to start reviews.

Step 4—Review information on the capabilities and limitations of human performance. Human factors texts and databases can provide basic information on important parameters such as reaction time and decision making.

7.4.2 Materials and equipment

Inexpensive tracing tissue can be used to quickly develop and revise tables. Word processing or desktop publishing software programs can be used to create smooth versions of tables, particularly for those that are complex.

Computing resources and software will be needed if simulations such as SAINT are to be used.

7.4.3 Procedure

Table 7.5 summarizes both static and dynamic approaches to FA.

7.4.3.1 STATIC

Chapanis (1996:100–5) offers this approach to FA:

Step 1—Identify all functions that must be allocated to either personnel or equipment and remove them from consideration.

Step 2—List the remaining functions and prepare descriptions of the alternate ways the functions could be accomplished.

Step 3—Establish weighting criteria.

Step 4—Compare alternatives against criteria.

Step 5—Select the alternative with the best performance and cost-effectiveness.

7.4.3.2 DYNAMIC

Clegg *et al.* (1989:175–90) propose a method, shown in Figure 7.8, that can be used to allocate functions in dynamic systems. Phases 1 through 3 are accounted for by other methods that are described in this text: operational analysis (Section 6.2) is used to specify system objectives and to account for

Table 7.5 Function allocation—assign functions to humans, hardware, and software.

Identify all functions that must be allocated to either personnel or equipment and remove them from consideration
List the remaining functions
Prepare descriptions of the alternate ways the functions could be accomplished
Establish weighting criteria
Compare alternatives against criteria
Select the alternative with the best performance and cost-effectiveness

requirements and functional analysis methods such as flow analysis (7.1), time line analysis (7.2) and link analysis (7.3) can be used to specify functions. Phases 4 through 7 contain the essence of the method which is listed here as five steps:

Step 1—Check against constraints and make any mandatory allocations.

In the sense of sorting out required assignments, the step mirrors Chapanis' static allocation approach.

Step 2—Determine provisional allocations using four groups of criteria:

- Technical feasibility
- Health and safety
- Operational requirements: physical, information processing, performance
- Function characteristics: criticality, unpredictability, psychological implications.

How well the development team can know these criteria and evaluate each option is open to question.

Step 3—Based on the review in the foregoing step, allocate functions to: human, machine or both. If neither human nor machine is appropriate, a reassessment of the intended 'human machine' product would be necessary.

Step 4—Check individual and cumulative allocations over time against:

- Requirements specification (including scenarios)
- Resources available (including financial, human)
- Sum total outcome.

Specifications and resources can be known. Presumably 'sum total outcome' would involve the collection of information on performance as well as the effects of the allocation (e.g. human subjective response).

Step 5—Allocate: the return to 'Phase 5' in Figure 7.8 indicates monitoring any changes that might occur in the criteria that had been used as the basis for allocation.

7.4.4 Result

FA identifies the means by which activities that are to be performed will be assigned to humans, hardware or software. It also identifies conditions under which those assignments will be changed and what the means for reallocation will be.

FA identifies needs for skills and information, staffing and training. It also provides the basis for the production of solution concepts based on elements that are identified to perform activities.

7.4.5 Example

Allocation is simplest for functions that lie at the extreme range of capabilities. Very simple functions can be too boring for humans and are easy to assign to hardware and software. Very complex functions remain beyond the ability of hardware and software and require a human to perform them. The greatest challenge in allocation decisions lies in requirements that lie between both extremes.

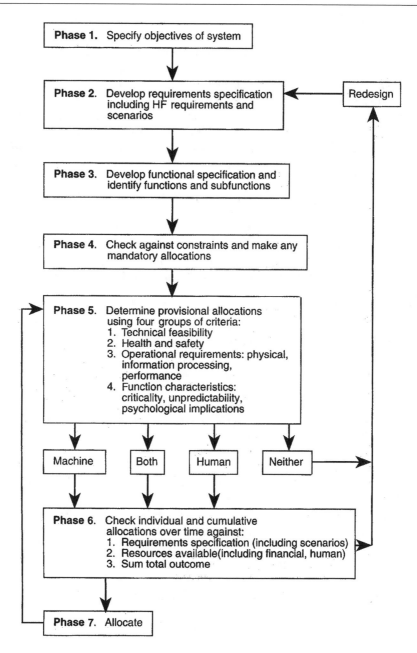

Figure 7.8 Broad-based methodology for allocation of functions—Clegg, Ravden, Corbett, and Johnson (1989) present an approach for the dynamic allocation of functions that is particularly germane to more sophisticated systems in which responsibility for performing functions can and should change with conditions.

Source: Copyright 1989 *Allocating Functions in Computer Integrated Manufacturing* by Clegg, C., Ravden, S., Corbett, M. & Johnson, S. Reproduced by permission of Routledge, Inc., part of the Taylor & Francis Group.

The development of an intelligent vehicle highway system (IVHS) driver interface poses such an opportunity. For years, drivers in many countries have spent long hours at the wheel of an auto to get from one place to another. Driving can afford greater convenience, cost-effectiveness and flexibility by comparison with public transportation. Driving long distances can also contribute to fatigue and discomfort that can cause safety problems.

Recent developments have made driving more convenient and safer through reallocating functions between driver and auto. Four-wheel drive originally required manual engagement. Now, 'all-wheel drive' engages multi-wheel drive without driver intervention. Cruise control made it possible for the auto to assume control over vehicle speed until overridden by the driver. This enabled the driver to forego paying constant attention to speed control under conditions such as long distance highway driving.

Efforts that are now underway to create guided highways may allow further allocation of functions from the driver to the auto. How would that be done and what are the implications? The following describes considerations for function allocation decisions.

A number of variables are involved, including weather and road conditions, traffic congestion, emergent situations, driver preference and the availability of IVHS guidance. The following considerations would make it possible to develop a model to allocate functions in such a system.

Mandatory allocations—manual control will be required:

- On any roadway that is not equipped with IVHS
- Upon driver demand

Provisional allocations:

- Technical feasibility—Is the technology necessary to create such a system available or attainable? Will systems and components that are necessary to accomplish the requirements do so within requirement parameters?
- Health and safety—Under what circumstances would driver or passenger safety be compromised?
- Operational requirements (physical, information processing, performance)—What are the steps that the system would need to accomplish?
- Function characteristics (criticality, unpredictability, psychological implications)—What aspects of system operation will influence allocation of tasks to the driver versus the hardware/software?

Allocate functions provisionally:

Developers might perform an analysis of similar systems (Section 6.3) including air traffic control systems to determine allocation strategies.

- Human—Control during transitions into and out of IVHS
- Machine—Control during operation on major long distance controlled access roadways
- Both—When conditions are compromised (e.g. rain, snow, ice, oil reduce traction).

Check individual and cumulative allocations over time against:

- Requirements specification (including scenarios)
- Resources available (including financial, human)
- Sum total outcome.

An IVHS system prototype (see Simulation, Section 8.1) would be used to develop data regarding system and human performance. Data that are collected can be used to review provisional allocation decisions. Final choices regarding allocation of function would be used in auto controller software development, control and display development and driver training.

7.5 Task description ('task identification')

Describe what humans are expected to do.

Task description (TD), or what David Meister (1985:29–31) has also called 'Task identification,' is a method that is used to list and describe tasks that are to be performed by means of a taxonomy (classification according to relationships). TD accounts for each aspect of human behavior that was derived through functional analysis and sets the stage for task analysis. Drury, Paramore, Van Cott, Grey and Corlett contend that task description resembles a flow chart process. The difference between TD and flow analysis is that what is being charted is what happens to an operator rather than what happens to a product (Sharit 1997:301–39).

Because TD is relatively easy to accomplish, most authors include it as a part of task analysis. Meister favors keeping it as a separate method that establishes design requirements in preparation for the more sophisticated method of task analysis. TD can, then, be considered the bridge between functional analysis and task analysis.

Many authors use the terms 'function' and 'task' interchangeably. One way to make the distinction is on the basis of continuous versus discrete. Functions tend to be continuous. Conversely, tasks are discrete, with a starting (trigger) point and an ending (goal) point and some element of feedback upon completion (Sharit 1997:301–39).

Meister (1989:181–219) admits that there is no absolute threshold for the distinction between function and task because 'At a certain level of detail—which is difficult to specify—the function shades almost imperceptibly into a task.' Meister distinguishes between functions and tasks on the basis of molar versus granular (i.e. more general versus more specific) description. 'Functions describe relatively molar behaviors—for example, to detect, to analyze, to repair. The individual tasks and behaviors needed to carry out the function are more detailed.' Which end should the investigator favor: molar or granular? Actually, either method is correct in certain circumstances. The investigator should remain more general (molar) when making a comparison of tasks across a variety of systems. A more molar approach makes it possible to avoid aspects that pertain to only one or a few instances. The investigator should favor a more detailed (granular) approach during the examination of tasks in an individual system design. The more granular approach is specific and thorough, which is essential to good individual system design.

7.5.1 Preparation

Review the results of functional analysis methods: flow analysis (Section 7.1), time line analysis (Section 7.2) and link analysis (Section 7.3). Arrange for the opportunity to interview subject matter experts (SMEs) regarding task description. Ensure that data are available on proposed concepts.

7.5.2 Materials and equipment

Refer to books or checklists for guidance on similar or related tasks.

7.5.3 Procedure

The following procedure for TD that Table 7.6 depicts is drawn primarily from Meister (1985:29–31).

Step 1—Examine each selected design alternative's described functions

- *Conduct unstructured interview*—Conduct unstructured interviews with subject matter experts. If possible, arrange for a '2 on 2' interview among two investigators and two experts. The use of two experts will offset the bias that is possible using a single expert and enable each to stimulate insights by the other. The use of two interviewers will broaden the scope of inquiry. It also allows one to question and the other to write, making it more likely that important comments by the experts will not be overlooked.
- *Conduct structured interview*—Conduct a series of structured interviews in order to establish difficulty, hazard levels, importance, learning problems, practice required to maintain proficiency, frequency of actual repetitions and training media that might be used.

Step 2—List all actions

List the actions that must be performed in order to accomplish each function that is necessary for the design alternative. Refer to books and checklists to supplement information that was gleaned from subject matter experts.

Step 3—Categorize the actions

Sort actions according to whether they are to be performed by an operator or maintainer and according to the hardware/software subsystem to which they belong.

Table 7.6 Task description ('task identification')—describe what humans are expected to do.

Review design alternatives
Examine each selected design alternative's described functions
List all actions
List the actions that must be performed in order to accomplish each function that is necessary for the design alternative
Categorize the actions
Sort actions according to whether they are to be performed by an operator or maintainer, and according to the hardware/software subsystem to which they belong
Describe each action
Use the construction of a behavioral verb, the equipment acted upon, the consequence of the action, the stimulus that initiates the action, feedback information that results from the action and task accomplishment criteria
Decompose tasks (if desired)
Break tasks down into subordinate tasks
Specify inputs and outputs from each task

Step 4—Describe each action

Use the construction of a behavioral verb (e.g. to turn on), the equipment acted upon (e.g. a motor), the consequence of the action (e.g. the motor operates), the stimulus that initiates the action (e.g. operator finger motion), feedback information that results from the action (e.g. the 'on' light illuminates) and the criteria of task accomplishment.

Figure 7.9 provides a formatted approach for task description. The upper portion of the format contains general task-related information. The lower portion is used to provide a detailed task description. The columns are arranged so that the terms can be read as a sentence. For example, subject (operator) performs an action (descriptive verb) addressing a parameter or component (object of the action) using the following component (means of action, such as a control or display) (Sharit 1997:301–39).

Step 5—Decompose tasks (if desired)

Break tasks down into subordinate tasks. Specify inputs and outputs from each task.

ACTION-INFORMATION REQUIREMENTS DETAILS (AIRD)   Sheet _____ of _____

PLANT: _____ UNIT: _____   REVIEWER: _____ DATE: _____

SEQUENCE: ___Alternate Shutdown Cooling_____ NO.: _____   INITIATING CUES: Cooldown requirements not met;
FUNCTION: ___Heat Removal_____ NO.: _____   all rods in past 06; RPV pressure LT 300 psig
TASK NAME: ___Initiate flow thru RHR heat exchgr___ NO.: _____   TERMINATING CUES:
TASK OBJECTIVE: ___To establish Suppression Chamber water cooling___   Flow through Hx in band (_____ – _____ gpm)
                                                        REMARKS:

NATURE OF TASK: Continuous_____ Discrete __X_____

Line No.	Verb	System/ Subsystem	Component	Parameter	State/ Direction	Value/ Range	Units/ Rate	Precision	Accuracy	Resp Time	Trend Req?	If Yes ··	If No ··	Comments	on AIRS Sh. No.	Seq Pkm
	positions	RHR	valve (Hx bypass)		close											
	observes	RHR	valve (Hx bypass)		closed											
	positions	RHR	valves (flow contr)		open (proper lineup to SC)											
	observes	RHR	valves (flow contr)		open											
	positions	RHR	pump		start											
	observes	RHR	pump		on											
	observes	RHR	pump	amps	in band											
	observes	RHR	pump	discharge pressure	in band											
	observes	RHR	Hx	flow (shell out)	in band											

Figure 7.9 Task description format—Drury, Paramore, Van Cott, Grey and Corlett provide a formatted approach to the description of tasks which is shown here. The upper portion of the format contains general task-related information. The lower portion is used to provide a detailed task description. The application in this instance is a task related to the control of a nuclear power plant.

Source: 'Task Analysis.' Drury, C., Paramore, B., Van Cott, H., Grey, S. and Corlett, E. In Salvendy, G., ed. *Handbook of Human Factors.* Copyright © 1987. This material is used by permission of John Wiley & Sons, Inc.

7.5.4 Result

TD produces categorized lists of what humans are expected to do in a system, sorted by role, hardware/software involved in task performance and task inputs and outputs. TD makes it possible to:

- Group tasks according to criteria such as purpose, common equipment or location, performance by the same operator
- Sort tasks into groupings of jobs or work roles
- Determine potential needs for control and display hardware
- Estimate requirements for human skills, knowledge and number of personnel
- Provide some of the input that is needed in order to perform task analysis (Meister 1985:29–31, 1989:181–219).

7.5.5 Example

Drury et al. (1987:370–401) illustrate the use of TD in control room operations at a nuclear power plant. While comparatively few individuals will work on nuclear power plants, many create workstations and control stations. Consider the similarities in task description methods between both types of applications.

Nuclear power plant control room operations were studied to provide task data in six areas: human engineering of control rooms (both retrofits and new construction), staffing, operation and training, procedures (normal, off-normal, emergency), job performance aids and communications. The operating sequence (a series of tasks) was chosen as the unit of analysis. Twenty-four such sequences were identified. Task descriptions were created through the collection of descriptive data (items that characterize the task and could be collected by the direct observation of worker performance). In case worker performance could not be observed, plant documentation and operators could be consulted.

Use of a format shown in Figure 7.9 made it possible to record descriptive information. Columns were arranged based on a model sentence format that encouraged content and construction consistency. For example, the element in the first line of Figure 7.9 could be read '(Operator) positions RHR heat exchanger bypass valve closed.' The deliberate control over terms facilitates the transition to other forms of analysis, such as task analysis.

Quantitative data would then be recorded in the fields to the right of the form. Such information can include the specific operating value or band of a parameter or units of change. The information provides the basis for task analysis, which assesses the tasks that are assigned to humans in terms of their abilities and limitations.

7.6 Task analysis

Create the tasks, training, equipment and information needed to perform a job.

Task analysis (TA) is used to refine the functional requirements of a system in order to describe the tasks that people will perform. Woodson (1960:965–6) points out that TA 'tells us not only whether the design is feasible, but also whether it is optimum from the user's standpoint.' TA also 'provides important information for the personnel subsystem specialist who must plan and devise personnel acquisition and training programs for the system.'

Woodson describes three areas in which TA influences a solution concept: task criticality, task duration and task difficulty.

Criticality—TA can discover potential errors in task performance that can be made, the effect each error would have and implications for possible loss or objectives achievement.

Duration—TA determines the time within which a task must be performed and the time within which an operator can perform the task.

Difficulty—TA can determine the potential for task performance requirements to be incompatible and possibly overload operator abilities. Difficulty can emanate from excessive precision (visual discrimination, fine motor responses, reaction speed, coordination among tasks), environmental stressors and information considerations (the integration of multiple sources, inadequacy, complexity, excessive demand on short-term memory).

Cushman and Rosenberg (1991:33–47) sort the kinds of tasks that need to be analyzed into three types: sequential, branching and process control.

Sequential (also referred to as a 'discrete' or 'procedural task')—Subtasks or task elements must be performed in a predetermined order (e.g. starting an auto).

Branching—Finite set of choices is available at each step in the task and tends to depend on previous choices made (e.g. word processing). TA begins by creating an operational sequence diagram (Section 8.4), then (time permitting) analyzing each branch as if it was a sequential task.

Process control (also referred to as a 'continuous task')—Involve continuous monitoring, user-initiated control movements in response to changes and feedback such as steering an auto.

Chapanis (1996:105–9) accounts for six categories of information that TA is used to identify:

- *Information requirements*—Information that is available to or required by operators, how operators evaluate information
- *Action requirements*—Actions the operator takes, movement frequency, precision, speed, feedback
- *Working conditions*—Available and required work space, required job aids, hazards
- *Staffing requirements*—Number of operators and their interaction
- *Training requirements*—Required training, experience
- *Environmental considerations*—Nature (e.g. temperature, noise, ice, sand) and location of work place.

Meister (1985:31–49) expands on the above list by provding questions that can be used to determine job design, staffing, training and test and evaluation.

Job design—What tasks are to be performed? How critical is this task? In what sequence will tasks be performed? What information does the task performer require? What control activations are required? What performance requirements are important? Is coordination with other tasks or personnel required? Are perceptual, cognitive, psychomotor or physical demands imposed by the task excessive? What errors are possible and how likely are they?

Staffing—How many people are needed to perform a task or job? What skill level is required to perform the task/job?

Training—On what behavioral dimensions are the tasks based? How difficult/complex is the task or job? What information does the operator need to perform the task/job? What criteria indicate that the task has been correctly performed? What are the consequences if the task or job is not performed or is performed incorrectly? How is the task related to other tasks, to the total job? How frequently is the task or job performed?

Test and evaluation—What are the performance criteria for the task or job?

At a minimum, the results of each analysis should include: task title, initiating stimulus, equipment acted upon, response to be made, feedback and output characteristics including performance criteria.

Drury *et al.* (1987:373) explain that a task is generally agreed to be 'a set of human actions that contributes to a specific functional objective and ultimately to the output goal of a system.' TA is the comparison of demands for performance (derived during task description) to human capabilities. The form of a TA depends on the purpose for which the analysis is to be conducted. Bailey (1996) offers two approaches that can be taken to analyze tasks: traditional and object-oriented. The traditional approach decomposes tasks and ends with the identification of work modules on the assumption that the user will follow a clearly understood process. The object-oriented approach focuses on the creation of user interface objects that are of most interest to typical users which can be used in any order while using the new product (most likely an intelligent machine such as a computer). In either case, the process involves task identification, task description and the synthesis of identified tasks into manageable work modules. Work modules make it possible to prepare facilitator materials (e.g. instructions, performance aids, training) and user scenarios.

Application subtleties and complexity have led to the creation of many refinements to TA. The large number of TA methods and the similarity among the methods can be confusing to any but those who are intimately familiar with the field.[1] For ease of understanding, the following description of TA provides the method's structure from Drury *et al.* (1987) and the specifics from Chapanis (1996) and Meister (1985).

7.6.1 Preparation

Collect and review the results of previous methods, including operational analysis (Section 6.2), functional analyses including flow analysis (Section 7.1), time line analysis (Section 7.2) and link analysis (Section 7.3) and function allocation (Section 7.4). Interview subject matter experts (SMEs) or others who can provide information on similar systems. Review (at least preliminary versions of) system documentation that describes hardware, software and operating procedures.

Decide why the TA is being performed, whether some or all tasks will be analyzed, the level of detail the analysis will include, questions that need to be answered and how they will be answered (Meister 1985:31–49).

7.6.2 Materials and equipment

No materials or equipment are unique to TA.

7.6.3 Procedure

Drury *et al.* (1987:389–95) describe four phases of TA, noting that there is no 'general purpose task analysis, only specific classes of application.' Each phase includes its own procedures that are shown in Table 7.7.

7.6.3.1 SYSTEM DESCRIPTION AND ANALYSIS

Step 1—System functional analysis

- Define potential modes of operation or system states (e.g. vehicle guidance)

Table 7.7 Task analysis—create the tasks, training, equipment and information needed to perform a job.

System description and analysis

System functional analysis
Define potential modes of operation or system states
Examine equipment configuration of the system
Describe the general process of function execution

Operational sequence analysis
Develop an operating sequence profile
Develop a narrative description of the sequence
Develop functional flow diagrams
Develop decision-action diagrams

Task description task list
Detailed task description

Descriptive data collection

Applications analyses

- Examine equipment configuration of the system—Determine how components and subsystems accomplish functions.
- Describe the general process of function execution—Include human and hardware roles. Identify performance criteria.

Step 2—Operational sequence analysis

- Develop an operating sequence profile—Describe sequence of major events.
- Develop a narrative description of the sequence—Define starting and ending conditions, major events or phases, functions invoked by the events or phases, major human or equipment actions.
- Develop functional flow diagrams—Depict the relationships and sequence of functions and performance requirements to account for the functions that the product is to perform and to portray their order. See Flow analysis (Section 7.1).
- Develop decision-action diagrams—(Can be in place of, or in addition to, functional flow diagrams.) This flow analysis method can be used to identify the information that is needed for each action or decision to occur.
- Develop operational sequence diagrams—Describe operator, equipment and software actions in time order sequence. See Operational sequence analysis (Section 8.4).

Step 3—Task description

- Task list—Account for the scope and sequence of human performance requirements. Include an alphanumeric code and/or title for mission/operating sequence, system function, behavioral function and task sequence (based on task start time); task title (what the human is expected to perform) and task purpose/objective (reason for the task, output that is expected).
- Detailed description—Woodson (1960:965–6) provides categories of information that can be used to organize a detailed description. Describe each task as a set of information require-

ments, including: location (where task is performed), task (equipment, support equipment and aids, displays, controls, detailed action sequence, feedback on performance), time (duration and how often it is performed), criticality (effect of failure on the purpose for the product) and performance criteria and standards. Drury *et al.* also recommend the inclusion of error potential, severity of error consequences, hazards/safety factors, subject matter expert estimate of task difficulty, tools and equipment, skills and knowledge and physical requirements for the individual (e.g. strength, dexterity, coordination). Environmental conditions also need to be taken into account.

Step 4—Descriptive data collection

Any of the following techniques can be employed to collect task description data.

- Documentation review—Prepare an initial task description from descriptions (e.g. specifications) and schematics (e.g. drawings) of existing or similar products. See task description (Section 7.5).
- Questionnaire survey—Collect information from subject matter experts regarding existing or similar systems by using questionnaires. See Questionnaires (Section 9.1).
- Interview—Collect information from subject matter experts regarding existing or similar systems by using interviews. See Interviews (Section 9.2).
- Observation—Observe performance of prototype (simulation, functional prototype) or similar system in order to collect initial or supplementary information on tasks.

7.6.4 Result

Task analysis results in formatted descriptions of tasks that include, at a minimum: task title, initiating stimulus, equipment acted upon, response to be made, feedback and output characteristics including performance criteria. Diagrams (e.g. functional flow diagrams, decision-action diagrams) can also be used to depict the elements that are involved in the task and their interaction through time. Task analyses can be used to build job descriptions, staffing/skill training requirements, human/system interface requirements and specifications.

7.6.5 Example

Drury *et al.* (1987:395–9) illustrate the use of TA in the context of aligning a lamp in a photocopier lamp holder.

Task description information was accounted for: task number, purpose, action and check (feedback information that indicated task completion). The information was organized into a tabular format as Figure 7.10 shows. Task analysis information also included problems that were related to the controls, the visual display and operator posture. Writing summaries of problems that were associated with the lamp holder forced the analyst to make clear statements about the requirements that were necessary to create an acceptable set of task conditions. At the beginning of the project, it was clear to the development team that a knee obstruction under the bench was the only real problem. However, the task analysis revealed that the visual display was a cause of awkward postures. The design requirements section at the bottom of the format specifies improvements to the displays, controls and operator posture that will facilitate task performance.

Construction of a mock-up made it possible to evaluate the use of video cameras as a replacement to the existing optical system and to substitute a deformable lamp holder for X and Y joysticks.

Task No.	Purpose	Action	Check	Control Problems	Display Problems
010–040	Insert lamp and holder	Not considered for analysis	Kinesthetic	—	
050	Adjust height	Estimate distances from top and bottom of coil to edges of rectangle. Use thumb wheel to set equal gaps.	Visual	Thumb wheel uses R hand which is also used as tilt joystick as 050, 060 are combined.	Contrast and brightness fall off at top and bottom of screen.
060	X tilt adjust	Estimate vertical from any line on screen (not dotted). Adjust RH joystick in X direction only. Compensate for position changes with LH joystick.	Visual	Must coordinate LH and RH joystick movements to keep image central. High "stiction" in tilt mechanism gives jerky action.	Too many vertical lines, including max tilt lines, cause confusion.
070	X tilt/height clamp	Tighten lamp clamp screws using torque driver No 1. Check still in alignment after tightening.	Kin (torque) visual	Must hold holder by LH while tightening clamp screws, no place to hold. Lamp house in way of screwdriver.	—
				Must now tighten X, Y position clamp screws to provide firm base for operation 080.	
080	Y tilt adjust and clamp	With X lamp off and Y lamp on, check Y tilt against verticals on screen. Adjust using Y direction only movement of RH joystick. Compensate for position changes with LH joystick.	Visual	As for 060, both points.	As for 060
		Tighten tilt adjust screws with socket wrench. Check alignment after tightening.	Kin (torque) visual	Must hold holder by LH while tightening screws. No place to hold.	Image obscured by socket wrench.
				Torque reaction moves chair on wheels—must grip bench between knees to compensate. Must reach over lamp house.	
090	X, Y position adjust and clamp	With both lamps on, check X, Y position against verticals on screen. Adjust X position with LH joystick and clamp RH position screw with screwdriver 2.	Visual	Extreme arm abduction with R arm to tighten screws. Must loosen X, Y position screws before adjustment. Must hold position with LH joystick while tightening screws.	Too many images on screen, both hard to see and hard to tell which is which. Image obscured by screwdriver. Images very low contrast. Two images here have two different standards on same screen.
		Adjust Y position and clamp LH position screw with screwdriver 2.	Kin		
		Check alignment after tightening.	Visual		
100	Unclamp holder, remove	Not analyzed			

General Problems	Design Requirements

Displays
"Exit pupil," of display is small and low down, forcing operator to bend head downward for maximum contrast and brightness.
 	
Dark work area in midst of lighter surroundings—should be reversed.
	1. *Displays*
	  a. Brighter and more contrasty.
	  b. Separate X, Y to prevent confusion
	  c. Raise display to reduce bending
	  d. Less restricted display viewing area
	2. *Controls*
	  a. Less static friction on tilt adjust
	  b. Less inertia on position adjust
	  c. Increase compatibility of controls/displays to make adjustments more natural
	  d. Provide for direct holding of lamp while tightening various screws

Posture
Knee position obstruction forces operator to sit too far back—hence extreme reaching and bending.
	3. *Posture*
	  a. Remove knee obstruction to get closer to task
	  b. Have smallest possible lamp housing to allow freedom of arm positioning
	  c. Remove need for extreme should abduction in task 090

Figure 7.10 Task analysis format—task analysis is the culmination of previous analysis steps after operational analysis (Section 6.2), flow analysis (Section 7.1) and task description (Section 7.5). A format such as the one shown makes it possible to thoroughly account for each consideration of requirements for human performance compared with human abilities such as those described in Chapter 2.

Source: 'Task analysis.' Drury, C., Paramore, B., Van Cott, H., Grey, S. and Corlett, E. *Applied Ergonomics*. Vol.14, No.1, 19–28. Reprinted with permission from Elsevier Science.

7.7 Cognitive task analysis

How do people solve problems and make decisions?

When Frederick Taylor developed time study and Frank Gilbreth invented motion study in the early 1900s, the work that they sought to quantify and optimize was primarily physical. A hundred years later, supervisory control work roles are being created at a greater rate than physical labor work roles. These supervisory control roles often involve decision making and the control of complex systems. The development and use of complex systems requires a much better understanding of human cognitive function and the tasks that humans are expected to perform than traditional task analysis methods can accommodate.

Cognitive task analysis (CTA) has evolved within the past twenty years as a variety of techniques that can be used to assist human performance in cognitive work (i.e. work primarily having to do with problem solving and decision making). CTA methods are used to determine specific information requirements. Requirements are used to develop products such as displays, procedures and training that will support operator decision and control and thereby improve problem diagnosis and solution (Potter, Roth, Woods and Elm 1998:395–9).

Hoffman *et al.* (1998:254–76) note that cognitive task analysis methods have been used since 1980 to learn about:

- Novice/expert differences
- Trainee and expert mental models in order to predict performance
- Decision making
- The assessment of mental workload and situation awareness
- The analysis of problems in 'ill-structured domains'
- The development of training and decision aids
- The preservation of corporate knowledge
- The identification of workstation and interface features.

The development of CTA parallels the evolution of dynamic function allocation (Section 7.4). Dynamic allocation is necessary to adjust function assignments, depending on changing conditions. That leaves many questions unanswered when it comes to the definition of tasks for humans. For example, when should the switch be made from rule-based behavior to knowledge-based behavior in order to enable the operator to engage an unusual set of circumstances? Do maintainers have access to the kind of information that will make it possible to adequately inspect complex conditions (e.g. building facade structural integrity)?

The programs that CTA methods are intended to produce, such as decision tools, are not universally accepted. Critics of decision tools consider them to be 'essentially misconceived' for a number of reasons. Formal models of knowing fail to capture human expertise. Detached, objective 'knowing what' cannot replace intuitive, involved and skilled 'knowing how.' Machines can only behave (follow rules) while humans can act (have intentions). Decision tool programs can only operate in very specific and confined areas and such tools can only be successful when the task does not involve 'knowing how' (Berg 1997:160).

7.7.1 Cognitive task analysis methods inventory

Because CTA is a collection of methods, it would be insufficient to describe just one approach. Thirteen of the better known CTA methods, drawn from Luczak (1997:362–75) are shown in Table 7.8 and briefly described in the following section. Critical decision method is then explained in detail.

7.7.1.1 DECISION LADDER: RASMUSSEN, 1994

Decision ladder uses general activity analysis to relate work environment requirements to staff member cognitive resources and subjective performance criteria. Structured interviews are conducted while an individual performs work to identify prototypical work situations and functions. These activities are defined in terms of control functions individuals carry out. The activity is described by identifying the mental strategies (skill-, rule- and knowledge-based) that individuals can use in exercising control. Their employment and short cuts that individuals will take to minimize effort are depicted in a ladder-like diagram. Section 3.4.3 provides further information on Rasmussen's decision ladder.

Table 7.8 Cognitive task analysis methods—Cognitive task analysis is a collection of methods that have been developed to elicit expert knowledge and decision patterns. Luczak (1997) provides an inventory of the more prominent CTA methods which are listed below.

Decision ladder Rasmussen, 1994	Used to identify prototypical work situations and functions through structured interviews while an individual performs work
Goals, operators, methods and selection rules (GOMS) Card, Moran and Newell, 1983	Used to predict human performance using proposed user interfaces
Recognition primed decisions (RPD) Klein, 1986	Combines situation assessment and mental simulation for the serial evaluation of options
Cognitive simulation model Cacciabue,1993	Emphasizes decision making in accident situations and emphasizes the selection and implementation of emergency procedures that are taken by an operator
Contextual control model Hollnagel, 1993	Describes the way actions are chosen in order to meet the requirements of a situation
Grammar techniques	Converts user knowledge into operational procedures by expressing them as grammatical statements that result in a subsequent sequence of related actions • Task-action-grammar (Payne and Greene, 1986) • BNF-grammar (Reisner, 1981)
Semiotic models Springer, *et.al.*, 1991	User interface concepts are developed at multiple (semantic, syntactic, lexical) levels that range from the most to the least abstract
Empirical framework Sebillotte, 1991	Uses informal and directed interview techniques to extract operator goals and the ways that operators achieve them
Concept mapping Glowin and Nowak, 1984	Uses an interactive interview method to formulate a user-oriented problem description, portrayed by a graphical concept map consisting of nodes (objects, actions, events) and links (relations between nodes)
Cognitive walkthroughs Nielsen and Mack, 1994	Relies on the generation of scenarios in order to simulate the problem solving process that an individual follows
Hierarchical task analysis Diaper, 1989	Describes complex tasks using the description of simple subtasks in a hierarchical organization and the presuppositions that are necessary to reach goals
Task analysis for knowledge description Payne and Greene, 1986	Describes the amount and context of knowledge to execute a task using the grammar of knowledge representation, mapped to a task description hierarchy
Verbal protocol analysis Newell and Simon, 1972	Provides a window on the user's mental model of what is going on while performing an activity by saying any and all thoughts that occur
Critical decision method (CDM) Klein *et.al.*, 1989	A case-specific multi-trial retrospection that combines forms of protocol analysis, case-based reasoning, structured interview and retrospection

7.7.1.2 GOALS, OPERATORS, METHODS AND SELECTION RULES (GOMS): CARD, MORAN, NEWELL, 1983

GOMS is a cognitive engineering model that is used to predict human performance using proposed user interfaces. There are four elements in the GOMS model: goals (symbolic structure that defines a state to be achieved), operator (elementary perceptual motor or cognitive acts that are necessary to change a user's mental strategy or the task environment), methods (procedures) for achieving goals and rules for choosing among competing methods for goals. Interfaces can be described in terms of methods sequences, operators and selection rules used to achieve goals. Goals are decomposed into sub-goals, time estimates (drawn from the literature in psychology) are made for mental operations and predictions are made about the time that is needed to complete tasks using alternative interfaces.

Heath and Luff (2000:10–11) describe GOMS as 'a model developed on several layers and utilising a framework based on explicit goals, operators, methods and rules for selecting between options.' Sociologist Lucy Suchman (2001a) has criticized 'goal-directed, plan-based models of human conduct' such as GOMS that diminish the importance of immediate human action, ignore their role as an assist to human work and ignore human agency.

7.7.1.3 RECOGNITION PRIMED DECISIONS (RPD): KLEIN, 1986

Recognition primed decisions combines situation assessment and mental simulation for the serial evaluation of options. RPD covers three phases: simple match (situation is recognized and obvious reaction implemented), developing a course of action (conscious evaluation in terms of a mental simulation) and complex RPD strategy (evaluation reveals flaws that require modification, or the option is judged inadequate and rejected in favor of the next most typical reaction). RPD is best suited to situations, such as aircraft operation by airline pilots, in which the user is experienced, time pressure is great and conditions are less than stable. Klein's naturalistic decision making approach is also described in Section 2.3.2.2.

7.7.1.4 COSIMO—COGNITIVE SIMULATION MODEL: CACCIABUE AND KJAER-HANSEN, 1993

Cognitive simulation model emphasizes decision making in accident situations and emphasizes the selection and implementation of emergency procedures that are taken by an operator. Three information processing activities are accounted for: activation, observation and execution. Emergency procedures are activated and chosen based on three elements: highly skilled operator heuristic knowledge, preset frames of cues regarding environmental data and actions to cope with emergency situations.

7.7.1.5 CONTEXTUAL CONTROL MODEL: HOLLNAGEL, 1993

Contextual control model describes the way actions are chosen in order to meet the requirements of a situation. Different control modes are used to match different performance characteristics. Variations in task complexity in different situations are considered to induce transitions between control modes.

7.7.1.6 GRAMMAR TECHNIQUES

Task-action-grammar (TAG) (Payne and Greene 1986) and Backus-Naur form (BNF) grammar (Reisner 1981) typify the methods that are used to convert user knowledge into operational procedures by expressing them as grammatical statements (e.g. 'delete a sentence') that will result in a subsequent sequence of related actions.

7.7.1.7 SEMIOTIC MODELS: SPRINGER, LANGNER, LUCZAK AND BEITZ, 1991

Semiotic models develop user interface concepts at multiple (semantic, syntactic, lexical) levels that range from the most to the least abstract. Both task description and system configuration are developed at each level (pragmatic, semantic, syntactic and physical) to complement each other.

7.7.1.8 EMPIRICAL FRAMEWORK: SEBILLOTTE, 1991

Empirical framework uses informal and directed interview techniques to extract operator goals and the ways that operators achieve them. Informal interviews allow operators to speak uninterrupted about their work. Directed interviews distinguish between the goals that operators intend to achieve and how they pursue them. Empirical framework elicits task-related information, decomposes them into sub-tasks, relies on formalism to represent tasks and relates the findings to cognitive models.

7.7.1.9 CONCEPT MAPPING: GLOWIN AND NOWAK, 1984

Concept mapping is an interactive interview method that is used to formulate a user-oriented problem clarification portrayed by a graphical 'concept map' consisting of nodes (objects, actions, events) and links (relations between nodes).

7.7.1.10 COGNITIVE WALKTHROUGHS: NIELSEN AND MACK, 1994

Cognitive walkthroughs relies on the generation of scenarios in order to simulate the problem solving process that an individual follows.

7.7.1.11 HIERARCHICAL TASK ANALYSIS: DIAPER, 1989

Hierarchical task analysis uses the description of simple subtasks and a hierarchical organization and the presuppositions that are necessary to reach goals, in order to describe complex tasks.

7.7.1.12 TASK ANALYSIS FOR KNOWLEDGE DESCRIPTION: PAYNE AND GREENE, 1986

Task analysis for knowledge description describes the amount and context of knowledge to execute a task using the grammar of knowledge representation, mapped to a task description hierarchy.

7.7.1.13 VERBAL PROTOCOL ANALYSIS: NEWELL AND SIMON, 1972

Verbal protocol analysis (Section 6.5) is also considered a cognitive task analysis method and uses verbalization to determine how subjects understand and pursue tasks. VPA provides a window on the user's mental model of what is going on while performing an activity by saying any and all thoughts that occur.

7.7.1.14 CRITICAL DECISION METHOD (CDM): KLEIN, CALDERWOOD AND MACGREGOR, 1989

Hoffman, Crandall, and Shadbolt (1998:254–76) describe critical decision method as a palette of knowledge elicitation methods such as retrospection, response to probe questions and time line construction that can serve as a framework which can be reconfigured. CDM is an adaptation of critical incident technique (Section 6.6). Sessions that last roughly two hours are structured and guided by probe questions that are posed by individuals who are referred to as information 'elicitors.' As a retrospective interview method, CDM is used to derive the goals, action options, indices and context elements that are related to making decisions. CDM is used in this text as the example for CTA because of its versatility across applications. It is also easier than many other CTA methods to grasp and use and has undergone substantial peer review.

7.7.2 Preparation

Organize the team of researchers who will interview experts (information elicitors). Recruit an adequate number of experts to be interviewed. Train elicitors in the CDM procedure. Elicitors need to become familiar with the information domain that will be investigated by reviewing documents and conducting preliminary interviews.

7.7.3 Materials and equipment

Reports, memoranda and similar materials are necessary for elicitors to become familiar with the information domain in the area of interest.

7.7.4 Procedure

Table 7.9 describes Klein, Calderwood and MacGregor's approach to CDM that is used here to describe one approach to the process of CTA.

Step 1—Select incident

Identify the incident in which the expert's actual decision making either altered the outcome, or produced an outcome that would have been different without expert intervention, or particularly challenged the expert's skills.

Step 2—Recall incident

The expert is asked to recount an incident in its entirety.

Step 3—Retell incident

The elicitor retells the incident in close fidelity to the original account. The expert is invited to attend to all details and usually adds details, clarifications and corrections.

Step 4—Verify time line, identify decision points

The expert is asked for the time of key events in order to capture salient events ordered by time and expressed in terms of the points at which decisions were made and actions taken.

Table 7.9 Cognitive task analysis: critical decision method—determine how people solve problems, make decisions. Hoffman, Crandall and Shadbolt (1988) describe the critical decision method as one of many methods within the approach known as cognitive task analysis that is used to determine how people solve problems and make decisions.

Select incident
Recall incident
Retell incident
Verify time line, identify decision points
Conduct 'progressive deepening' probe
Conduct 'what if' inquiries

Step 5—Conduct 'progressive deepening' probe

The elicitor focuses attention on each decision making event by asking probe questions. Probes include such questions as cues (What were you seeing, hearing, smelling?), knowledge (What information did you use in making this decision and how was it obtained?) and analogues (Were you reminded of any previous experience?).

Step 6—Conduct 'what if' inquiries

The elicitor poses hypothetical changes to the incident account and asks the expert to speculate on what might have happened differently in order to specify pertinent dimensions of variation for key features that are contained in the account of the incident.

7.7.5 Result

Critical decision method can be used to develop training and remediation including illustrative stories to aid skills recall, taxonomies of informational and diagnostic cues and skill level assessment. CDM can also be used to develop products that are of benefit in complex situations such as decision aids and information processing workstations.

7.7.6 Example

Crandall's 1989 study explored the effect of probe questions as a means to elicit knowledge about perceptually based assessments such as critical cues. Twenty firefighters were asked to play the role of fireground commander, the individual who would be responsible for both strategic and tactical decisions at the scene of a fire. Each was exposed to a fire scenario that included decision points.

Two approaches were used to elicit knowledge. The first approach was similar to verbal protocol analysis (Section 6.5). Subjects were asked to describe what they noticed and to describe the basis for judgments and choices of action each time a decision was required. In the second, participants were asked probe questions at each decision point about cues, judgments, goals and actions.

Transcripts were coded according to remarks about:

- Salient or important cues
- Cues the expert inferred on the basis of information that was available
- Goals the expert hoped to accomplish through a particular action
- Actions that were under consideration or were being implemented
- Goal-action links—Statements in which goals were connected to a plan of action
- Remark specificity—The nature of comments in a range from abstract (e.g. referring to the 'look of a fire') to concrete (e.g. referring to 'spurting orange flames').

Crandall's study demonstrates the nature of CDM as an information elicitation method. The comparison between unstructured descriptions and response to probe questions showed that the latter approach was more effective. Cue information in answers to probe questions was greater in both quantity and in quality than unstructured statements by experts, demonstrating that 'you will learn what you ask about' (Hoffman *et al*. 1998:254–76)

> 'Murphy's Law': If something can go wrong, it will.
> (Anonymous)

7.8 Technique for human error rate prediction (THERP)

Determine how vulnerable assigned tasks are to human error.

The technique for human error rate prediction (THERP) applies engineering reliability analysis to the task of predicting human error. That prediction makes it possible to determine how vulnerable a proposed solution is to human errors that will inevitably occur.

Chapter 2 provided a basic orientation to human behavior, including human error. Park (1997:990–1,004) notes 'If the variability in human performance is recognized as inevitable, then it is easy to understand that when humans are involved, errors will be made, regardless of the level of training, experience or skill.' Park depicts THERP as an engineering and reliability-based approach that 'models events as a sequence of binary decision branches' based on probability. THERP can be used to estimate the probability of task success and failure. That estimate can be used in the study of design trade-offs as well as risk assessments.

THERP was developed by architect Alan Swain from the 1960s through the 1980s (along with colleagues Bell, Guttmann, and Weston). Miller and Swain (1997:229) describe this 'oldest and most widely used' human reliability technique as:

> A method to predict human error probabilities and to evaluate the degradation of a man-machine system likely to be caused by human errors alone or in connection with equipment functioning, operational procedures and practices, or other system and human characteristics that influence system behavior.

The method originally focused on nuclear power plant (NPP) operator behavioral error and evolved later to include higher level cognitive behaviors such as diagnostic errors.[2] The core of THERP is represented in twenty-seven tables of human error probabilities that are included in Part IV of Swain and Guttmann (1983). The handbook's tables provide human error probabilities that are based on expert opinion. The tables also provide data on activities that are similar to those that are performed by nuclear power plant (NPP) operators. The authors consider the data appropriate to other industrial settings. For error probability estimates on activities that are unlike those of NPP operators, the investigator must rely on other sources including subject matter expert (SME) opinion(s).

Wickens (1992:389–439) outlines three components of THERP: human error probability (HEP), the human reliability analysis (HRA) event tree and performance shaping factors (PSF).

Human error probability (HEP)—Estimates are obtained from databases of actual human performance (when available) or expert estimates (typically within a +/– 95 percent confidence bracket).

HRA event tree—The procedures that are required to complete a task can be mapped out in a tree-like diagram in order to determine the likelihood that the task will be completed successfully. For example, reading a form, entering keystrokes and reading a display are all procedures that are a part of data transcription. The event tree diagram can be used to determine the likelihood that the task will be completed.

Miller and Swain (1997:219–47) recommend the HRA event tree as the basic tool to model tasks and task sequences. The HRA event tree is not a fault tree diagram. A fault tree (Section 8.2) diagram starts with an undesirable outcome (fault) and works backward to construct a series of events that contribute to the occurrence of that fault. By contrast, the event tree portrays steps that result from task analysis (Section 7.6). Construction of an event tree such as the example in Figure 7.11 can start at any time that is convenient in the stream of tasks that are

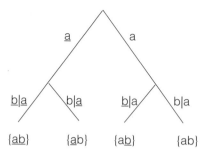

Serial systems

Succeed only if all human activities are performed correctly

$$R_{ser} = \text{Prob \{a\} Prob \{b|a\}}$$

Parallel systems

Succeed if either task is performed correctly. Individual paths are summed to represent probability for more than one path.

$$R_{par} = \text{Prob \{\underline{a}\} Prob \{\underline{b}|\underline{a}\} + Prob \{\underline{a}\} Prob \{b|\underline{a}\} + Prob \{a\} Prob \{\underline{b}|a\}}$$

Figure 7.11 Human reliability analysis event tree—Park (1987) explains fault tree logic by representing the performance of two tasks (A and B). Underlined lower case letters indicate desirable states or success while plain letters indicate failure. Task A is performed first, making task B dependent on the outcome of A. This interdependence is represented by the series of pairs using the vertical stroke (|) symbol. Once conditional probabilities of success or failure are assigned to each tree limb, the probability through each path can be calculated using the formulae shown.

Source: Reprinted from *Human Reliability*, Park, K., copyright 1987, with permission from Elsevier Science

performed. Starting at any convenient point in time, a tree-like series of branching structures is constructed to represent tasks that are performed correctly or incorrectly.

Performance shaping factors (PSF)—Multipliers can be used to modify human error probability (HEP) estimates according to the effects of influences such as expertise or stress. Miller and Swain account for two types of performance shaping factors: external and internal. External PSFs include inadequate work space and layout, poor environmental conditions, inadequately designed controls and equipment, inadequate training and job aids and poor supervision. Internal PSFs include stress and inexperience. For example, the authors report that an extremely high stress level (an internal PSF) can increase the probability of error among skilled workers five-fold and novice workers ten-fold.

Wickens (1992:429–36) enumerates four shortcomings of THERP when it comes to the prediction of human error due to its origins as an engineering reliability analysis method: lack of data base, error monitoring, nonindependence of human errors and integrating human and machine reliabilities.

Lack of data base—While data are available for simple tasks, little or no data are available for more complex activities such as diagnosis or problem solving.

Error monitoring—Humans can monitor their own performance and may catch errors and correct them before overall performance is affected.

Nonindependence of human errors—Reliability analysis makes the assumption that hardware components can be evaluated independently (i.e. one component failure does not cause the failure of others). However, human actions are dependent. An error can induce increased caution or further error. Multiple human operators are also likely to interact, causing increased or decreased likelihood of error based on a complex set of considerations.

Integrating human and machine reliabilities—Human error data and hardware reliability data are essentially different. The combination of both in order to represent some statement of overall human-machine reliability poses a problem.

Meister (1985:150–66) points out that it is necessary to perform four constituent tasks in order to implement THERP and all require a good deal of judgment:

- Determine all possible errors
- Choose error rates that are appropriate to those possible errors
- Determine the degree of dependence among the errors
- Determine the factors that affect the performance shaping factors (PSFs)

Reason (1990:229–31) finds that all human reliability analysis (HRA) techniques suffer from problems with validation. For instance, human factors practitioners have produced widely varying probabilities of failure when using THERP in controlled studies. Even so, THERP offers what Reason finds 'the most accessible and widely used' among all of the available human reliability analysis techniques.

7.8.1 Preparation

Obtain information on human error (e.g. data in the behavioral literature, test reports, the Data Store data base on human error probability and experts on human error estimation) (Meister 1985:153–5). Review the results of operational analysis (Section 6.2), functional flow analysis (Section 7.1) and task analysis (Section 7.6).

7.8.2 Materials and equipment

No materials or equipment are unique to THERP.

7.8.3 Procedure

Five steps from Meister (1985:150–66) and Miller and Swain (1997:219–47) are used to perform THERP and summarized in Table 7.10. The third and fourth steps (7.8.3.3 and 7.8.3.4) form the core of the method.

7.8.3.1 SYSTEM DESCRIPTION

Describe the system goals and functions, situational and personnel characteristics, particularly those that can be influenced by human error. See operational analysis (Section 6.2).

Table 7.10 Technique for human error rate prediction (THERP)—determine how vulnerable assigned
tasks are to human error.

Describe the system

Describe system goals and functions, situational and personnel characteristics, particularly those that
can be influenced by human error.

Describe the job

Identify, list and analyze human operations that are performed and relate to the system and functions
of interest.

Describe jobs and tasks that are performed by individuals.

Identify error-likely situations by performing a function/task analysis oriented around error identification.

Estimate error

Estimate the likelihood of each potential error in each task and the likelihood that the error will be
undetected.

Estimate consequences

Determine the effects of human error on events of interest.

Recommend changes

Suggest changes to the product in order to mitigate the potential for error to affect task performance.

7.8.3.2 JOB DESCRIPTION

Identify, list and analyze human operations that are performed and relate to the system and functions of interest. Describe jobs and tasks that are performed by individuals. See task description
(Section 7.5) and work design (Section 7.9).

Identify error-likely situations by performing a function/task analysis that is oriented around error
identification. See flow analysis (Section 7.1) and task analysis (Section 7.6).

7.8.3.3 ESTIMATE ERROR

Estimate the likelihood of each potential error in each task and the likelihood that the error will be
undetected. Refer to sources that contain information on error probability such as data banks,
behavioral literature (books, journals, proceedings, study reports). Alternately, consult with experts
on the tasks that are under study in order to obtain estimates of error probability.

7.8.3.4 ESTIMATE CONSEQUENCES

Determine the effects of human error on events of interest.

7.8.3.5 RECOMMEND CHANGES

Suggest changes to the product in order to mitigate the potential for error to affect task performance.
Steps 7.8.3.2 through 7.8.3.5 may be repeated in order to evaluate recommended changes.

7.8.4 Result

THERP is used to produce what Park (1997:990–1004) describes as 'estimates of task success and
failure probabilities that can be used in design trade-off studies or in probabilistic risk assessments.'

The probability tree diagram that results from THERP can be used as an analysis tool to represent
anticipated conditions, assess the probability of human error and propose solutions.

7.8.5 Example

Swain performed THERP at a Department of Energy contractor's facility when it became apparent that an assembler-technician (AT) had made an error in the handling of electronic programmer (EP) units that were installed into a junction box of a major assembly. THERP was used to estimate the probability of other defective units among the 284 assemblies that had already been completed.

The EP units had two rack-and-panel plugs that had to be protected from exposure to static electricity by the installation of foam protector pads. One AT would insert black three-eighths inch thick carbon-filled foam pads inside the two identical-looking connectors, then install a yellow plastic dust cover with a rubber band to compress the pads onto the plug pins. When brought to another area, a different AT had to remove the pads so that the two EP plugs could connect to the junction box before inserting the programmer into the rigid foam support in the major assembly. Inspection of one unit showed that the second AT had failed to remove the pads from one of the plugs.

A detailed task analysis was performed in order to construct an event tree. Major cognitive aspects of behavior (e.g. diagnosis and decision making) were negligible. Stress level was not an issue and could be ignored. Only one individual was involved so individual differences in performance could be ignored. Three human reliability analysts who served as subject matter experts (SMEs) independently assigned error probabilities, taking task dependencies into account. Most estimates differed by less than a factor of two and the analysts reached a consensus for all estimates. Figure 7.12 shows the event tree, with probabilities that describe the likelihood that the two pads (labeled P1 and P2) were removed. The tree accounts for each possible occurrence: a pad is removed (recovered), it sticks inside the connector and may or may not be detected, or it falls out or dangles outside of the connector and is assumed to be recovered. Twelve limbs ('a' through 'k') represent the permutations of what could happen. As in Figure 7.12, the limbs represent a series of steps that occur through time. Underlined lower case labels ('a') indicate successful task performance while plain characters ('a') indicate failure. The rationales for some of the limbs in Figure 7.12 are described below.

a—Limb '<u>a</u>' states that the pad fall out or dangles. It is assumed that it will be detected and removed. Limb 'a' states the pad sticks, meaning it would remain inside the connector, undetected. SMEs estimated a likelihood of 0.8 for '<u>a</u>' and 0.2 for 'a.'

c—The human error probability (HEP) is lower than limb d, on the assumption that the AT would be more likely to take out a stuck pad (P2) after having just seen another pad (P1) fall out as indicated by limb '<u>a</u>.'

d—'P1 missed' indicates the AT failed to remove the pad. The 0.05 probability estimate is a downward adjustment of the normal inspection error rate for the detection of defects that occur at a rate of about 0.01 or less based on McCormack's work in the early 1960s.

g—The high likelihood of successful performance at limb 'g' is based on the completion of '<u>d</u>' where one pad (P1) was successfully removed, providing a cue to the successful removal of its companion pad (P2).

h—Pad P2 falls out (or dangles). It is assumed that it will be detected and removed.

i—Limb 'i' represents the recovery of an AT error that was made at limb '<u>a</u>.' Because one pad (P2) falls out or dangles at limb '<u>h</u>,' it reminds the assembler to check for the presence of the other pad (P1) that had been overlooked at '<u>a</u>.'

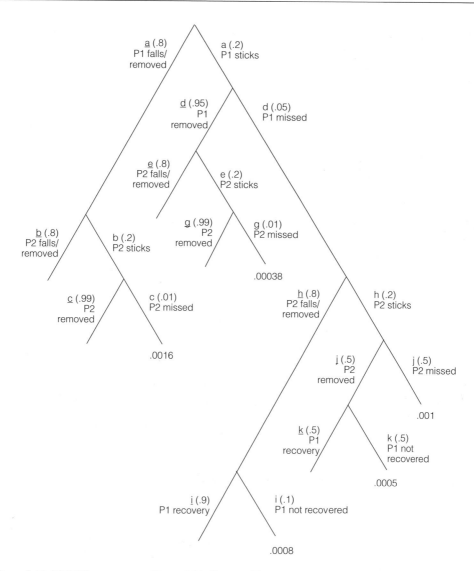

Figure 7.12 THERP event tree—Figure 7.11 illustrated how an event tree can be built to evaluate reliability. The event tree here illustrates a case study conducted by Swain during the analysis of an industrial fabrication process. After nearly 284 electronic programmer units had been installed in junction boxes, it was discovered that the assembler technician responsible had failed to remove one of a series of foam pads used to protect the programmer from static electricity damage. A reliability study was conducted to estimate the probability that one or two of the pads had been left in other assemblies.

Source: Reprinted from *Human Reliability*, Park, K., copyright 1987, with permission from Elsevier Science

Addition of the HEPs shown in Figure 7.12 showed that the joint probability that an EP will have one or two stuck pads and that the AT would fail to remove them is about equal to 0.004 (0.0016 + 0.00038 + 0.0008 + 0.0005 + 0.001). However, HEP estimation is not certain. A sensitivity analysis was performed to develop conservative (pessimistic) and optimistic estimates. Calculations showed that in a lot of 284 units, 1.14 EPs could be expected to be defective (within a range of 2.84~0.14).

In addition to the one EP that was found to have an undetected pad, 64 additional EP's were disassembled in order to verify the accuracy of the 0.004 estimate. No additional pads were found, making the fraction of defective units one in 65, or 0.015. A significance level was calculated to check the agreement between the data and the probability estimate. The result demonstrated that there was no reason to reject the assessment of 0.004. Swain's conclusion was that even though the process relied on subjective assessment, analysts who are trained properly can perform THERP accurately enough for industrial applications (Park 1987:222–32).

7.9 Work design

Assemble tasks into work for individuals to perform.

Functional analysis methods were used to decompose, or break apart, what needs to be done to fulfill the product's purpose. Function allocation was used to sort functions among system elements and those that were allocated to humans were described by task identification. Task analysis specified what is to be done and what was needed (e.g. tools, decision aids, training) in order to accomplish it.

Work design (WD) is used to assemble the tasks into meaningful modules of work that can comprise a job. While each of the previous methods decomposes what is required, WD is the step that composes tasks into work modules and work modules into jobs.

Medsker and Campion (1997:450–89) account for a variety of methods that can be used in order to design jobs: mechanistic, motivational, perceptual/motor and biological.

Mechanistic—The classical industrial engineering approach based on early work by Frederick Taylor (1911) and Frank Gilbreth (1911). Mechanistic design seeks to maximize efficiency in terms of productivity and the use of human resources.

Motivational—Based on organizational psychological work of Herzberg (1966) and Hackman and Oldham (1980), motivational design seeks to fit work system and job design to the external environment. It also seeks to optimize both social and technical systems in the internal environment of the organization. Motivational design advocates greater flexibility, employee involvement and training, decentralized decision making, less hierarchical structures and more formalized relationships and procedures.

Perceptual/motor—Based on the body of knowledge in experimental psychology and human factors, the perceptual/motor approach seeks an efficient and safe use of humans in the human machine system. The approach considers human mental and physical capabilities and ensures that requirements do not exceed the abilities of the least capable potential worker.

Biological—Based on the physiology and ergonomics knowledge developed by such authors as Tischauer (1978) and Grandjean (1980), the biological approach deals with equipment and work place design as well as task design. The biological approach seeks to maintain the employee's comfort through improvements to the physical surrounding and elimination of work-related aspects that may threaten employee well being.

Work design can blend the approaches but Medsker and Campion caution that not all approaches are equally compatible. For example, the motivational approach seeks to make jobs more stimulating by increasing mental demands while risking reductions in productivity. On the other hand, the mechanistic and perceptual/motor skill approaches seek to lessen mental demands in the interest of protecting productivity. Trade-offs may be needed if jobs are redesigned using more than one approach. The reader is invited to consult Medsker and Campion for a comprehensive review of various alternatives to work design for individuals and teams.

Bailey (1989:195–205) considers work module design to be the third part of process-oriented task analysis after task identification and task description. Tasks are identified, then analyzed, then selected for assignment to work modules. Finally, work modules are assembled into the set of activities that comprise a job. One or more work modules may be combined to create a job for a specific type of person. Figures 7.13 and 7.14 illustrate the assembly of tasks into modules and modules into a job. Bailey discusses task selection and module assignment in terms of task complexity. Of the four approaches that Medsker and Campion describe, Bailey's concern for individual abilities would appear to put it well within the perceptual/motor approach.

7.9.1 Preparation

Identify tasks through flow analysis (Section 7.1). Diagram tasks using a task level functional flow chart. Describe tasks through task description (Section 7.5).

7.9.2 Materials and equipment

No unique materials or equipment are needed for work design.

7.9.3 Procedure

The steps to perform work design, based on Bailey's (1989:195–205) work module design approach, are shown in Table 7.11.

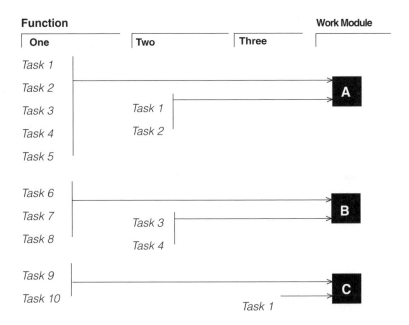

Figure 7.13 Relationship of functions to work modules—work modules can be made up of tasks from several different functions. Modules are assembled into jobs that can be evaluated according to work load. Figure 7.14 shows how work module configuration can affect performance.

Source: Adapted from Bailey, R. *Human Performance Engineering.*

7.9 Work design

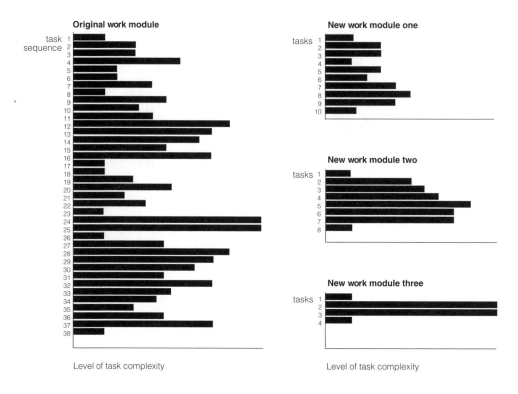

Figure 7.14 Work module design—two diagrams demonstrate how task complexity and sequence can be used to create work modules that produce varying results. The diagram at left shows a work module for the preparation of a service order. Thirty-eight tasks of widely varying complexity resulted in degraded human performance. The set of diagrams at right shows how the single module was broken into three separate modules in order to resolve problems with longer-than expected manual processing time and operator learning.

Source: Adapted from Bailey, R. *Human Performance Engineering*.

Step 1—List all of the tasks that are required according to function.

Step 2—Review tasks

Review each task according to five considerations: data relationships, skill level for a task, task relationships and sequence, time dependencies and special human–computer interaction considerations.

Step 3—Identify tasks that share similar traits

Aggregate tasks that share traits. When possible, subdivide groupings. For example, twenty tasks may relate to customer information (a data relationship) and twelve may be clerical in nature (e.g. data entry) while eight are supervisory level considerations (e.g. related to credit approval). The subdivision based on a second trait, skill level for a task, makes it possible to cluster tasks according to the skills and aptitude of those expected to perform them.

Table 7.11 Work design—determine how vulnerable assigned tasks are to human error.

List tasks

List all of the tasks that are required according to function

Review tasks

Consider each task according to five considerations:

    Data relationships
    Skill level for a task
    Task relationships and sequence
    Time dependencies
    Special human-computer interaction considerations

Identify tasks that share similar traits

Aggregate tasks that share traits

When possible, subdivide groupings

Assemble 4 to 9 tasks into a work module

Continue until all tasks have been assigned

Assess the work module throughput requirements

Determine the volume of work that is expected to flow through these tasks during a specified length of time

Specify staffing for work modules

If less than a full job, combine with other work modules

Confirm that one work module is the equivalent of one job

If more than a full job for one person, specify more than one individual to perform the work module

Complete work design

Draft job descriptions

Assemble decision aids, training and other materials

Step 4—Assemble four to nine tasks into a work module.

This initial module will be reviewed and adjusted if necessary. Why use four to nine tasks? Fewer than four tasks may afford insufficient variety to make a job interesting. More than nine tasks may make the job vulnerable to overload. Continue until all tasks have been assigned.

Step 5—Assess the work module throughput requirements.

Determine the volume of work that is expected to flow through these tasks during a specified length of time (e.g. per hour, per shift, per workday). The staffing that is necessary to perform the work module will be either less than, equal to or more than a job for one person. For guidance on work-load assessment, see Section 8.5.

Step 6—Specify staffing for work modules.

Select among three options:

- If less than a full job, combine with other work modules.
- Confirm that one work module is the equivalent of one job.

- If more than a full job for one person, specify more than one individual to perform the work module.

Step 7—Complete work design

Draft job descriptions. Assemble decision aids, training and other materials related to the performance of each job.

7.9.4 Result

Work design produces complete descriptions of jobs that can be used to recruit, train and assign individuals to perform tasks.

7.9.5 Example

Medsker and Campion (1997:450–89) present an example of the application of work design (which they refer to as job design) in a large financial services company.

A unit in the company processed paperwork on behalf of other units that sold the company's products. Jobs had been designed using the mechanistic approach and that approach had become unsatisfactory for the firm. Individuals prepared, sorted, coded and translated the hard copy flow into data files. The firm chose a motivational approach to enlarge jobs and obtain three objectives: improve employee motivation and satisfaction, increase current employee feelings of work ownership in order to improve customer service and maintain productivity in spite of the decrease that implementing a motivational approach tends to bring.

The study was conducted at five different locations. During evaluation, data was collected on job design and enlarged jobs' potential benefits and costs. During the research phase, jobs that had undergone enlargement were compared with those that had not. Questionnaires were administered and incumbent workers, managers and analysts were convened in focused team meetings.

The focus of the study was task enlargement (the redesign of jobs based on changes to task assignment). Results of the study indicated that jobs in which tasks had been enlarged showed benefits and some costs that those benefits incurred. Benefits included improvements to employee satisfaction, decreased boredom and improvements to work quality and customer service. Costs included some increases in training, skill and compensation requirements (although it appeared that not all anticipated costs were actually incurred).

A follow-up evaluation study that was conducted two years later showed that job enlargement benefits and costs change over time, depending on the type of enlargement. Task enlargement had incurred long term costs such as lower employee satisfaction, efficiency and customer service along with increases in mental overload and errors. Knowledge enlargement (which had evolved in the time since the original study) appeared to yield benefits such as improved satisfaction and customer service and decreases in overhead and errors.

7.10 Decision analysis

Select the preferred alternative among a number of potential solutions.

Research and development teams typically create a variety of solutions to problems and all have some potential to succeed. Which is the right solution to implement? The answer is not as apparent as it may seem. Because systems are often complex and can be difficult to evaluate, their value is not always evident. There are also interpersonal issues at play. Team members, management and other parties may prefer one alternative or another on the basis of their own agendas and perceptions. Decision analysis (DA) allows a fair comparison and choice of the best solution by basing the choice on a consensus among those involved in the decision process.

Kirkwood (1999:1119) describes five steps that should comprise a systematic approach to quantitative decision analysis:

1   Specify objectives and scales for measuring achievement with respect to objectives
2   Develop alternatives that potentially might achieve these objectives
3   Determine how well each alternative achieves each objective
4   Consider trade-offs among the objectives
5   Select the alternative that, on balance, best achieves the objectives, taking into account uncertainties.

A broad literature has evolved to analyze how decisions are made. Rather than account for all approaches, the intent of this chapter is to provide a straightforward means to structure comparative decision making when no other method is preferred or available.

DA uses a structured, weighted set of objectives to evaluate each alternative in a consistent manner. In simple examples, the value is less apparent when comparing complex solutions against a large number of objectives. Internal inconsistencies can arise if reactions to one choice influence subsequent reactions to other choices. This crossover effect can erode the method's validity. To prevent such an effect, create different versions of the comparison matrix (e.g. version A, B and C) that have different orders of objectives and alternatives. When each version is administered, marks should be fairly consistent among similar groups of respondents such as engineers or managers. If a particular version of the matrix (e.g. the 'A' version, completed by engineers) shows a significant variation from the values that are assigned using other versions, check the values in the 'A' version across different sets of respondents (e.g. managers). If the pattern is also evident among 'A' versions completed by managers, it is possible that the 'A' version is not valid and results collected using it may need to be discarded.

Fogler and LeBlanc (1995:98–101) caution that there are potential pitfalls in DA that need to be avoided: consistency and level of detail.

Consistency—The assessments of the weights that are assigned to the wants in a decision analysis must be checked for internal consistency if the result is to be valid. To ensure consistent evaluation, change the order of criteria. Then, reevaluate the solutions in order to determine whether order has an effect on rating. If a rating changes in the newly ordered matrix, it is possible that the first rating was influenced by an inconsistency such as a positive or negative association with the criteria or the solution preceding it.

Level of detail—Criteria should all be written at the same level so that they are neither too global nor too particular. Adding a number of detailed requirements can 'pack' the matrix and skew an otherwise balanced comparison.

A comparison matrix can be used to make sure that each objective is accounted for fairly and consistently. Solution ratings and objective weights can both be adjusted until the research and development team can agree on the results. How this is accomplished must be done by authentic consensus. The value of DA is diminished if adjustments are influenced by criteria other than project objectives such as political influence.

7.10 Decision analysis

7.10.1 Preparation

Step 1—Compose a statement that specifies what will be decided

Step 2—Collect and verify information that is available through documents, interviews with individuals and first hand observation.

Step 3—Review data on preferences and performance that have been collected for each of the solutions that are under consideration.

7.10.2 Materials and equipment

Any prototypes or simulations that can be used to represent the alternative solutions.

7.10.3 Procedure

The following approach to decision analysis from Fogler and LeBlanc (1995:98–101) is summarized in Table 7.12.

Step 1—Set criteria by establishing the objectives of the decision

Step 2—Sort objectives by importance. Charles Owen (Institute of Design, IIT, Chicago, IL) has found that three levels of constraints on requirements are effective: those that must be met (would

Table 7.12 Decision analysis—select the preferred alternative among a number of potential solutions.

Set criteria
Establish the objectives of the decision

Sort objectives by importance
Sort according to criteria that:
    Must be met (would result in failure if not satisfied)
    Should be met (will degrade performance if not met)
    Ought to be met (would lend a certain character or quality that is desirable)

Assign a weight
Assign a numerical rating to give each importance category (a weight)

Create a table
List each solution under consideration

Evaluate solutions
Rate each solution concept according to how well it meets the objective
Multiply each solution rating by the assigned objective weight
Add the multiplied ratings for each solution

Select the preferred solution
The solution with the highest cumulative rating can be considered to best meet solution requirements

Qualify the preferred solution
Check the comparison results against the results of the failure modes and effects analysis that may have been performed earlier
Verify the prospects for and potential cost of failure for the top-ranked solution

result in failure if not satisfied), should be met (will degrade performance if not met) or ought to be met (would lend a certain character or quality that is desirable).

Step 3—Assign each importance category (i.e. must, should, ought) a weight by giving it a numerical rating. The 'must' category would be likely to receive the highest numerical value on a scale of 1 to 10.

Step 4—Create a table and list each solution under consideration, as Table 7.13 depicts.

Step 5—Rate each solution concept according to how well it meets the objective.

Multiply each solution rating by the assigned objective weight (e.g. a solution that fully meets an objective with a 'must' level of importance would receive a rating of 100 (i.e. received a 10 rating in a requirement that is rated as 10 in importance).
   Add the multiplied ratings for each solution. The solution with the highest cumulative rating can be considered to best meet solution requirements.

Step 6—Check the comparison results against the results of the failure modes and effects analysis (Section 8.3) that may have been performed earlier. Verify the prospects for and potential cost of failure for the top-ranked solution. It may be necessary to choose a 'second best' alternative if the potential for loss is unacceptable.

7.10.4 Result

Decision analysis identifies the solution alternative that best satisfies product objectives.

7.10.5 Example

Fogler and LeBlanc (1995:98–101) demonstrate the use of decision analysis in making a choice among three electrostatic paint gun alternatives for use in an auto manufacturing plant.

Table 7.13 Solution comparison through decision analysis—decision analysis provides a structured approach to the consideration of alternative solutions. In the example below, Fogler and LeBlanc (1995) show how three kinds of paint spray guns are compared to choose one for use in an industrial facility. The table accounts for requirements, requirement importance ('weight') and ratings for each alternative. Gun 3's failure to provide adequate flow control eliminated it from further consideration.

		Gun 1		Gun 2		Gun 3
Musts						
Adequate flow control		Yes		Yes		No
Acceptable appearance		Yes		Yes		Yes
Wants	*Weight*	*Rating*	*Score*	*Rating*	*Score*	
Easy service	7	2	14	9	63	
Low cost	4	3	12	7	28	
Durability	6	8	48	6	36	
Experience	4	9	36	2	8	
	Total		110		135	

Gun 1 is considered the industry standard. While its performance meets requirements, its manufacturer is overseas which affects service speed and efficiency. Its price is also considered inflated, possibly due its long-time market dominance. Two other alternatives (Gun 2 and Gun 3) are newcomers to the market. The decision is to select a paint spray gun and the alternatives are Guns 1, 2 and 3.

Mandatory performance objectives (musts) have to be measurable. Alternatives were evaluated in a lab setting according to two mandatory objectives: adequate flow control and acceptable paint appearance. During the testing, Gun 3 could not be adjusted to control the flow of paint as needed and was dropped from further consideration.

Paired comparisons among preferences ('wants') can be used to verify the consistency of assigned weights and avoid order bias (order bias is explained in Section 9.4). Interviews with plant personnel indicated that four preferences existed: ease of service, low cost, durability and experience using the product. Weights assigned to the four preferences favored ease of service and durability. Ratings were collected among the decision makers and adjusted according to the assigned weights. While Gun 1 rated highly in terms of durability and experience, Gun 2 fared much better in terms of ease of service and cost. Total ratings for the product identified Gun 2 as a better choice. A follow-on analysis was performed to examine the potential risks (in terms of probability and criticality) that were associated with each alternative.

7.11 *Summary*

This chapter has addressed what is often considered the core of human factors involvement in research and development: methods to determine what humans are expected to do. They are called 'design guidance' methods here because each uses knowledge about the capabilities and limits of humans in order to guide the design of products, systems and services.

Functional analysis is used to account for the series of events, or actions, that are necessary for a product to fulfill its purpose. Functions are expressed in concise statements of what is (or needs to be) done in the product or system. Three approaches to functional analysis (flow analysis, time line analysis and network analysis) were reviewed and three methods explained: flow analysis, time line analysis and link analysis.

Flow analysis (Section 7.1) is used to account for the functions that the product is to perform and to portray their order.

Time line analysis (Section 7.2) is a functional analysis method that combines functions or tasks with time-related information in order to predict workload levels for a solution that is under development.

Link analysis (Section 7.3) is a network analysis method that is used to assess the relationships among a selected number of elements. Link analysis uses interaction frequencies to determine affinities among elements and to configure arrangements based on those relationships.

Function allocation (Section 7.4) is used to assign the performance of each function to the element (among humans, hardware, or software) that is best suited to perform it.

Task description (Section 7.5), also called 'task identification,' is a taxonomic method that is used to list and describe tasks that are to be performed. Task description accounts for each aspect of human behavior that is contained in the functions that were derived through functional analysis methods.

Task analysis (Section 7.6) is used to refine the functional requirements of a system in order to describe the tasks that people will perform.

Cognitive task analysis methods (Section 7.7) are used to determine the specific information requirements that will support operator decision and control activities, in order to improve problem diagnosis and solution.

The technique for human error rate prediction (Section 7.8) applies engineering reliability analysis to the task of predicting human error. That prediction makes it possible to determine how vulnerable a proposed solution is to human errors that will inevitably occur.

Work design (Section 7.9) is used to assemble the tasks that are assigned for human performance into meaningful modules of work that can comprise a job.

Decision analysis (Section 7.10) allows a fair comparison and choice of the best solution by basing the choice on a consensus among those involved in the decision process. Decision analysis uses a structured, weighted set of objectives to evaluate each alternative in a consistent manner. Caution needs to be taken, though, to ensure that the assigned weights are valid.

The evaluation methods in Chapter 8 make it possible to see how well the decisions that were made during the design process turn out.

8　Evaluation methods

How will an intended concept appear or perform?
Simulation (8.1)

What could go wrong?
Fault tree analysis (8.2)

How can loss be minimized? Avoided?
Failure modes and effects analysis (8.3)

Do events happen in the order that they need to occur?
Operational sequence analysis (8.4)

Can individuals handle the tasks that we expect them to perform?
Workload assessment (8.5)

The analysis methods that were presented in Chapter 6 are used to learn about the nature, cause, scope and composition of a problem or opportunity. Methods that were covered in Chapter 7 are used to determine what humans are expected to do. The evaluation methods that are presented in this chapter make it possible to see how well the decisions that were made during the design process actually turn out.

Research and development teams typically generate many solutions to a problem. Which solution is best? How can we know whether the design of a concept is adequate? Cushman and Rosenberg (1991:17–31) define adequacy by comparing objective performance data that have been collected through tests with performance requirements that have been developed and written into a product requirements document. Evaluation is the point at which the clear and measurable requirements that were produced during problem definition come into play. Simply put, evaluation asks 'are we getting the results we need?' Evaluation methods are best suited to learning how well a solution concept performs, whether it is in actual use or is under development.

Three approaches are typically used to evaluate systems.

- *Theory*—Performed in early design and development stages
- *Simulation*—Performed by either of, or a combination of, two methods:
  - Computer—Computer-based simulation is the faster, more accurate, at times less costly approach
  - Mock-up—Making a physical prototype is preferable when a family of related systems is to be evaluated
- *Actual operation*—Similar to mock-up simulation, except the evaluator has less control over test conditions

All three approaches are covered by methods in this chapter.

As in Chapters 6 and 7, each section in this chapter follows the same format so that information is easy to find and understand: introduction, preparation, materials and equipment, process, results and example.

- *Introduction*—Explains the background of how the method was developed and different approaches that have been taken to perform it.
- *Preparation*—Describes what needs to be done before performing the method.
- *Materials and equipment*—Accounts for physical items that are necessary including materials and equipment.
- *Process*—Enumerates the steps that are involved in performing the method.
- *Results*—Describes the outcome that performing the method produces.
- *Example*—Leads the reader through a brief discussion to illustrate the method's use.

8.1 *Simulation*

Demonstrate a concept in order to collect human preference and performance data.

Simulation is used to predict systems/parts of systems that do not exist and to allow users to experience systems/parts of systems that are dangerous, complex and expensive. Through simulation, potential solutions can be examined, evaluated and optimized at minimal cost and least risk to people. Simulation can be used to compare alternatives, on the assumption that an alternative that performs better than others during evaluation will be best in actual use. Some uncomplicated specific purpose products can be evaluated with a simple 'yes' or 'no' according to whether the concept does what it is supposed to do. Section 4.4 described the use of simulation through rapid prototyping in order to develop a consensus among project stakeholders.

Simulation bridges research and design. The mock-ups, models and prototypes made by designers and engineers are used to simulate a concept. In that sense, simulation is a design method. However, each technique is also a means to collect information on human preference and performance. In that sense, simulation is also a human factors research method.

While Ulrich and Eppinger (1995:218–32) refer to all three-dimensional visualization methods as prototypes, they classify prototypes along two dimensions: physical/analytical and comprehensive/focused. The information that each dimension represents can be used to better explore and understand a solution concept.

Physical—Approximate the product in order to evaluate its feel, demonstrate the concept, or demonstrate how it functions.

Analytical—Represent product traits for analysis (e.g. solids modeling software to analyze the expected performance of materials under stress).

Comprehensive—Full-scale operation version of the product (e.g. beta version of software that is made available prior to production release).

Focused—Implement one or a few of the product's attributes (e.g. the form of a hand held object or the function of an interface). Choosing to explore a few traits is much more efficient than exploring all traits at once. For that reason, studies can be created to examine a few attributes such as form and composition (a 'looks like' prototype), or function (an 'acts like' prototype).

The more complete a simulation is, the closer it is to reality and the more reliable its results will be. Yet, the more complete a simulation is the most costly it will be. The trade-off facing the development team is to select the simulation approach that derives the greatest learning for the least cost.

Virtual reality (VR) systems have evolved in recent years as a method to simulate environments. Knott (2000:6–7) describes VR systems as a sophisticated user-computer interface that provides a user with the illusion of being immersed in a computer-generated (virtual) environment. VR systems employ one or more interface devices to provide the user with sensory cues: head-mounted visual displays or large projection screens, spatially distributed audio and acoustic signals, tactile force feedback and inertial displays. Input devices range from a mouse or joystick to devices that are used to sense the body's position in three-dimensional space. The computer-generated environment consists of a virtual world database, software to dynamically render the database in real time and in three dimensions and software that manages user input. VR systems find use in applications that benefit from the simulation of a real environment including data visualization, design, training, medicine, education and entertainment.

A number of considerations can inhibit the creation of a reliable simulation: available time, available funding and presumptions the investigator may have about what would be learned. To mitigate such limitations, the investigator should test for conditions that are expected to occur most often as well as conditions that could result in failure.

Simulations can be conducted through informal observations or more formal demonstrations.

Observations—In informal simulations, a quick series of inexpensive mock-ups (often made of foam, glue and available objects) are made and evaluated. Their purpose is to observe what the concept looks like. Human factors knowledge can be quickly brought to bear to shape the evolution of these fast studies. In such informal simulations, the investigator uses professional discipline to guide evaluation of the quick study mock-ups. The development team acts as a group of expert evaluators. Meister (1985:237) describes placing equipment in a full-size mock-up and using a human factors checklist to make a record of judgements regarding their location. The demonstration version of this arrangement would be to have subjects role play to show how they can, or would, use the set-up.

Demonstrations—Demonstrations can be developed for others who are outside of the development team to use the simulated solutions. In these more formal simulations, more rigorous procedures need to be followed. The investigator must create an environment for fair and consistent evaluation that other team members and the subjects will follow. Meister (1991:489) warns against 'seat of the pants' intuitive measurement while conducting formal simulation.

> To be acceptable, rapid prototyping must be initiated with the development of test objectives, the specification of standards of satisfactory performance, and of criteria and measures, the selection of a sufficient number of representative users as subjects, a description of the training which subjects may require, and an adequate data analysis following the test.

Measures should collect both subjective and objective data. Both performance and acceptability (which includes safety) should be evaluated. Failure to follow such procedures may cause the results of a development project to suffer from problems with reliability and validity. Chapters 5, 10 and 11 provide guidance on how to ensure that the simulation will produce valid, consistent results.

8.1.1 Preparation

Step 1—Review information generated during task analysis (Section 7.6).

Step 2—Review the hardware, software and operating procedures that are intended to be used to demonstrate the solution. If there is a script or set of responsibilities required to operate the simulation, be sure to understand assumptions the team may be making.

Step 3—Set up the mock-up in a physical situation that is as close to actual intended operating conditions as possible. If evaluating a snow blower mock-up, performing the evaluation in a northern climate in January would be much more likely to provide a consistent evaluation than inside a garage or at a more convenient, inexpensive locale.

Step 4—The level of abstraction in a simulation may allow for ambiguity or misinterpretation. This may call for subjects to perform a certain degree of interpretation or role playing. Ensure that the subjects who participate in the simulation understand the simulation's goals, procedures and situation (environment).

8.1.2 Materials and equipment

Choice of materials and equipment depends on two considerations: the state of concept development (i.e. computer simulation or physical mock-up) and the type of information expected to be collected while evaluating the simulation. Computing simulations typically rely on a dedicated platform (CPU/display workstation) using a particular application software program to demonstrate a proposed product. Web service concepts developed by a web-authoring program are an example. Physical mock-ups can be either 'looks like' or 'acts like' mock-ups. 'Looks like' mock-ups can be used to obtain qualitative response from subjects. Subjects can operate 'acts like' mock-ups, making it possible for the investigator to collect subjective responses and, in some cases, quantitative information. Interview guides and questionnaires that are described in Chapter 9 can be used to collect qualitative data.

Which means to use for quantitative data collection depends on the simulation. Computer-based simulation can often be accomplished by data capture and analysis programs. Such programs monitor the type, frequency and rate of input for later review. Usability labs routinely use such programs that combine user operation with time line information. Compact video cameras are often used in combination with software analysis programs to capture subject comments and other information that the software cannot.

Quantitative data collection for physical 'acts like' models is more of a challenge and depends on the kind of data that would be most informative. For example, retinal eye tracking is used to monitor eye scan across a field of vision. The resulting information can be used to understand what a subject pays attention to. Back strain and injury can be minimized through studies using electronic monitors to measure muscle activity (e.g. electromyography) and skeletal movement (e.g. external strain gauges). Refer to the collection of data on physiological measures during workload assessment (Section 8.5) for more information.

8.1.3 Procedure

Table 8.1 summarizes two approaches to simulation that can be used to either explore concepts within a team (observational) or to collect information using subjects (demonstrational) (Meister 1985:233–241).

Table 8.1 Simulation—demonstrate a concept in order to collect human preference and performance data

---

Informal simulations (observational)

Develop evaluation criteria based on project goals

Determine what aspects of the concept need to be explored

Select the feature(s) that need to be examined

Develop one or more 'looks like' or 'acts like' prototypes for evaluation

Have the evaluators discuss the prototypes in open conversation

Formal simulations (demonstrational)

Develop test objectives

Specify standards of satisfactory performance

Specify criteria and measures

Select a sufficient number of representative users as subjects

Describe the training which subjects may require

Conduct the evaluation and collect data

Analyze the data

8.1.3.1 OBSERVATIONAL

Step 1—Develop evaluation criteria, based on project goals.

Step 2—Determine what aspects of the concept need to be explored (e.g. form or composition, function, assembly).

Step 3—Select the feature(s) that need to be examined (e.g. a display, a control device). Determine the objective and the subjective measures for each feature.

Step 4—Develop one or more 'looks like' or 'acts like' prototypes for evaluation. A 'looks like' control prototype might explore the concept's composition and might be used to determine how others think it works. An 'acts like' prototype might be used to determine the results the mechanism produces when it is operated.

Step 5—Have the evaluators discuss the prototypes in open conversation, stating reactions to the alternatives in terms of the project criteria. Early observational simulation often involves subject matter experts (SMEs) who can provide highly qualified subjective evaluations. Record the observations and suggestions for change for reference when further prototypes are created.

8.1.3.2 DEMONSTRATIONAL

Meister (1991:489) provides a series of seven steps that are appropriate to the collection of data during demonstrational simulation.

Step 1—Develop test objectives.
Step 2—Specify standards of satisfactory performance.
Step 3—Specify criteria and measures.
Step 4—Select a sufficient number of representative users as subjects. More advanced demonstrational simulations usually involve new users who have recently been trained.

Step 5—Describe the training that subjects may require.
Step 6—Conduct the evaluation and collect data.
Step 7—Analyze the data.

A control panel project serves as an example. The development team would choose to evaluate reach and accessibility for a user who would use the panel in a confined space. Standards for satisfactory performance might be to reach all controls and to avoid accidental activation. Criteria and measures might include hand movement distance, hand movement frequency and activation. Subjects might require some training to understand aspects of the mock-up. Data that are collected for one mock-up might be compared with data collected from other subjects who used a different control arrangement in the same mock-up.

8.1.4 Result

Simulation results can be used to predict the performance of a product before it is produced. Simulation can be used to predict aspects of the solution that will be necessary to complete its development, including assessments of workload and operating procedures and training. Alternative solutions or configurations can be evaluated and selected through the use of simulation. Simulation also makes it possible to identify situations that might provoke incidents, errors or accidents.

8.1.5 Example

Four different types of simulation can be used to answer questions about the design of an intelligent system user interface. They are ranked below in terms of lowest to greatest cost and greatest to least versatility.

8.1.5.1 PAPER HIERARCHY

Each node in the interface is written on a small piece of paper (e.g. the popular 3M brand 'Post It' self-adhesive memo notes). The collection of paper notes is then sorted into clusters that have similar traits and organized into a hierarchy. Nodes can be easily added, moved, removed and given representative codes and symbols. Verbal protocol analysis (Section 6.5) can be used to determine how prospective users react to terms and hierarchical structures.

8.1.5.2 PAPER OR HTML SIMULATION

Two approaches can be taken to what Meister (1985:233–41) refers to as facading, or the quick and inexpensive simulation of the external appearance of a system interface: paper physical model or Web-capable software (e.g. HTML) simulation on a monitor.

Nemeth (1995) describes the development of a concept for a math and science learning tool software program for middle school students. The object of the learning tool was to teach the relationship between natural resources and finished goods (e.g. tree, to lumber, to frame, to house). Fome Bord (a commercial product made of a thin foam center with Bristol Board paper cladding) was used to make a mock display and controls. Hand drawn images on paper were used to simulate those that would be displayed by the program on a display screen. Members of the development team described the interface operation for their middle school student subjects, then presented a choice of options. When the student would choose, the team member would swap the current image by placing another sketch in response to the choice. Student enthusiasm and willingness to role-play produced a rich evaluation experience for a nominal cost of materials.

Programs designed to create Worldwide Web-capable hypertext mark-up language (HTML) files can be used to represent controls and displays. The development team creates a hierarchy of system nodes, then develops screens for each node that are connected by hypertext links. Users can be asked to operate controls by pressing buttons (clicking a JAVA-scripted button) that change the screen that is displayed within the browser window. Developers can observe quantitative (e.g. error, time) and qualitative (e.g. satisfaction, learning) data through a sample of subjects using one or more concepts.

8.1.5.3 'WIZARD OF OZ'

Two personal computers are set up in a lab environment. One, which the test subject uses, is installed at a normal workstation. The other computer, unknown to the subject, is behind a partition and is operated by a test monitor. The test subject is asked to operate the interface that is under evaluation. The test monitor, observing the subject's input, quickly enters inputs that make the subject's interface respond as if it were an actual operating product. The Wizard of Oz technique has been used successfully in recent development work on automatic speech recognition systems.

8.1.5.4 'VERTICAL SLICE'

Actual code is written to demonstrate how the interface would work. However, only a limited set of options is fleshed out at each level of the hierarchy. The result shows how the interface works without the cost of writing all code for the finished product. By using actual code, it serves as a pilot run for the concept.

8.2 Fault tree analysis

Discover what conditions may combine to create a hazard and possibly cause loss.

Fault tree analysis (FTA) is one of the three methods presented in this book that can be used to answer the questions 'what might go wrong?' and 'what can we do about it now?'

Management oversight and risk tree analysis (Section 6.7) looks for possible threats to performance that may lead to failure. Failure modes and effects analysis (Section 8.3) reviews hardware, software and human elements and the effect their failure might have on performance.

Park (1997:150–71) defines FTA as 'a method of system reliability/safety analysis that … shows a logical description of the cumulative effects of faults within the system.' The method is in the same family of predictive techniques for error probabilities as the technique for human error rate prediction (Section 7.8).

Safety is the protection of humans from the consequences of either their own errors or machine and material failure. Safety is a primary responsibility of those who work in human factors. Chapter 2 covered human capabilities and limits and the consequences of exceeding them. The research and development team member with human factors responsibilities needs to ensure each solution concept is safe. This can be accomplished by making a best attempt to anticipate possible situations that may create a hazard. As an evaluation method, FTA can be used to assess a solution concept with regard to conditions that might create a hazard.

FTA uses symbols in a logic diagram as a graphical technique. Typical symbols, shown in Figure 8.1, represent cause and effect relations among events. Events are paired and assigned a logical state that is represented by a graphical symbol that indicates AND or OR, as Figure 8.2 shows. The use of a tree diagram 'allows us to see where our problems lie and to place our thoughts on paper to produce a logical sequence of causal events' (Ferry 1988:80–1, 150–1).

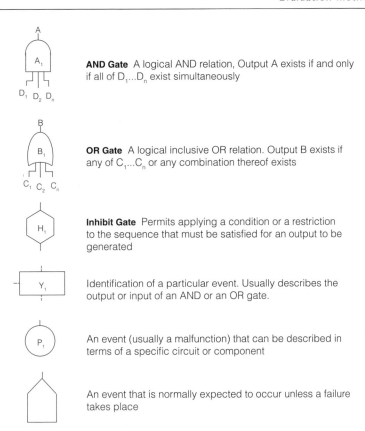

AND Gate A logical AND relation, Output A exists if and only if all of $D_1...D_n$ exist simultaneously

OR Gate A logical inclusive OR relation. Output B exists if any of $C_1...C_n$ or any combination thereof exists

Inhibit Gate Permits applying a condition or a restriction to the sequence that must be satisfied for an output to be generated

Identification of a particular event. Usually describes the output or input of an AND or an OR gate.

An event (usually a malfunction) that can be described in terms of a specific circuit or component

An event that is normally expected to occur unless a failure takes place

Figure 8.1 Fault tree logic symbols—fault tree analysis uses symbolic logic to structure circumstances that may result in an undesirable outcome. Six symbols are used to create a logical tree diagram. Two symbols, AND and OR, are used most frequently.

Source: Adapted from *Product Safety Management and Engineering*. Hammer, W.

Bahr (1997:136–8) observes that the more AND gates a tree contains, the more fault tolerant and safer a system is. Conversely, a proliferation of OR gates depicts a failure-prone situation.

Chapanis (1996:109–13) depicts FTA as an accident investigation method that 'starts with an undesirable event (for example, a train or aircraft collision, injury to personnel) and attempts to determine those combinations of events and circumstances that could lead to it.' It is the combination of events that this method relies on to sort out potentially hazardous outcomes, although there are two significant limitations to the method:

- Each event has to be described in terms of only two possible conditions: either it succeeds or it fails.
- Probability data on human activities is limited. The same issue affects function allocation (Section 7.4) and the technique for human error prediction (Section 7.8).

Stephenson (1991:274–5) describes event tree analysis as a variant of FTA that can be used to 'explore both success and failure alternatives at each level.' Event trees are portrayed in a

8.2 Fault tree analysis

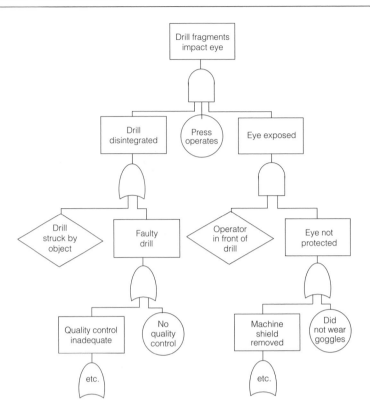

Figure 8.2 Fault tree diagram—industrial facilities are inherently hazardous. The design of workstations for handling equipment and materials requires careful consideration of what might go wrong. Fault tree analysis can be used for that purpose. Figure 8.1 shows logic symbols that are used to construct a fault tree. In the example above, fragments strike a drill operator's eye. Note that three events must all occur for the result to occur. The fault tree shows the uppermost conditions that could result in substantial loss if not prevented. The tree can extend as far back as necessary to account for contributing events. The OR logic gates marked 'etc.' at the bottom of the tree indicate links where foregoing contributing events might be added.

Source: *Modern Accident Investigation and Analysis*. Ferry, T.S. Copyright © 1988. This material is used by permission of John Wiley & Sons, Inc.

tree that branches from left to right and uses only OR gate logic, in contrast with fault tree analysis' top to bottom hierarchy and use of both AND and OR gate logic.

8.2.1 Preparation

Understand the people, hardware and software that have been chosen to comprise the product, as well as the thought processes that were used to develop the product. Factors that could cause the undesired result can be inherent in either the elements or the development logic (Vincoli 1994:184–5).

Perform, or review the results of, methods that account for functions and tasks that are expected: management oversight and risk tree analysis (Section 6.7), flow analysis (7.1), task analysis (Section 7.6). Interview (Section 9.2) experts can provide background on its development.

8.2.2 Materials and equipment

Information on product elements and their specifications is important to the construction of a fault tree.

Inexpensive tracing tissue can be used to quickly develop and revise fault tree diagrams. Illustration or desktop editor software programs can be used to create smooth versions of tree diagrams, particularly for those that include multiple and extensive branches.

8.2.3 Procedure

The following approach that is summarized in Table 8.2 is drawn in part from Hammer (1980:204–5) and is used to perform FTA.

Step 1—Identify the top event for which the probability will be determined. The results of management oversight and risk tree analysis can be used to identify undesired events.

Step 2—Add the events or conditions that can contribute to the occurrence of the top event.

Step 3—Assign the appropriate logic gate to the combined events indicating whether both events must occur at the same time and place (AND) or either event may occur (OR). Figure 8.1 shows symbols used in fault trees. Progression down the branches that are shown in Figure 8.2 indicates causes. Moving up the branches indicates effects.

Step 4—Continue to identify contributory events and assign logic symbols to link higher events to causes. While some conditions will present pairs of events, any contributing event can be placed on the tree.

Step 5—Determine the probability that each event will occur by figuring its likelihood based on the probability of each event in its contributory pair. Figure 8.3 provides two equations from Chapanis (1996:112–3) that can be used to figure the probability of two of the most often used logic gates: AND and OR. Even if probability events cannot be assigned, FTA can still serve as a valuable evaluation method.

Step 6—Examine the events and conditions to determine whether the malfunction is due to: primary fault (component failure), secondary effect (another component's failure, or conditions), input or command (erroneous signal, error, or input).

Step 7—Develop strategies to remedy the presence or combination of events in order to prevent the top event from occurring.

Table 8.2 Fault tree analysis—discover what conditions may combine to create a hazard and possibly cause loss

Identify the sources of the potential hazard (threat to life or property)
Identify the initiating events that could lead to this hazard
Establish the possible sequences that could follow from various initiating events using event trees
Quantify each event sequence
Determine the overall risk

gate	equation	situation
AND	$p_A = p_B p_C$	Where A is the result of both B and C occurring simultaneously
OR	$p_B = 1 - (1 - p_D)(1 - p_E)$	Where B is the result of either D or E occurring

Figure 8.3 Fault tree probability—logic gates are used in fault tree analysis to refine the structure of events and their likely occurrence. Chapanis (1996: 112–13) provides equations to figure the probability of the two most popularly used gates: AND and OR.

8.2.4 Result

Fault tree analysis produces a diagram that represents the collective faults, or impairments to performance, that can result in dysfunction. It can also spell out how a series of events may lead up to an undesirable event. The portrayal can be used to anticipate and rectify potential sources of failure. It can also be beneficial in accident analysis.

8.2.5 Example

In industrial accidents, it is not unusual for a number of hazards to exist on a site. Some work sites (e.g. steel mills, foundries) are inherently hazardous. Exposure to intense heat, molten metal, noxious gasses, the impact of a heavy moving object and more require steps to create an environment that is appropriate for workers and compliant with work place regulations (e.g. in the U.S., Occupational Safety and Health Administration (OSHA)). Research and development teams that are responsible for the creation of a work station in hazardous industrial settings need to identify circumstances that can expose workers to hazard, then take steps to remove the hazard or protect workers from exposure. Solutions that flow from such analyses can include improved plans for facilities and work areas, new kinds of protective apparel and new equipment.

Begin with the identification of top events, such as molten metal on unprotected worker. Events or conditions that would contribute to the event would include worker unprotected and molten metal spill. An unprotected worker could result from three events: unavailable, inadequate or unused protection. A molten metal spill could result from improper handling or implements. Figure 8.2 shows an example of a fault tree that describes a set of conditions and the logic gates for each set. Probability for each of the events to occur can be estimated, then figured using the formulas in Figure 8.3.

Construction of a fault tree and the assignment of probabilities for the events create a model that can be used to assess the causes for the occurrences. Causes can be combined with actions that can be taken to remedy them. Both causes and possible remedies are shown below.

- Protection unavailable
    - Failure to procure correct items—Order and issue protective gear
    - Failure to issue items—Assign issuance to work center supervisor tasks
- Protection inadequate
    - Equipment inadequately maintained—Assign maintenance responsibilities
- Protection unused
    - Worker unaware of hazard—Develop proficiency training
    - Worker not motivated—Set clear hiring standards
- Improper handling
    - Failure to follow procedure—Make training clear, memorable
    - Worker not trained—Provide worker training

- Improper implements
  - – Failure to select correct implement—Design workstation and training to ensure implements are available and understood by work crew
  - – Failure to maintain workstation—Assign maintenance to work center supervisor tasks.

Industrial safety programs strive to maintain good work habits and avoid employee exposures to hazards. FTA can assist operational risk management programs. If an accident does occur, FTA can be a helpful tool to review what factors combined to make it possible. Results of the study may be used during litigation and to inform future development work and safety programs.

8.3 Failure modes and effects analysis

Discover potential, implications and remedies for possible failures.

Failure modes and effects analysis (FMEA) is used to examine the potential for the concept under development to fail. Failure is the inability to fulfill the purpose that was determined during operational analysis (Section 6.2). FMEA leads the investigator to develop recommendations for remedial action that may avert (or at least minimize) the potential for failure to affect the operation of a product.

The method can be performed using either a 'top-down' or 'bottom-up' approach to reveal possible sources for failure. A top-down FMEA is used to review all major portions of a product, break them into components and constituent parts and speculate what might go wrong with each. A bottom-up FMEA is used to review all individual parts and speculate what might go wrong with each component. Individual possible failures among parts can then be aggregated into a larger picture of component, subassembly or product potential failure. Either approach can be successful. The choice of which to use depends on the investigator's preference and the information that is readily available.

In contrast with fault tree analysis' focus on conditions, FMEA reviews each hardware and software element that may fail and each human that may err. A tabular format, shown in Table 8.3, is used to organize information for each part of the product. Categories include: the condition that item failure causes, the source for the item failure, how failure affects product performance, the likelihood that the failure will occur, the degree to which item failure affects accomplishment of the product's purpose and changes that can be made to the product in order to avert failure.

FMEA accounts for both aspects of product dysfunction: hardware/software failure and human erroneous acts. Recommendations for changes are made with regard to the product, not the human operator, maintainer or user. If humans actions can cause a malfunction, it is incumbent on the development team to change the hardware and software so that it does not induce errors and can tolerate as many errors as possible.

Depending on the erroneous action, changes might allocate control over the function to other elements (e.g. hardware/software) or provide decision aids (e.g. displays with timely and clear information). Hardware and software reliability can be reinforced through a number of techniques, such as redundancy. Human erroneous acts can be minimized by providing assists at each of the three levels of experience based behavior (skill, rule, knowledge) that were described in Chapters 2 and 3. What matters in FMEA is that potential product dysfunction is foreseen and that loss is avoided or at least lessened.

8.3 Failure modes and effects analysis

Table 8.3 CD player failure modes and effects analysis—failure modes and effects analysis (FMEA) can be used to determine the potential causes of failure, resulting effects, and remedies. In the example below, the researcher identified 19 failure modes for the Sony D-125 CD player, concluding that only errors in set-up and CD handling were critical. Remedies included dividing functions into separate controls and offering more control states for the user to select through keypad entry.

Task	Item	Failure mode	Erroneous action	Error type	Cause of failure	Possible effects	Probability	Criticality	Remedy
1 Remove disc from player	CD, CD player	a) CD stuck on player hub	a) Attempt to remove disc from hub	a) Error of commission	a) Dropping, manual dexterity fault	a) Scratching, breakage	a) Medium	a) Medium	Wider grip area 0.25–0.5" around perimeter of CD to facilitate handling
2 Press Open button to open lid	Open button	a) CD lid will not open	a) Pry lid open	a) Error of commission	a) Button b) Unaware of how to open	a) Lid damage b) High	a) Low	a) High	Push-push lid, push-pull lid may simplify operation
3 Clean disc	CD, cloth	CD surface gets dirty	a) Use abrasive cleaning agent b) Wipe brusquely	a) Error of commission b) Time error	a) Lack proper materials b) User rushing	a) Degrade CD surface b) Scratch CD	a) Medium b) High	a) High	Change material to abrasion-resistant resistant polymer

FMEA is also a valuable method to investigate accidents. Malaskey (1974:107–41) describes an approach to management oversight and risk tree analysis that follows a procedure that is similar in logic and format to FMEA. A twelve column format is used to account for the:

- Hazard
- Element or procedure that could give rise to the hazard
- System element's function or purpose
- Hazard category (attempts to indicate cause: hardware, personnel)
- Nature of injury/damage
- Personnel affected
- Consequence, primary (hazard's effect on system)
- Consequence, secondary (cascading effects)
- Hazard classification (safe, marginal, critical, catastrophic)
- Cause of hazard
- Action taken (corrective steps)
- Probability of occurrence.

Ferry (1988:150–1) views a malfunction as part of a system of complex and interrelated components and traces its effects through the system while evaluating the effect on task performance. Due to FMEA's singular focus, other failures and interactions may be overlooked.

8.3.1 Preparation

FMEA relies on the review of information that has been collected through thorough observation of people operating or using the product or similar products. Functional analysis methods (Sections 7.1 to 7.3) and activity analysis (Section 6.4) are both helpful methods to perform prior to FMEA. Critical incident study (Section 6.6) can also provide insights into patterns of performance that may create near incidents or accidents or actual incidents or accidents.

8.3.2 Materials and equipment

The existing product or a functional 'acts like' prototype simulation of the solution concept will be required. While a visual 'looks like' model or mock-up can be used, important aspects such as function will be unavailable to the investigator.

Desktop publishing or text editor software can be used to develop a FMEA table.

8.3.3 Procedure

Failure modes and effects analysis consists of the following steps that are shown in Table 8.4.

Step 1—List all noteworthy product components (i.e. element, subassembly). Set the level of detail for the list to be particular enough to accurately assess each, yet general enough to make the method practically manageable. For a review of system definition, see Chapter 1.

Step 2—Add each component name to the first column of a table that is organized according to seven columns:

Table 8.4 Failure modes and effects analysis—discover potential implications and remedies for possible failures. Table 8.3 provides an example.

List all significant elements

Set the level of detail for the list to be particular enough to accurately assess each element, yet general enough to make the method practically manageable.

Create a seven category table

Create a table that is organized according to: Item, Failure mode, Cause, Effects, Probability, Criticality, and Remedy

Enter elements

Enter names of each element under 'Item' column in table

Rate each element

Evaluate elements according to each consideration

Enter evaluation ratings and comments in table

Identify 'critical' elements

Review evaluations for elements that may compromise performance and have probability and criticality ratings that are anything other than 'low'

Apply remedies

Determine how remedies can be incorporated into the concept's hardware or software in order to avert possible failure

---

- Item (the part of the product that could fail)
- Failure mode (the condition that item failure causes)
- Cause (the source for the item failure)
- Effects (how item failure affects product performance)
- Probability (the likelihood that the failure will occur)
- Criticality (the degree to which item failure affects accomplishment of the product's purpose)
- Remedy (changes that can be made to the product in order to avert failure).

Step 3—Review each component according to each consideration and complete the table.

Probability can be rated and assigned a representative code from frequent ('A') to probable ('B'), occasional ('C'), remote ('D') and improbable ('E'). Criticality can be similarly represented as catastrophic ('I'), critical ('II'), marginal ('III') or negligible ('IV') (Ferry 1988:80–1, 150–1). None of the ratings need to be expressed in absolute terms. They serve as a means to sort among implications for failure. For example, an event that is estimated to have high criticality but low probability would merit less immediate attention than one with medium criticality and high probability. Ultimately, it is management that decides where to set the threshold with regard to failures that are or are not acceptable.

8.3.4 Result

The use of FMEA directs design and development efforts to the correction of potential problems, making the product safer and more reliable. Specifications for the product design can be adjusted to eliminate options that might cause dysfunction.

FMEA can be of significant benefit in the event that a legal claim is brought against the product's producer. Demonstrating the process that was followed, remedies that were recommended and actions that were taken to minimize loss is evidence that the producer was not negligent in development efforts. Retain FMEA records for possible future use in legal investigations.

8.3.5 Example

Medication is a necessary aspect of health care and injection is one means of providing medication. While injection can be safe and effective, problems can occur. For a team that is assigned to develop a new approach to medication injection, FMEA would be a helpful method to use. The following example describes how FMEA could be used to foresee and remedy potential problems.

Begin by listing all noteworthy product components (i.e. element, subassembly). Set the level of detail for the list to be particular enough to accurately assess each, yet general enough to make the method practically manageable.

Add each component name to the first column of a table that is organized according to seven columns:

- Item (the part of the product that could fail)
- Failure mode (the condition that item failure causes)
- Cause (the source for the item failure)
- Effects (how item failure affects product performance)
- Probability (the likelihood that the failure will occur)
- Criticality (the degree to which item failure affects accomplishment of product mission)
- Remedy (changes that can be made to the product in order to avert failure by either reducing the probability or the consequences).

Review each component according to each consideration and complete the table. Determine probability and criticality based on the description in Section 8.3.3. In this case, examination of the injector would note the potential for the item that comes into contact with the subject to become contaminated. Medication needs to be delivered at a specific location (e.g. below the skin, in muscle tissue) to have the correct therapeutic effect. Failure would compromise the medication's effectiveness.

Failures that have great impact and are highly likely to occur will normally be obvious and will garner immediate attention. An event that is estimated to have high criticality but low probability would merit less immediate attention than one with medium criticality and high probability. Thorough examination through the use of FMEA can enable the team to uncover events that can escalate above low impact and low probability. For example, failure to inject the correct type or amount of medication may have significant effects.

Review each component that merits attention and explore remedies that will make each less likely or unlikely to result in failure. Some elements that are related to people (e.g. hiring, training, decision aids) may mitigate the potential for failure. Remember, though, that it is changes to hardware and software that will improve system reliability.

8.4 Operational sequence analysis

Determine the interaction among elements through time.

Operational sequence analysis (OSA) combines the events, information, decisions and actions involved in product use into a single diagram that is organized along a timeline. Chapanis (1996:131–5) considers the method to be 'one of the most powerful techniques available to the human factors specialist.' As a sophisticated extension of task analysis combined with time line data, OSA can be used to develop a picture of system operation, indicate the functional relationships among multiple system elements, trace the flow of materials and information, show the physical or sequential distribution of operations, identify subsystem inputs and outputs and make the results of alternative design configurations explicit (Chapanis 1985).

OSA can be a challenge to perform. Because it is comprehensive, OSA can take a significant amount of time to complete. OSA requires a fairly substantial amount of knowledge about a concept. That knowledge may or may not be available. In addition, inferences about workload or interface evaluation must come from cues that either may not yet be specified or may require special expertise. OSA is an excellent way to depict the sequential flow of information and activity and makes it easy to classify activities. These two advantages make OSA useful in interface description (Wickens 1992:383–402).

Figure 8.4 shows how the diagram that OSA creates portrays every element and its interaction. Standard symbols are used to represent typical events and conditions. Reviewing the diagram makes it possible to ensure that information, decisions and results occur in the desired order and in sufficient time.

8.4.1 Preparation

OSA is a comprehensive representation of one or more solution alternatives in diagrammatic form. In order to be of value, the diagrams must reflect accurate information. Many of the issues and methods that have been described so far in this text can contribute to the information that is necessary for OSA:

- What the solution should include and how it should be organized: operational analysis (Section 6.2)
- How others have dealt with this problem: analysis of similar systems (Section 6.3)
- What people actually do and how frequently: activity analysis (Section 6.4)
- How users perceive the use of a product: verbal protocol analysis (Section 6.5)
- What is causing difficulties or losses: critical incident study (Section 6.6)
- What might go wrong and how it could be avoided: management oversight and risk tree analysis (Section 6.7)
- Nature and order of functions: flow analysis (Section 7.1)
- The order and duration of events: time line analysis (Section 7.2)
- How elements that comprise the product are related: link analysis (Section 7.3)
- Which person, hardware or software will perform each proposed function: function allocation (Section 7.4)
- What people will do in the proposed solution concept: task description (Section 7.5)
- Tasks, training, equipment and information will be needed to perform a job: task analysis (Section 7.6)
- What could go wrong: fault tree analysis (Section 8.2)
- How loss can be minimized or avoided: failure modes and effects analysis (Section 8.3)

8.4.2 Materials and equipment

Inexpensive tracing tissue can be used to quickly develop and revise the operations analysis diagram. Desktop publishing software programs can be used to create smooth versions of the diagram.

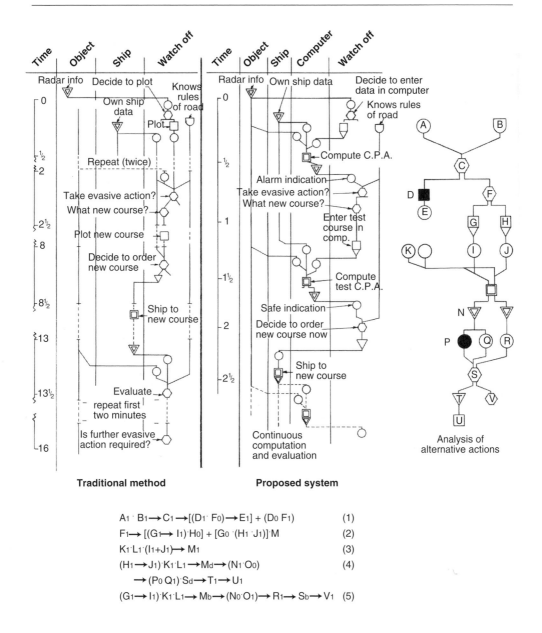

$$A_1 \cdot B_1 \longrightarrow C_1 \longrightarrow [(D_1 \cdot F_0) \longrightarrow E_1] + (D_0 \, F_1) \tag{1}$$

$$F_1 \longrightarrow [(G_1 \longmapsto I_1) \cdot H_0] + [G_0 \cdot (H_1 \cdot J_1)] \cdot M \tag{2}$$

$$K_1 \cdot L_1 \cdot (I_1 + J_1) \longrightarrow M_1 \tag{3}$$

$$(H_1 \longrightarrow J_1) \cdot K_1 \cdot L_1 \longrightarrow M_d \longrightarrow (N_1 \cdot O_0) \tag{4}$$
$$\longrightarrow (P_0 \, Q_1) \cdot S_d \longrightarrow T_1 \longrightarrow U_1$$

$$(G_1 \longrightarrow I_1) \cdot K_1 \cdot L_1 \longrightarrow M_b \longrightarrow (N_0 \cdot O_1) \longrightarrow R_1 \longrightarrow S_b \longrightarrow V_1 \tag{5}$$

Figure 8.4 Operational sequence diagram—Woodson demonstrates the use of operational sequence Analysis in the operation of sensor equipment (radar, computer and ship) and a human (watch officer). Two approaches are compared in order to assess traditional and computer-supported decision making. Activities are laid out along a time line. By allocating rule-based activity to the computer, the watch officer is able to review information and make decisions much faster.

Source: Woodson, W. *Human Factors Design Handbook*. 1981. McGraw-Hill. Reproduced by permission of the McGraw-Hill Companies

8.4.3 Procedure

Follow these steps, summarized in Table 8.5, to perform operational sequence analysis.

Step 1—Create strip chart that is similar in format to Figure 8.4 including columns for elements that are of interest to the problem: objects (e.g. equipment), individuals, elapsed time.

Step 2—Create key for special symbols that represent typical activities. Figure 8.4 shows examples of such symbols. New symbols can be added as long as they are used consistently and are explained in a key.

Step 3—Plot each action. Using the order of events from earliest to latest, write a label to describe the action. Select a symbol to represent the type of action. Place the symbol/label in the column of the system element that performs the action. Use a single line to indicate the flow of activity from earlier to later items.

Step 4—Review and evaluate the diagram. Identify discontinuities in timed relationships. Look for opportunities to cut redundancy.

Step 5—Recommend changes to remedy problems and optimize interactions.

8.4.4 Result

Operational sequence analysis creates a time-based chart that shows functional relationships among elements, the flow of materials and information, the sequence of operations that are to be performed and the input and output of subsystems that must be coordinated for adequate performance. Review of the results can demonstrate the consequences among alternative designs as well as potential human/system difficulties (Chapanis 1996:131–5).

Table 8.5 Operational sequence analysis—examine the interactions among elements. Figure 8.4 provides an example

---

Create strip chart

Include objects (e.g. equipment) and individuals that are of interest to the problem
Include a column for elapsed time

Identify symbols

Select symbol set for use in diagram

Create key to describe symbols

Create symbol(s) if standard set is insufficient

Plot each action

Using the order of events from earliest to latest, write a label to describe the action

Select a symbol to represent the type of action

Place the symbol/label in the column of the system element that performs the action

Use a single line to indicate the flow of activity from earlier to later items

Review and evaluate the diagram

Identify discontinuities in timed relationships

Look for opportunities to cut redundancy

Recommend changes

8.4.5 *Example*

Many transactions such as purchases and identification are managed through the use of automated systems. These systems require the user to insert and pass a coded identification card through a reader. Both user and system need to interact in certain ways, in a certain order and within a certain length of time. Operational sequence analysis can be a useful method to map out the interactions that need to be accomplished.

A strip chart such as the chart in Figure 8.4 would include the reader, display, remote systems that will perform activities (e.g. communication, authentication, authorization) and the individual who is using the product. Symbols would indicate activities such as an operator action, system action, system decision, operator decision or interaction between the operator and equipment. Plot each step in the process. Begin with the operator producing the transaction card (an operator action). Continue for each step in the process, sorting the activities according to the correct column on the diagram. The final diagram will depict the order of each activity, necessary relationships among activities, the duration of each activity and the cumulative time that will be required for the sequence.

A review of the diagram might reveal information that development team members did not anticipate. The entire sequence may take more time than users prefer to spend. Remote systems (e.g. databases) may not be able to conduct the transaction in the intended order. Users may require a delay in order to allow for such activities as making a decision or consulting information.

Further methods that can be used to determine answers to these questions include simulation (Section 8.1), interviews including focus groups (Section 9.2) and questionnaires (Section 9.1).

8.5 *Workload assessment*

Determine load of work required and individual's ability to perform it.

A century ago, the effort that was needed to perform a task had a great deal to do with physical capacity and reserves. The lethargy from physical work left little doubt when those reserves had been exhausted. As work roles and their tasks evolved to incorporate a greater proportion of mental challenges, the question of capacity and reserves has become far less obvious.

Workload assessment (WA) is used to evaluate the number and type of tasks that humans are expected to perform in terms of both anticipated conditions and the amount of time that is available to perform them. Workload increases when there are more things to be done, when there is less time to complete them, when tasks are more difficult and when conditions under which tasks are to be performed are more stressful (Chapanis 1996:135–8). The issue is not tasks alone, but the fluctuation of tasks, conditions and time that may produce dysfunctional situations. Dysfunction can result from too little work (producing inattention) or too much work (resulting in stress).

WA focuses on an individual's mental resources that are available to perform tasks. The method makes it possible to determine whether the work that is expected fits those resources. Figure 2.1 illustrates the concept of limited attention resources and its effect on perception, decision and response selection and response execution.

WA can also be used to compare two or more concepts. In some circumstances, workload may be the only satisfactory criterion to use when choosing between alternatives (Wickens 1992:383–402). Meister (1989) faults WA approaches for ignoring the actual issue: the amount of work load that is associated with a particular piece of equipment or a task configuration.

There is no consensus on the one best way to determine workload. Four approaches have evolved in practice: primary task, secondary task, physiological measures and subjective rating. The first three are summarized here from Wickens (1992:383–402).

Primary task—Primary task WA directly evaluates task performance. While it is the most direct approach, it may not lend itself to accurate analysis for a number of reasons. An individual may not reach maximum performance limits, making it difficult to know when reserve capacity has been exhausted. Two primary tasks may vary in their measures or meaning. It can be impossible to obtain good primary task performance measures. Also, primary tasks may vary by differences in data limits.

Secondary (dual) task—The subject is asked to perform a task in addition to the primary task. The intention is to use any mental resources that may be left over from performing the primary task. Under changing conditions, degradation in performance of the secondary task heralds a degradation in the performance of the primary task. That is because the performance of a secondary task is assumed to be inversely proportional to the resource demands that are imposed by the primary task. Examples of secondary tasks include rhythmic toe or finger tapping, random number generation and estimation of the time that has elapsed during the problem. Nakayama *et al.* (1998) described a study to evaluate driver steering performance while also performing additional activities. For the secondary task, a light was illuminated at random intervals and the subject was instructed to push a switch upon noticing the illuminated light. The time from light illumination to the switch being pushed was measured with the reaction time serving as an index of workload.

The secondary task approach has a high degree of face validity (the extent to which a measure looks as though it measures what is intended). It will also yield workload measures in the same units when applied to very different primary tasks. However, secondary tasks may not always be appropriate for certain primary tasks. Secondary tasks can also be obtrusive by interfering with primary task performance.

Physiological correlates of workload—The investigator monitors selected aspects of the subject's central nervous system (e.g. evoked brain potential, pupil diameter, respiration rate, heart rate). Variability in a physical measure is monitored for a correlation with responses to increased workload demand. Physiological measures can be used to develop a continuous record of data over time and do not impede primary task performance.

Monitoring is accomplished through the use of devices (e.g. electrodes, pupil measurement equipment) that can make the method more obtrusive than others cited here. Measurement apparatus may be initially intrusive. They tend to become less intrusive as the subject gets accustomed to them.

Because the causal relationships between the physiological variable and workload are not always clear, test results may only be used to infer predicted workload performance. Use of physiological methods requires particular caution when it comes to individual differences. Each individual has a primary physiological response to changes in workload. It may be a change in heart rate, or in perspiration or in the rate of eye blink. Without first identifying what the typical response is for each subject, the data collected for physiological measures of workload is of little value.

Subjective rating—The subjective rating method uses questionnaires to collect data on each subject's estimate of workload. Estimates can be collected along a single variable (univariate) or along a variety of variables (multivariate). Methods may be used to either predict workload (projective estimation) or measure it after the completion of a task (empirical evaluation).

The subjective approach methods do not interrupt the performance of primary tasks and are relatively easy to derive. The validity of the methods depends on the degree to which an operator's spoken or written statements truly reflect available reserves of mental processing resources and demands on them. Subjective evaluations can require substantial time to prepare and compile and can also be vulnerable to subjects who may harbor preconceived notions about task ease or difficulty. Subject reactivity (Section 5.1.2.1) can cause subjects to modify their responses based on demand characteristics. Subjects may also modify responses based on peer or organizational influence (e.g. prefers to demonstrate the ability to withstand difficult conditions because that is what the organizational culture expects).

Charlton (1996a:181–97) describes two widely accepted techniques for subjective workload assessment: subjective workload assessment technique (SWAT) and NASA task load index (TLX).

Subjective workload assessment technique (SWAT)—Reid *et al.*'s (1981) subjective workload assessment technique (SWAT) evaluates workload on three 3-point scales, as shown in Figure 8.5. SWAT has also been adapted for use as a projective technique ('Pro-SWAT') that can be used if a new concept is in development. Through Pro-SWAT, subjects are given extensive briefings on tasks and can use SWAT to project themselves into tasks one at a time using equipment or mock-ups, then complete estimates and receive extensive de-briefs. Conditions and equipment/systems need to be well defined, as 'the higher the fidelity of the mock-up and the task procedures used the better the predictive utility of the workload results.'

Figure 8.5 Subjective workload assessment technique—Reid, Shingledecker, Nygren and Eggemeier's (1981) subjective workload assessment technique (SWAT) can be used to examine workload on three 3-point scales and has found wide acceptance as an evaluation method. Subjects are asked to mark a choice for each of three areas: time, mental effort and stress.

Source: *Development of Multidimensional Subjective Measures of Workload.* Reid, G.B., Shingledecker, C.A., Nygren, T.E. & Eggemeier, F.T. Proceedings of the International Conference on Cybernetics and Society. Copyright © 1988 IEEE.

National Aeronautical and Space Administration (NASA) task load index (TLX)—Hart and Staveland's 1988 NASA TLX method assesses workload on five 7-point scales as shown in Figure 8.6. Increments of high, medium and low estimates for each point result in 21 gradations on the scales.

Figures 8.5 and 8.6 show the questionnaires that participants are asked to complete.

The choice of which subjective method to employ is up to the investigator. SWAT is typically employed during task performance. Because of the length of evaluations, TLX is often employed following task performance while participants watch a video recording of the event.

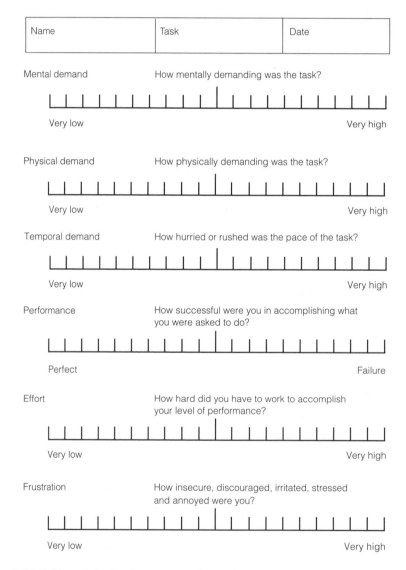

Figure 8.6 NASA task load index—Hart and Staveland's NASA task load index (TLX) method assesses workload on five 7-point scales. Increments of high, medium and low estimates for each point result in 21 gradations on the scales.

Source: Reprinted with the permission of Sandy Hart.

8.5.1 Preparation

Review information that has been collected on task time frequency and precision data from forego-ing analyses such as observation, analysis of similar systems (Section 6.3), task analysis (Section 7.6), time line analysis (Section 7.2), controlled studies (Chapter 11), simulation (Section 8.1) and databases that contain information on human performance.

Subjects will need tasks to perform (for empirical estimates) or to consider (for projective esti-mates) during the use of this method. The work may result from the performance of tasks from a current job, or tasks that have been developed through the performance of task analysis (Section 7.6). Ensure that both tasks and performance objectives are clearly defined.

8.5.2 Materials and equipment

The requirement for accurate and reliable data makes it necessary to use equipment for the measurement of primary and secondary task measures, particularly of physiological measures. The use of equipment to measure physiological functions requires specialized knowledge and experience in how to install and operate it. Development teams that need to collect physiological measures for workload assessment should either obtain qualified staff from another corporate department, or retain an outside consulting firm or university-based research group that is famil-iar with such work. NASA has developed a computer-supported version of TLX that runs on a PC.

8.5.3 Procedure

Subjective rating is presented here as an efficient, direct means to perform workload assessment. It may be helpful to administer a brief preliminary questionnaire to sample participants on attitudes about task ease or difficulty in order to minimize the effect of preconceived notions about workload assessment.

Table 8.6 summarizes the steps to perform subjective workload assessment using either SWAT or NASA TLX.

Table 8.6 Workload assessment—determine load of work required and individual's ability to perform it. Figures 8.5 and 8.6 provide examples.

Subjective rating method
Identify tasks of interest
Recruit subjects who will be asked for their assessment
Select rating scale:
Subjective workload assessment technique
NASA task load index
Develop questionnaire
Normalize sample participants
Administer questionnaire
Compile data
Plot workload estimates for task segments
Identify under- or over-loading
If necessary, adjust task assignments

Step 1—Identify tasks of interest in the problem.

Tasks are a function of the work that is to be accomplished. Select tasks that will have the most direct bearing on the evaluation of a process, solution concept or the comparison of multiple solution alternatives.

Step 2—Recruit subjects who will be asked for their assessment.

Participants in the sample can be individuals who are performing a current related work role or who are experts with in-depth knowledge of the tasks and circumstances.

Step 3—Select rating scales.

Figures 8.5 and 8.6 portray the most commonly used rating scales for subjective workload assessment: SWAT and NASA TLX.

Subjective workload analysis technique

Subjects are asked to choose from among three discrete ratings of three kinds of load: time, mental effort and stress, as shown in Figure 8.6.

Twenty-seven cards, representing all possible combinations of the three factors, are given to the subjects. Subjects are asked to sort the cards in order of decreasing workload. The sorted cards are used to interpret the subjects' ratings of a task in accordance with one of six profiles. SWAT has a prescribed method for scale development:

1   Write the nine levels of time, mental effort and stress on separate cards. Review Figure 8.5 for the three 3-point scales that provide the nine levels.
2   Have subjects rank order the cards (among 27 possible combinations).
3   Perform a series of axiom tests to identify a rule for combining dimensions for each subject.
4   Apply conjoint scaling to develop an appropriate unidimensional workload scale (range from 0 to 100, representing no workload to highest workload).
5   Compare the results of the unidimensional workload to the original rank ordering. The subject is deemed valid and can be used to collect data if the correlation between the workload and rank order achieves a Kendall's Tau coefficient of 0.79 or greater.

Figure 8.7 provides a description of how to use Kendall's Tau coefficient in order to determine correlation between rank ordered sets, based on Bailey. Review rank ordered sets to determine how many pairs are possible. Compare the pairs graphically in order to determine the number of inversions that exist. Determine the value of Tau by the formula [Tau = 1—2 * (number of inversions) / (number of pairs)]. In the instance of SWAT scale development, each individual's rank order for SWAT scales is compared with the individual's single dimension rank order. Charlton sets a threshold of 0.79 as the acceptable correlation between card rank order and the single dimension workload. If the correlation is 0.79 or higher, then the subject's individualized workload scale is valid and can be used to collect data (Bateman, 2000c).

Time, mental effort and stress ratings include the following descriptors.

Example—Nine subjects (with the names 'A' through 'I') fill out a single scale workload estimate.

Pairs determination—There are 36 pairs (the total number of links that can be made among all of the subjects in the set).

Compare pairs to determine inversions—The subjects' workload estimates place them in this order: ACBFDEGIH ('A' gave the highest estimate and 'H' the lowest). Connect like letters. Each line that crosses is an inversion. There are four inversions.

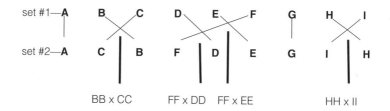

Kendall's Tau coefficient formula

Tau = 1 - 2 * (number of inversions) / (number of pairs).

Tau= 1 - 2 * (4) / (36)

Tau = 0.78

Figure 8.7 Kendall's Tau coefficient—Bailey (1971) explains how Kendall's Tau coefficient is a measure of the correlation between two rank ordered sets. Using Kendall's Tau coefficient makes it possible to know whether test subject use of SWAT scales to determine subjective workload are high enough to use with confidence. High correlation demonstrates that subjects use the scales in a consistent manner.

TIME LOAD

1   Often have spare time. Interruptions or overlap among frequent activities occur infrequently or not at all.
2   Occasionally have spare time. Interruptions or overlap among activities occur frequently.
3   Almost never have spare time. Interruptions or overlap among activities are very frequent, or occur all the time.

MENTAL EFFORT LOAD

1   Very little conscious mental effort or concentration required. Activity is almost automatic, requiring little or no attention.
2   Moderate conscious mental effort or concentration required. Complexity of activity is moderately high due to uncertainty, unpredictability, or unfamiliarity. Considerable attention required.
3   Extensive mental effort and concentration are necessary. Very complex activity requiring total attention.

STRESS LOAD

1   Little confusion, risk, frustration, or anxiety exists and can be easily accommodated.
2   Moderate stress due to confusion, frustration, or anxiety noticeably adds to workload. Significant compensation is required to maintain adequate performance.

2   High to very intense stress due to confusion, frustration, or anxiety. High to extreme determina-
     tion and self-control required.

Actual rating of workload during the test (such as '1, 3, 3') is interpreted as points on a univariate
scale.

NASA task load index

Subjects are asked to indicate an estimate on a scale for each of six aspects of workload, as Figure
8.6 shows. Elaboration on each of the variables from Wickens (1992) that are shown below provides
further background for subjects to use in developing an estimate.

   Mental demand—How much mental and perceptual activity was required (e.g. thinking, deciding,
calculating, remembering, looking, searching)? Was the task easy or demanding, simple or complex, exacting or forgiving?

   Physical demand—How much physical activity was required (e.g. pushing, pulling, turning, controlling, activating, etc.)? Was the task easy or demanding, slow or brisk, slack or strenuous, restful
or laborious?

   Temporal demand—How much time pressure did you feel due to the rate or pace at which the tasks
or task elements occurred? Was the pace slow and leisurely or rapid and frantic?

   Performance—How successful do you think you were in accomplishing the goals or task set by
the experimenter (or yourself)? How satisfied were you with your performance in accomplishing
these goals?

   Effort—How hard did you have to work (mentally and physically) to accomplish your level of performance?

   Frustration—How insecure, discouraged, irritated, stressed and annoyed versus secure, gratified, content, relaxed and complacent did you feel during the task?

   Usually the least important of the six measures is eliminated. Importance is based on a pairwise
selection before conducting the test.

Step 4—Normalize the sample

Step 5—Administer questionnaire

SWAT can be administered at predetermined times during an event or after it. Convert ratings to
the normalized workload scale using the rule that was identified during the procedure used in
scale development.

The investigator needs to observe consistent and appropriate procedures to ensure that data are valid
and reliable. Section 9.1 provides guidance on how to administer a questionnaire.

Step 6—Plot workload estimates for task segments.

Compile responses to the questionnaire. Portray the results in a diagrammatic form to make it possible to see patterns and compare task elements. A bar graph can be used to depict estimates for
workload as a function of time or task. Organize the graph according to workload on the vertical axis
and task elements along the horizontal axis. NASA TLX can portray weighted workload graphically.
A rating of each factor by importance weight can be shown in a bar graph along with overall workload rating.

Step 7—Identify under- or over-loading and adjust task assignments. Estimates that indicate moderate workload confirm current or intended design requirements. Estimates that are noticeably high or low present opportunities for reassignment. If work overloads are discovered, tasks can be reallocated, possibly deleted, or shifted elsewhere in the system. If an individual does not have enough work, tasks can be added or can be reallocated (Bahr 1997:136–8). Reallocation can be done among other team members (e.g. adding tasks from another job that may have too many tasks) or by returning to function allocation (Section 7.4) to revisit the assignment of tasks between hardware/software and operator.

8.5.4 Result

Workload assessment estimates the amount of work that is expected of individuals and determines whether the potential exists for that load to be inadequate, adequate or excessive.

8.5.5 Example

Remote communication is an increasing activity among professionals and consumers. Convenience and ease of access are benefits that users can enjoy through wireless access to sophisticated information systems such as web sites and sophisticated databases. However, the location and condition of use can have a significant effect on performance. For example, a complex interface may receive positive ratings when operated in a benign environment. It may have a much different result when operated in the tumult of the operational environment.

An information-based interface can be configured in nearly endless variations. Which variation is best depends on a variety of criteria that are related to user work role and conditions.

The development team can create a working prototype using simulation. Focus group interviews can be used to demonstrate the prototype and perform a subjective WA method such as SWAT to predict the effect on workload. The results of the sessions can be used to revise the prototype.

Alpha and beta usability assessment (Chapter 10) can be conducted to determine how the prototype fares in actual conditions. Subjective WA can then be used after the field trials as an assessment tool. The WA might find that subjective sense of workload changed significantly due to the distractions, delays and complications that remote use imposes. Discoveries such as these can make it possible to offer interface options that the user might select or overhaul the main interface to avoid problems in the final release version of the product.

8.6 Summary

The evaluation methods in this chapter make it possible to see how well the decisions that were made during the design process turn out. Evaluation methods are best suited to learning how well a solution concept performs whether it is under development or in actual use.

Simulation (Section 8.1) is used to predict systems/parts of systems that do not exist and to allow users to experience systems/parts of systems that are dangerous, complex and expensive. Through simulation, potential solutions can be examined, evaluated and optimized at minimal cost and least risk to people. Simulation can be used to compare alternatives on the assumption that an alternative that performs better than others during evaluation will be best in actual use.

Fault tree analysis (Section 8.2) uses the symbolic logic diagram as a graphical technique to represent cause and effect relations among events.

Failure modes and effects analysis (Section 8.3) is used to examine the potential for each hardware, software element that may fail and each human that may err. A top-down FMEA is used to review all major portions of a product, break them into components and constituent parts and speculate what might go wrong with each. A bottom-up FMEA is used to review all individual parts and speculate what might go wrong with each component. Individual possible failures among parts can then be aggregated into a larger picture of component, subassembly or product potential failure. Remedial actions are directed at solving potential problems before they are put into production.

Operational sequence analysis (Section 8.4) combines the events, information, decisions and actions involved in product use into a single diagram that is organized along a time line. As a sophisticated extension of task analysis combined with time line data, OSA can be used to develop a picture of system operation, indicate the functional relationships among multiple system elements, trace the flow of materials and information, show the physical or sequential distribution of operations, identify subsystem inputs and outputs and make the results of alternative design configurations explicit.

Workload assessment (Section 8.5) is used to evaluate the number and type of tasks that humans are expected to perform in terms of both anticipated conditions and the amount of time that is available to perform them. The issue is not tasks alone, but the fluctuation of tasks, conditions and time that may produce dysfunctional situations. Dysfunction can come from too little or too much work. Four methods have evolved to assess workload: primary task, secondary task, physiological measure and subjective rating.

In addition to the above methods, interviews (Section 9.2) and questionnaires (Section 9.1) can also be used for the purpose of evaluation.

9 Surveys

Interviews and questionnaires

How can I collect attitude and preference information?
Questionnaires (9.1)

How can I collect information on attitudes and preferences in person?
Interviews (9.2)

How can I collect information on attitudes and preferences from distant subjects?
Self-administered questionnaires (9.3)

What obstacles can compromise self-administered questionnaires?
Bias, presumption and non-response (9.4)

Chapters 6 through 8 provided human factors research methods that can be used to perform thorough observation. As broad as they are, observation methods may not fit some occasions when it is necessary to obtain information from people. For example, it may be necessary to learn about the ways that members of a group understand or value a routine way of doing something, or how individuals perceive a product or institution. In such cases, it can be useful to employ survey methods such as conducting interviews or administering questionnaires.

Surveys are methods that require involvement and an active response from subjects. Questions are the means to elicit what needs to be learned. The questionnaire is the basis for collecting the information. This chapter discusses questionnaire development and two approaches to administration: directly by subjects and by others in interviews. Both can be used singly such as a phone interview or in combination such as a focus group that includes the completion of a brief self-administered questionnaire.

Thomas Malone (1996:101–16) considers questionnaires and interviews helpful tools to obtain opinions, attitudes and preferences of those who have experience in a situation such as the hands-on use of equipment. Both methods can be used to identify and determine the size of use problems, to elicit information on causal factors and to test users' perceptions of implications and effects of use problems. John Chris Jones (1992:214–34) notes that the questions can be used to discover why people do what they do and to reveal either routine or out of the ordinary patterns of activity. They can be used to find critical information and sources for uncertainty. Questions can be used to probe how users make up for inadequate products, to learn about conditions to which users cannot adapt and to identify content that might be included in a questionnaire. Jones finds that the questionnaire is often the only practical way to gather factual information from members of a large, dispersed population.

Chapter 1 showed how a business plan guides departmental and project plans including budget considerations. Sections 4.2 and 4.3 described the connection between need or opportunity and research and development activity. Section 5.1.1 showed how research spanned the practical considerations of the moment with the possibility of future change. Before launching a survey project, weigh the cost of time and effort involved in collecting information against the value of the information that is collected (White 2001:2).

This chapter provides an orientation to the nature of interviews and how to develop, conduct and interpret them. It will also address how to develop, administer and evaluate information through the use of questionnaires.

> *Long-standing acquaintance is one of the favorite hiding places for the unexamined assumption.*
> (G. McCracken)

9.1 Questionnaires

Human factors research relies on the use of questionnaires to collect a variety of data in the development process. This makes the questionnaire a helpful guide for a research assistant to use while conducting an interview or collecting information through either interview or self reports.

While interviews require the presence of an interviewer, written questionnaires do not. Salvendy and Carayon (1997) cite four reasons why questionnaires are advantageous: their ease of use, their facility as a tool to collect quantitative data, their relatively low cost and the potential to collect information from many respondents who may be in separate locations or organizations. The authors also point to shortcomings for this method. Questionnaires have a low response rate compared with interviews. How individuals respond to them can result in biases (from such causes as unanswered questions and selective responses). In contrast to interviews, questionnaires offer no ability to ask follow-up questions and no opportunity to open a new line of inquiry. They also allow subjects to interpret (or misinterpret) the question and provide answers according to their own (not the investigator's) needs and interests.

Charlton (1996b:81–99) has found that test and evaluation questionnaires are best used to supplement and explain objective data that already exists. Questionnaires are, however, a tricky tool to develop. Poorly crafted questionnaires can allow 'irrelevant factors that may contaminate questionnaire responses, including the respondent's mood, degree of attention, how he or she interprets the questions, etc.' Irrelevant factors may also induce measurement errors and erode the reliability of results. To avoid such difficulties, use the steps outlined in Section 9.1.2 to improve questionnaire reliability and validity.

9.1.1 Questionnaire types

Two types of questionnaires are used in human factors research: standardized and custom.

9.1.1.1 STANDARDIZED QUESTIONNAIRES

Standardized questionnaires are used to collect information on more general traits of products and performance such as standard measures of workload and comfort at workstations measured across different applications. The Cooper Harper standardized questionnaire on flying qualities is an example of a standardized questionnaire. These instruments have been developed and validated through repeated use among members of the scientific and technical communities. Standardized questionnaires do not, however, collect information that is specific to individual products and systems.

9.1.1.2 CUSTOM QUESTIONNAIRES

Custom questionnaires are developed in order to inquire about aspects of a specific product. Unlike standard questionnaires, custom questionnaires are created to collect information that is germane to a particular application or product. This unique aspect makes the custom questionnaire efficient as a collection tool, yet more costly and difficult to validate than standard questionnaires.

Logs can be considered a custom questionnaire that provides a formatted means to record answers to what amounts to simple implicit questions (e.g. What did you do? See? Hear? Use? When did that occur?). Those who complete logs can make entries according to items of interest such as type of activity, time and equipment used. If specific time intervals are of interest, the subject can be prompted by a signal (e.g. timer alarm) to make a log entry at pre-determined time periods. When entries are made at timed intervals, log completion can amount to a self-administered version of activity analysis.

Checklists serve as a reminder for both research and development work as well as operational settings regarding variables of interest in any given situation.

9.1.2 Process

Table 9.1 summarizes the steps that are drawn from Jones (1992:221–34) and from Charlton (1996b:81–99) to develop accurate and reliable questionnaires: identify goals, develop a first draft, conduct review/pre-test, adjust first draft, retest, distribute and administer, collect and interpret responses and report on the results.[1]

Step 1—Identify goals

Determine information that is needed and how results will be analyzed. Jones (1992:221) recommends that the investigator first conduct a non-directive 'pre-pilot investigation' interview in order to compare the investigator's expectations with the knowledge and attitudes of potential respondents.

Step 2—Develop first draft

Use a first draft to develop the approach to information and to determine whether there are any sources of ambiguity or bias that need to be corrected before actual use.

Select a questionnaire type—Choose the type of data that is needed and the five means to obtain it: rating scales, hierarchical, semantic differential, multiple choice and open-ended questionnaires.

Charlton (1996b:84) accounts for the kinds of questionnaires, shown in Table 9.2, which are commonly used in test and evaluation.

- *Rating scales*—Presenting graduated linear scales can be used to collect easily quantifiable data from many subjects.
- *Hierarchical*—Hierarchical questionnaires combine several types of questionnaires and levels of data. The hierarchical structure guides the subject through levels of information, or around areas of information that do not pertain to the respondent.
- *Semantic differential*—The semantic differential presents a respondent with a series of word pairs in order to measure values, attitudes, or complex relationships.
- *Multiple choice*—Multiple choice questionnaires that present a set of predetermined answers for respondent selection can be used to screen respondents and to collect demographic data.
- *Open-ended*—Open-ended questionnaires that ask respondents to provide information in response to prompting questions can be used to collect information from small samples of qualified subjects such as experts.

9.1 Questionnaires

Table 9.1 Creating a questionnaire—questionnaires are easy to use, help in the efficient collection of quantitative data, are relatively low in cost, and assist the collection of information from many respondents who may be in separate locations or organizations. Charlton (1996b) and Jones (1992) recommend these steps to create questionnaires that are accurate and reliable.

Identify goals

Determine information that is needed, and how results will be analyzed

Develop first draft

    Select a questionnaire type

    Select a response scale and descriptor set
    Determines the form of answers

    Word the questions
    Invites information from respondent

    Assemble questionnaire elements

Conduct review/pre-test

Adjust first draft

Retest (if indicated)

Distribute and administer

Collect and interpret and/or tabulate responses

Report on the results

Table 9.2 Types of questionnaires—a questionnaire is a set of questions that are used to obtain information. Salvendy and Carayon (1997) consider questionnaires an advantage in the light of their ease of use, their facility as a tool to collect quantitative data, their relatively low cost, and the potential to collect information from many respondents who may be in separate locations or organizations. Charlton (1996b) offers five approaches to questionnaires, each having advantages as well as drawbacks.

Method/purpose	*Advantages*	*Drawbacks*
Rating scales—Collect easily quantifiable data from many subjects	Can be answered quickly. Data reliable and amenable to summary statistics	Little room for unanticipated answers. Questions must be well thought out
Hierarchical—Combine several types of questionnaires and levels of data.	Answers are detailed and specific. Speeds time to complete. Only important questions are asked.	Requires time and familiarity with content in order to prepare
Semantic differential—Measure values, attitudes, or complex relationships	Provides data on relative similarity of attributes. Easy to prepare and administer	Correct analysis and interpretation requires experience
Multiple choice—Screen respondents, collect demographic data	Answers easy to summarize	Questions must be simple and may force a choice
Open-ended—Collect information from small samples of qualified subjects	Easy to write. May discover unanticipated information	Difficult to summarize. Require more time to answer.

Select a response scale and descriptor set—The scale and descriptors determine the form of answers that respondents will provide. The response scale can be balanced (having a neutral option at the mid-point) or asymmetrical. While any number of intervals can be used, providing up to seven alternatives makes it possible for subjects to discriminate among available choices. Descriptors (the labels for each choice along the scale) can be of particular help if symmetrical wording is used such as 'substantially agree, agree, disagree, substantially disagree.'

Word the questions—Questions invite information from respondent. Bias, which Hawkins (1993:217–18, 231–5) defines as 'the distortion of answers through prejudice, suggestion or other extraneous influences,' can invalidate a questionnaire. Avoid jargon or overly technical terms. State choice options positively. Separate issues so that each can be considered individually. Keep questions short. Ask only relevant questions. Bailey (1989:510) recommends the use of the most simple and familiar words possible and using words that have precise meanings. Avoid terms that have significant emotional value, words that suggest a desired response, or words that that might embarrass respondents. Ask for information that the respondent might reasonably remember. Use a familiar context for questions or to define question context. Order questions from most general to most specific.

Assemble questionnaire elements—Combine the questionnaire with a cover sheet that includes information on the respondent and session date/time, instructions, description of how data will be used and requests permission from the respondent.

Step 3—Conduct review/pre-test

Administer the questionnaire to a small sample to detect problems such as ambiguous terms, presumptions and typing errors. Pre-testing can also detect order bias. Order bias is the influence that one question or a series of questions can have on the minds of respondents which may influence responses to other questions that follow. The sample can include members of the group that will receive the final questionnaire, or other experienced researchers who were not involved in questionnaire development.

Step 4—Adjust first draft

Revise the questionnaire to account for what was learned during the pre-test.

Step 5—Retest

If significant revisions are made, it is a good idea to again pre-test the questionnaire. Use different subjects from those who served as subjects for the first pre-test.

Step 6—Distribute and administer

If the questionnaire is administered multiple times, use the same version for each. Conditions for data collection also need to be consistent. Be aware of conditions that can influence responses. They include, among others: the environment (e.g. ventilation, light, heat), the amount of time made available, the time of day and day of the week and the attitude and delivery of the individual who administers the questionnaire. Related events can also exert an influence. For example, respondents to a questionnaire on transportation safety will react differently if it is administered the day after a major airline crash.

Step 7—Collect and interpret and/or tabulate responses

Read qualitative responses, interpret them and collect them into discrete categories. During this phase, the investigator looks for patterns in the collected information. Those patterns provide the investigator with information that can be used to guide the development of a solution and to produce a statement of findings for others who are interested in what was learned.

Step 8—Report on results

Chapter 13 discusses the methods of communicating results of research activities, including questionnaires.

9.2　Interviews

Wolcott (1995:102–17), defines a fieldworker as an individual who works 'in the field' instead of in an office. Interviewing includes any situation in which a fieldworker seeks information by asking for it, however casually. This deliberate choice to seek information separates interviewing from observation. 'In the simple act of asking, the fieldworker makes a 180-degree shift from observer to interlocutor … In the first case, one takes what one is offered; in the second, one makes personal preferences known.' In so doing, the interviewer creates a structure for the transaction by the choice of particular information to seek, creation and ordering of questions and pursuit of a line of questioning. Structuring the interaction between questioner and subject makes the interview a convenient, efficient way to gather data.

Interviews use the strength of interpersonal interaction in order to build a base of knowledge. The interviewer can probe further to understand the answers a subject gives, can observe the subject and ask questions in response to non-verbal signals, can observe the context (e.g. room) in which the subject answers the questions and can encourage the subject to be forthcoming. Interviews require both interviewer and subject to be in the same place at the same time, which can be particularly difficult when interviewing experts.

9.2.1　Purpose

Salvendy and Carayon (1997:1458–65) value interviews as a means to measure human factors and ergonomic (HFE) outcomes. Interviews are used most frequently to collect information on comfort (e.g. measures of musculoskeletal discomfort), physical and mental health and attitude (e.g. job satisfaction). Among the six types of interviewing that Wolcott finds are of interest to the social scientist, three are particularly helpful in human factors research:

- Casual/conversational interviewing
- Open-ended interviewing
- Interviewing that includes formal eliciting techniques

Table 9.3 shows the types of interviews of interest to human factors research and how they vary according to advantages and drawbacks.

9.2.2　Data reduction

Data reduction is the translation of rough data into expressions that lend themselves to analysis. Efficient survey methods lend themselves to reduction and are often favored because they take less time to tabulate. That can help an investigator to accommodate time and budget constraints. Efficiency refers to how easy it is to tabulate results, which is most often accomplished by the use of computer software. For example, tabulating responses from one hundred subjects can take just minutes if the question requests a simple 'yes' or 'no' answer. On the other hand, interpreting an in-depth interview with one person can take days depending on the content. Efficiency aids practical considerations but also has drawbacks. Efficiency makes it possible to complete the interviewing task more quickly and to convert its results into a usable form. However, important insights may be lost that a less efficient interview might capture.

9.2.3 Informal interview

Casual/conversational interviews are a popular and an inexpensive way to collect information. They are a friendly, yet professional, exchange in which the degree of interaction between interviewer and subject can vary. The informality of the casual interview can lead a less experienced interviewer to overlook the structure that must exist in order to make it different from a freewheeling conversation between friends. The interviewer needs to follow a few guidelines in order to maintain structure.

Table 9.3 Types of interviews—surveys are research activities that are used to elicit information from subjects that observation or controlled studies cannot obtain. The questionnaire is the basis for information collection and can be administered via remote means as a self report or in-person as an interview. The interview is a flexible means to obtain information, and is often employed during the early stages of a project (e.g. while conducting an audit). Which interview approach to choose depends on the kind of information that the researcher needs, and practical constraints such as respondent availability, time and budget.

Type/example	Data reduction	Advantages	Drawbacks
Casual	Depending on length, can require extensive analysis	Inexpensive, flexible	Can be inconsistent if administered by more than one interviewer; may not have the benefit of notes and other records that the interviewer can use for later reference
Intercept	Efficient	Respondent can offer opinions, ask questions, and explain answers	Limited in depth
Long interview	Can require extensive analysis	Can discover otherwise unknown attitudes and motivations	Can take significant amount of time to summarize, interpret
Focus group	Significant cost and time to prepare and analyze	Can collect attitude, preference and qualitative data; well suited to obtaining reactions and comparative preferences	Can require substantial sample development and recruiting
Phone	Efficient	Rapport more direct than self-report methods	Limited opportunity to ask clarifying questions or to open new lines of inquiry; limited in depth

- Be the one who conducts the interview, rather than allow the subject to slip into the role of the questioner.
- Stay with the topic in order to collect information in a consistent and objective manner.
- Use a simple set of notes that includes topics of interest to make sure the session covers necessary ground.
- Use a tape recorder to ensure an accurate record is available for review after the discussion.

For example, a researcher might need to learn background information while getting started on a new project to develop a new infant feeding container. The researcher will eventually conduct more extensive secondary and primary research, yet it would help to get a quick orientation to infant feeding. Visiting a relative who has a six-month-old child could be such an opportunity.

Casual group interviews can also be a rich source of information. Their interaction can generate insights that individual sessions might not discover. Casual group interviews can be used to become familiar with a particular area of content, attitudes or norms. For example, an investigator working on a new interface for databases may happen on three graduate students who have recent experience using a number of different databases. This chance meeting provides an opportunity to quickly learn useful insights. Even in an informal setting, though, group sessions need to be structured in order to account for differences in how individuals behave when they are among others. For example, individuals in a group may withhold comments that may be controversial. This is particularly true when dealing with content that may reflect on others in a less than positive way (e.g. review of critical incidents). To manage such issues, a moderator needs to follow a purposeful line of thought in the same manner as an individual interview.

9.2.4 In-depth interview

In-depth interviews, what McCracken terms the long interview, are a frequent tool of social scientists and market researchers. Conducted in either a brief or an extended period of time such interviews also have a place in human factors work in research and development. For McCracken (1991:21) qualitative methods such as the in-depth interview are 'most useful and powerful when they are used to discover how the respondent sees the world.' Lasting up to two or three hours, the in-depth interview makes it possible to develop an understanding about the way that culture influences how individuals view the world.

Culture consists of the customary beliefs, social forms and material traits of a racial, religious or social group. Research into culture may be thought of as exotic but can just as easily be familiar. For example, in-depth interviews can be used to understand social groups at work, such as truck drivers. The results from such interviews can yield insights into what happens on the road in distant locations that can have an effect on safety and reliability in the design of highways, rest facilities, semi tractor cabs and work schedules.

Results from in-depth interviews are not neatly organized. Instead, the investigator interprets, then explains, the meaning of what the subject has said during the session. In so doing, the interviewer must observe what McCracken describes as symptoms of truth. These symptoms are the conditions that the explanations must satisfy in order to demonstrate the traits that typify good intellectual craftsmanship. These characteristics include:

- exactness (no unnecessary ambiguity)
- economy (requiring the minimum number of assumptions)
- mutual consistency (assertions do not conflict)

- external consistency (conforms to what is known independently about the subject matter
- unity (explanations form a set of interrelated ideas)
- power (explains the data simply yet comprehensively)
- fertility (suggests new opportunities for insight).

External consistency is the most challenging condition to manage. It requires the investigator to distinguish between inconsistencies and new information that may open the way to new understanding.

McCracken suggests that the interviewer follow a questionnaire during an in-depth interview to ensure the same material will be covered in the same way for each interview subject. The use of a questionnaire provides the same type of prompts for information for each interview. Use of a questionnaire establishes the direction and scope of the interview. It frees the interviewer to concentrate on what the subject is saying rather than on what should be covered next. It also provides a way for the interviewer to get back on the topic after pursuing an unplanned line of inquiry.

Two formal interview methods can be used to elicit information from experts: nominal group technique and Delphi technique.

Nominal group technique—Individuals (e.g. experts) are interviewed individually and summaries of responses and ideas are provided to other members. The method is a way to avoid the influence of group opinion on individual opinions.

Delphi technique—Delphi technique is specialized application of nominal group technique used to develop forecasts of future trends, particularly with regard to technology. A panel of experts is polled for their opinions on future trends. Summaries are distributed, a second round of opinions collected, and so on until a consensus is obtained.

9.2.5 Intercept interview

Intercept interviews may be the polar opposite of the long interview and an example of what Salvendy and Carayon term a structured interview. Often conducted by phone or in a public space such as a market or shopping area, oral questionnaires are a quick way to collect many individuals' reactions. Their imposition on a subject, unannounced, makes it important to be efficient with the subject's time.

In this method, the interviewer begins by asking the individual for permission to take a limited amount of time (e.g. five minutes) to conduct a research session. With permission, the interviewer may invite the subject to consider an object or image and answer questions related to it. Even in this structured situation there is still value in having an interviewer present rather than asking an unsupervised subject to simply fill out answers on a questionnaire. The interviewer can observe a reaction or hear a nuance or 'throwaway' comment that can contribute to further understanding.

9.2.6 Phone interview

Interviews by phone can be accomplished in much less time than in-person interviews. In fact, Bailey recommends that phone interviews last no longer than fifteen minutes, using an 'attitude of friendly neutrality' to ensure the subject understands instructions and questions. Phone interviewers can qualify subjects by confirming they meet interview requirements, establish a rapport that would not be possible using written means and can prompt the respondent with explanations and follow-on questions. Ask for information that is readily available to the respondent so that there is no need to check elsewhere to find the answer.

9.2.7 Focus groups

Stewart and Shamdasani (1990:9–32) define the contemporary focus group as an interview conducted by a moderator for among eight to twelve participants over one and a half to two hours. The purpose is to collect qualitative data that is expressed in the participants' own words. Focus groups are one of many formal interview techniques that also include brainstorming and synectics.

Brainstorming—Brainstorming is a group technique used to facilitate the generation of new ideas and encourage creative expression. Group members are invited to generate large numbers of speculations about how to solve problems.

Synectics—Synectics is a more structured version of brainstorming. A moderator leads the group using a formal body of methods in order to optimize the statement and solution of problems.

9.2.7.1 USES

The greatest benefit of focus groups is in discovery and exploration. Focus groups can be used to gather background information, generate research hypotheses, stimulate new ideas, diagnose the potential for problems, generate impressions with regard to objects of interest (e.g. products, services), learn how respondents discuss a topic (for use in research instrument development) and interpret quantitative results. Rogers *et al.* (1998:111–25) demonstrate the use of focus groups in human factors research to assess constraints on daily living of 59 adults 65 to 88 years of age. The study identified problems in activities of daily living and how amenable the problems are to remediation by human factors solutions.

9.2.7.2 ADVANTAGES AND DISADVANTAGES

Focus groups make it possible for the investigator to interact directly with respondents and for respondents to interact and build on each other's responses. The method produces easily understandable results. It is flexible and it is one of the few tools that is suited to data collection from children and those of low literacy.

There are also limitations to the focus group approach. Participants may report one behavior in a focus group while performing a different behavior in reality. Because there are often differences between what an individual says and actually does, Jakob Nielsen (1997b:1) cautions that 'direct observation of one user at a time always needs to be done to supplement focus groups. Some marketing research organizations have developed what has been termed observational research to perform this function. Osborne (2002:29) explains that the approach 'borrows the techniques of academic enthnography and anthropology' to question users about their experience with consumer products while using them. Whether observational research is a reliable source of unbiased information is open to discussion.

Observational methods such as those in Chapter 6 can be used to overcome focus group shortcomings. For example, a certain type of equipment may have been involved in accidents. Activity analysis (Section 6.4) could be used to document what actually happens when individuals operate the equipment. A video recording could be made during activity analysis to anonymously record the equipment operation. If the video recording shows errors that occur, a focus group might be asked to view the tape. Follow-on discussion would then make it possible for the group to provide information on such topics as motivation and circumstances.

An opinionated member may dominate focus group interaction. If a member is reticent to talk in a group, his or her opinion may be unavailable. In-person opinions are more dramatic than impersonal means (e.g. surveys). As a result, these in-person opinions may be given more weight than they deserve.

Focus group results cannot be generalized to a larger population. Summarization and results interpretation can be difficult.

9.2.7.3 PROCESS

The method of focus group development and execution is similar in many ways to other qualitative methods: define the problem, identify the sampling frame (subjects whom the investigator believes may represent the larger population), identify a moderator, create and pre-test an interview guide, recruit the sample, conduct the group, analyze and interpret data and report on the results.

9.2.7.4 LOGISTICS

Sessions are normally staged in a facility that is equipped for staff and clients to observe and record the proceedings. There may also be computer support available for individuals to enter responses to questions of interest. In this instance, the use of computer terminals encourages anonymous inputs. The system allows individuals to contribute to the record and react to comments by others without many of the behavioral issues that attend face-to-face sessions.

Internet-based services have been developed to make it possible for individuals to view and comment on options that are presented at a web site. Individual viewing and response submission approximates nominal group technique. Participating in a moderated interactive text or voice 'chat' session to exchange opinions has more of the character of a focus group.

9.2.7.5 ISSUES

Stewart and Shamdasani (1990) note that an effective focus group moderator is essential to obtaining rich and valid insights. The moderator sets the tone and ground rules, guides the discussion and must be ready to employ one of a variety of strategies to get the desired outcome. Firms often turn to agencies that are experienced in the conduct of focus groups in order to obtain the expertise necessary for successful sessions.

9.2.8 Interview process

Whether brief or long, casual or formal, interviews rely on a methodical approach in the same way as the observation methods in Chapters 6 through 8 followed a deliberate process. Table 9.4 summarizes the four steps that Riordan and Pauley (1996:112–27) recommend to develop an interview: prepare, develop questions, conduct the interview and follow-up.

Step 1—Prepare

Become familiar enough with the problem area to understand what interview subjects can provide that other sources cannot. Establish a structure of what needs to be learned from the interviews. That structure will make it easier to stick with the topic during the interview and to ensure that information that is needed is collected.

Table 9.4 Creating an interview—Riordan and Pauley (1996) recommend following four steps to create interviews that are informative and efficient. During the session, pay attention to the subject's comments, delivery and demeanor in order to guide the course of questions.

Prepare

Become familiar with the problem through preliminary research

Establish a structure of what needs to be learned

Prepare a simple form to include both essential subject information and interview permission

Develop questions

Write closed questions to seek information that will be quantified

Write open-ended questions to seek more subtle information and to open the way to new discoveries

Conduct interview

Review the basic information and permission form and, if it is acceptable, ask the subject to sign it

Conduct informal conversation initially to establish rapport and to orient subject to interview purpose and content

Use questions to inquire about information of interest

Follow-up

Review interview record as soon as possible after session

Draw conclusions, make comments, and extract significant information while the event is fresh

Send correspondence to the subject, if appropriate

Agree on a simple set of ground rules with the subject. Agree on an appointed time, interview duration, topic and scope of discussion. If the interviewee is an expert it may be necessary to provide a fee. If so, ensure the amount and terms of the compensation are suitable.

Before the session, prepare a simple form to include both essential information and interview subject permission. One side of the formatted sheet can include the interview subject's name, the date, location and name of the project. If there is additional information that is meaningful with regard to the project (e.g. years of experience at a job, type of equipment operated), include it as well. The other side of the sheet can include a simple description of what will be covered, how the information will be used, a statement indicating that the subject gives permission for the interview as described and a signature line.

Step 2—Develop questions

Ask basic questions to get at essential information. Ask questions about the significance of things for the subject in order to understand the context (e.g. 'Is this important to you? Please tell me why.'). Any individual who has experience with a situation can answer such questions, which can be either closed or open-ended. Closed questions seek information that will be quantified (e.g. 'How many times a week do you take the bus to work?'). Subjects may be asked to indicate their answer on a scale to indicate their response. Open-ended questions seek more subtle information and open the way to new discoveries ('Please tell me about using the bus for commuting to work.').

Review printed information such as reports, brochures and publications to learn more about the topic beyond what the experience of individuals offers.

Step 3—Conduct the interview

Not every individual is equally comfortable in interviews. Informal conversation with the interview subject will establish some amount of rapport and will provide a brief period of orientation. Ask the subject to review the basic information and permission form and, if it is acceptable, to sign it. Retain all such forms in project records.

Following a brief discussion of the topic, the purpose for the interview and the amount of time the session should take, ask if it is all right to proceed. Use the questions that were developed during the previous step to guide the session.

How active the interviewer should be during the interview is a topic of some discussion. Riordan and Pauley encourage the interviewer to take an active role and to be flexible in order to elicit needed information from the interview subject. Because the subject may not know whether the interviewer is interested in what he or she has to say, they recommend that the interviewer listen first, then change to an active role. That includes having the interviewer ask probing open-ended questions, use the echo technique (repeat significant words and wait for a more specific answer), repeat in his or her own words what the subject just said (reformulate) and ask the subject to describe the process which he or she uses to perform a task. McCracken, on the other hand, favors a more passive, nondirective approach in order 'to discover how the respondent sees the world' and to avoid the imposition of interviewer bias. Prompts for information are necessary. Offering answers, though, runs the risk that the interview results will reflect the interviewer's point of view and not the respondent's. Responses can be used to encourage the interview subject to provide more information. A response by the interviewer can be as simple as an 'eyebrow flash' of recognition, repeating a word the subject used in anticipation of a clarifying response (e.g. after a statement 'I found it confusing' the interviewer simply says 'Confusing?') or a question to obtain more information (e.g. 'What do you mean when you use that term?').

The interviewer should:

- Keep track of elapsed time during the interview.
- Take brief single phrase notes and write down the elapsed time when each comment was made. Such notes help to keep track of the flow of thought. They also make it easier to locate the content on a tape recording and can make editing and interpretation easier.
- Let the interview subject know that his or her time and thoughts are appreciated and helpful before, during and after the session.
- Provide some form of compensation at the conclusion of the session. Subjects who have given ten minutes of their time to respond to an oral questionnaire might merit a gift certificate to a local ice cream or coffee retail store. Expert interview subjects will typically receive a fee for the time they have spent in an interview that is commensurate with their professional fees.

Step 4—Follow-up

The questioner should review the interview record as soon as possible after the session is completed. Draw conclusions, make comments and extract significant information while the memory of the event is fresh.

If the session was long, it may be appropriate to send a letter of thanks to the subject within a few days of the interview. If the subject was to receive a fee for the interview and did not receive at it the end of the session, include a check for it in the letter.

The data obtained is as good as the measurement tool.
(G. Salvendy and P. Carayon)

9.3 Self-administered questionnaires

Like interviews, self-administered questionnaires are used to seek information from subjects. Unlike interviews, respondents provide the answers on their own. Self-reports have a number of advantages. They can be used to collect information from a large numbers of subjects. They can be sent to a sample that is spread over a large geographic area. They do not incur the cost of an interviewer, as the subject is invited to fill out the report.

Self-reports are used to determine basic facts or conditions and assess the significance or importance of the facts. Riordan and Pauley recommend a standard approach to their development and use:

- Use a context-setting introduction to explain why the person was chosen as a subject for the survey, what the goal is in collecting the information and how the information will be used.
- Ask the respondent about information that can be easily remembered.
- Use closed questions to ask for easily tabulated discrete answers (e.g. yes/no, a number, a day).
- Use open-ended questions to ask for comments that may provide insight, such as the kind of problems the respondent has experienced.
- Design a format that can be easily read and tabulated.

Three ways to self-administer questionnaires include use of the mail, the Worldwide Web, and having a moderator administer the reports to a large group.

9.3.1 Mail

Mailing questionnaires can save on the cost of in-person or phone interviews. Even though less costly, a number of considerations affect the success of self-reports sent by mail, whether hard copy, CD or diskette. The mailing list may not be current, containing the names of those who are no longer associated with organizations. Response quality relies on the interest and available time of the respondent. Those who complete the form are not necessarily qualified. For example, if respondents are supposed to have a certain work role, the role may no longer exist. Reports can be referred to others (such as administrative assistants) to answer. As a result, the respondent may not necessarily be the person for whom the questionnaire was intended. Reports that are mailed can also be subject to the same issues as self-administered reports: questions can be misunderstood or subjectively interpreted and answers can be omitted.

9.3.2 Internet/intranet

Most large firms have a web-based Intranet that links its members in a secure and homogenous environment. Interaction via the media via web-based media through the Internet provides a relatively low cost, high speed means to reach a wide potential range of respondents. In either internal (intranet) or public (Internet) venues, an investigator can invite prospects to complete a web-based self-report form that is available at a specific web site address (universal resource locator, or URL).

As pervasive as the Internet is, not all subjects will necessarily have Internet access. In addition, subjects will not necessarily feel inclined to respond to an Internet self report any more than traditional approaches such as face-to-face interviews. Web-based surveys share many of the same traits as mail surveys. Participation is self-motivated, as the respondent needs to find the web site and complete the questionnaire. Because of this, the number and kinds of responses to questionnaires depends on who is interested in responding.

To create a web-based survey, develop a form, post it on an available web site and send e-mail to potential respondents inviting them to visit the site and fill out the form. The approach minimizes intrusion on those who are not interested. Well-designed web-based forms can also perform tabulation automatically, saving time on data collection and tabulation.

A self-administered questionnaire survey can also be directed to members of a group that share similar interests via the Internet. For example, broadcast e-mail services called 'listserv' groups exist to serve a particular interest such as design research. Participants exchange correspondence, notices and comments on their topic. To correspond with a listserv group, send correspondence to its moderator to request permission to correspond with the participants. With the moderator's permission, invite group members to complete a web-based self-report at a specific URL.

9.3.3 Facilitator

Large assemblies of people offer an opportunity to collect a large amount of data in a short time. A facilitator can distribute surveys to a group and ask for participants to fill out the questionnaires. Advantages of the approach include a higher likelihood of response. It also lessens the potential for confusion, as respondents can ask clarifying questions. The sample will not represent the general population because attendance at a meeting is not random and membership of a meeting group does not match society. In special interest studies, this may not be a concern.

9.4 Bias, presumption and non-response

The flexibility of interview and questionnaire methods makes them versatile and useful tools. Because the methods are so flexible, they can lend themselves to presumption, bias and non-response errors if they are not used correctly.

9.4.1 Presumption

Presumption is the act of taking for granted. The following is an example of what presumption can cause, although the context is not what one might expect in a human factors book. On a trip in Italy in 1975, the author sat at a restaurant table among Italian friends anticipating a luscious lunch. As our group finished ordering, the waiter nodded at me and said in Italian to one of my table mates 'En l'stile Americano, si?' His question ('In American style, yes?') implied that I would not prefer food prepared in the style that is normal for that region. If I had not intervened to change the waiter's presumption, I would have been served food that was modified to 'American style' instead of the authentic Piedmont-style cuisine they normally served (and that I was looking forward to). In this way, tourists who travel the world continually run a risk: the imposition of '-style' on whatever they encounter in order to suit perceptions others have about them. From food to entertainment, authentic local culture is often readily adapted to the tastes of visitors, without the visitors' awareness. The result is a change to their experience that the visitors may not even perceive.

Individuals who perform behavioral research can experience the same result. The investigator can unwittingly ignore what actually occurs by imposing a personal view on any situation or group. Or, those who are being asked for their opinion can offer responses that are intended to please the investigator. Such situations can produce research results that reflect the view of the investigator, not the subjects.

9.4.2 Bias

Questionnaires and interviews can alter responses by virtue of question order and question content.

Order—The order in which questions are presented to the subject can present a source of bias. To minimize the potential for order bias, Bailey (1989:513–5) encourages the interviewer to organize questions according to three levels of activity:

- Order questions according to topical groups so that all of the questions about a particular content area are asked while the subject is thinking about it.
- Sort questions within topical groups in order to achieve a logical flow of thought from question to question and avoid order bias (previous question predisposes a respondent to an attitude that may influence a question that follows).
- Order response alternatives within individual questions. Start with questions for neutral information, then address easier, simple topics and move eventually into more challenging, complex material.

Content—Hawkins (1993: 217–8, 231–5) cautions against biasing respondents through content via leading questions ('Are you against giving too much power to the police?'), loaded words ('Do you think union bosses properly reflect their members' views?'), invitations to please the questioner ('Do you find our product enjoyable?'), social or professional expectations ('Do you avoid using alcohol within the prescribed period prior to flying?'), multiple questions ('Do you enjoy traveling by road and rail?'), acquiescence ('Do you agree or disagree that individuals rather than social conditions are more to blame for crime in society?'), imprecision, complexity or ambiguity.

9.4.3 Non-response

Both response and lack of response are significant. For example, a survey may receive a very strong indication from those who respond to it. However, if the number of respondents is small in proportion to the total number of those sampled, the number who did not respond may be even more significant. To control for low response, questionnaire tabulation should also include analysis of the number and nature of responses compared with expected norms. How do we know what the norms are? Responses collected during the pre-test and initial test, as well as interviews that may have been conducted on the same topic can provide a basis for comparison.

9.5 Summary

Methods that are particularly helpful in conducting human factors research include: interviews (casual/conversational, intercept, phone, long, casual group and facilitated group) and self reports (mail, web-based and facilitated group). Interviewing includes any situation in which a an investigator seeks information by asking for it. Interview development involves four steps: preparation, question development, conducting the interview and conducting the follow-up. Three interview approaches are particularly helpful in human factors research: casual/conversational interviewing, open-ended interviewing and interviewing that includes formal eliciting techniques.

Self-reports are easy to use and can help in the efficient collection of quantitative data. They are relatively low in cost and they assist the collection of information from many respondents who may be in separate locations or organizations. Self-report surveys are best used to supplement and explain objective data that already exists. To develop accurate, unbiased and reliable custom reports it is necessary to identify goals, develop a first draft (select a questionnaire type, select a response scale and descriptor set, word the questions, assemble questionnaire elements), conduct review/pre-test, adjust first draft, retest (if indicated), distribute and administer, collect and interpret and/or tabulate responses, report on the results. Particular care should be taken to ensure that bias and other potential errors do not invalidate the responses.

The purposeful selection, ordering and pursuit of information can be accomplished by structuring the interaction between questioner and subject. The more structured a technique is, the more efficient it is as a means to collect data. Efficiency in data reduction makes it possible to complete a survey more quickly and to convert its results into a useable form. However, emphasis on ease of data reduction risks the loss of important insights that a more thorough, sensitive interview or report might provide.

Awareness of the respondents and their context will help to avoid presumption. Attention to the order of questions and their effect on respondent answers will help to avoid order bias. Comparison between expected norms and the number of questionnaires received can help to correct for potential problems that are related to non-response.

10 Usability assessment

How can we ensure this product fits what users need to do their job?
Usability (10.1)

What techniques are available?
Usability assessment methods (10.2)

How can I describe and compare user and product performance quantitatively?
Usability testing (10.3)

How can we 'sell' usability to others in our organization? What are its shortcomings?
Usability benefits and issues (10.4)

As personal computers came into general use in the late 1980s, a much-Xeroxed cartoon could often be seen taped near office workstations. It was Cosimo, the overweight bald eagle cartoon character from Jeff MacNelley's syndicated series 'Shoe.' Wearing glasses and rumpled sport coat in his role as newspaper editor, he was poised with a sledgehammer at the peak of its swing over a personal computer. The droll caption underneath read 'Press any key to continue.' Its humor pivots on an issue that pertains to all products, including computers: usability. Products are supposed to make things better. All too often, though, the products and systems we hope to make life better actually create problems. That is why a sledgehammer solution made sense to many users then. It still does to many users today.

Usability is the condition of a product, system or service being suited for human use. In fact, usability is often the way that organizations think of and use human factors in product development. The concept of usability underlies most chapters of this text. A number of the usability assessment methods in this chapter logically belong in other chapters. So, why is there a separate chapter on usability? Frankly, it is because a reader would expect to see 'usability assessment' somewhere in these pages. Rather than fragment the topic across a number of chapters, it is treated here on its own. References to other sections help the reader to make connections with related content.

Chapters 6, 7 and 8 provided a variety of methods for the analysis, design support and assessment of human-centered products, systems and services. Chapter 9 offered guidance on how to develop and use interviews and questionnaires as a human factors research tool. This chapter describes usability and the nature and methods of usability assessment. It also includes a review of the simple statistics that can be used to describe the performance of individuals and groups of users. Chapter 11 describes the nature and method of controlled studies, including experiments.

10.1 Usability

A useful product is a good fit between need and solution, between human user and artifact. A product that is useful makes it possible for an individual to perform a task more reliably and with fewer errors. Useful products are easier to learn and more satisfying to use (making them more acceptable). Usability pertains to all aspects of a product including hardware, software, menus, icons, messages, manual, quick reference, online help and training.

Usability engineering is the practice of accounting for usability traits during product development. Usability assessment is the approach that is used to ensure that a product, system or service is useful.

Usability has become well known as a result of the development of software user interfaces since the 1980s. In 1985, Gould and Lewis (1985:300–11) described four needs to ensure usability: to establish an early and continuous focus on users, to integrate consideration of all aspects of usability, to test versions with users early and continuously and to iterate the design (assess a concept, adjust it and retest it). The methods in the following section can be used to meet each of those needs. The cognitive task analysis (Section 7.7) method that emanated from software development efforts during the 1980s can also be used to assess usability.

10.2 Usability assessment methods

Usability assessment follows a three-step pattern. Present actual users with a prototype or actual product. Invite them to perform actual tasks with it. Observe what the subject says (or does not say) and does (and does not do).

As in all evaluation methods, the closer test conditions are to actual conditions the more valid it will be. Actual users who perform actual tasks will yield valid results.

10.2.1 Purpose

Usability assessment can be performed either to determine whether there are difficulties with using a product or to uncover opportunities for improvements. In the process, those who are making observations often make new discoveries about how users behave while performing tasks. Products that people operate and use are all candidates for usability assessment, including software programs (e.g. Internet tools, operating systems, application programs, games), training and information materials (video and audio recordings, publications, periodicals, CDs, signage) and equipment (e.g. control/display interfaces for intelligent machines such as appliances, vehicles).

Products that are easy to use may not be easy to maintain. Products may be easy to maintain but their documentation may not be easy to use. Knowledge that is gained during usability assessment can be used to make products easier to maintain and to make maintenance documentation and training materials easier to use. Ongoing programs to design and test products according to inspection and repair needs are referred to as design for maintenance.

10.2.2 Types of usability assessment methods

Table 10.1 lists a selection of methods that can be used within the umbrella of usability assessment. Many of the methods were used in design and human factors work long before the notion of usability emerged as a practice. Even so, it can be useful to consider how each method can be used to contribute to an understanding of user needs and product opportunities. Brief comments on method strengths and drawbacks also accompany the descriptions. The methods and elements of usability testing are drawn in part from Rubin's (1994) practical and informative work. Readers who are interested in further usability testing information are encouraged to consult it.

10.2.2.1 PARTICIPATORY DESIGN

Participatory design is the practice of incorporating a number of representative users into the design team. Users bring current knowledge about actual practice directly to bear on the development of a new concept. If users remain on the development team for an extended time, though, it can lead to them adjusting their performance and their point of view to suit the team. See Rapid prototyping (Section 4.4).

10.2.2.2 EXPERT EVALUATIONS

An expert evaluation is the review of a product by a human factors specialist (considered to be the expert in usability). Retaining a human factors specialist who has knowledge of the product's particular application or technology is likely to result in a more qualified opinion. Fu *et al.* (1998:1341–5) report that human factors experts are more effective than users when identifying problems of usability in skill-based and rule-based performance. However, they find users excel in problem identification when dealing with knowledge-based performance. The best results can be realized through a collaboration between users and human factors specialists.

See Chapter 6, 7 and 8 for methods that incorporate subject matter expert opinion.

10.2.2.3 WALK-THROUGHS

During a walk-through, an investigator leads a test subject through actual user tasks while another team member records information on the user's performance. See Simulation (Section 8.1).

10.2.2.4 PAPER AND PENCIL EVALUATIONS

In a paper and pencil evaluation, a test subject is shown an aspect of product on paper (e.g. software program desktop, menu bar) and is asked questions about how to operate it based on task requirements. The method is inexpensive, simple to produce and can be done quickly.

Table 10.1 Usability assessment methods—Rubin (1994:19–23) describes eleven approaches that can be used in order to assess the usability of products, systems, or services. The methods range from informal, low cost approaches (e.g. paper and pencil) to formal methods that can be fairly costly (e.g. research study).

Participatory design	Incorporation of a few representative users into the design team
Expert evaluations	Review of product by human factors specialist
Walk-throughs	Investigator leads test subject through actual user tasks
Paper and pencil	User shown an aspect of product on paper and is asked questions about how to operate it based on task requirements
Focus group	Discussion among 8–10 users, managed by a skilled moderator
Survey	Collection of user preferences and opinions by mail or phone
Alpha and beta tests	Early release of product to a few users
Follow-up studies	Surveys, interviews, and observations to collect information on usability while product is in use
Usability audit	Compare design against human factors standards and checklists
Usability test	Informal iterative series of tests
Research study	Formal usability test

Paper and pencil evaluation requires the subject to interpret what is seen, because the simple labels and arrangements are an abstraction by comparison with an actual product. It also must be transposed into another medium if it is to be developed as a prototype. See Simulation (Section 8.1).

10.2.2.5 FOCUS GROUPS

A focus group is a structured discussion among roughly eight to ten users that is managed by a skilled moderator. The focus group can be used to collect qualitative information such as participant opinions on topics of interest to a client or reactions to alternatives that are presented for consideration.

Examples that are abstract require intuitive talent among participants and can yield vague opinions. Participant opinions may vary from what they would actually do and can be affected by a number of influences. Focus groups do not generate quantitative performance data as usability tests do. See Focus group interviews (Section 9.2.7).

10.2.2.6 SURVEYS

The survey is the collection of user preferences or opinions by mail or phone in order to understand the attitudes of a broad sample of subjects.

Surveys require extensive validation through pilot tests and improvements. There is no opportunity to collect performance data as there is in usability tests. As in focus groups, opinions expressed in surveys do not necessarily reflect what participants actually think or would choose to do. See Interviews (Section 9.2) and Self-administered questionnaires (Section 9.3).

10.2.2.7 ALPHA AND BETA TESTS

Alpha and beta tests (also known as field trials and user acceptance tests) are the early release of a product to a small group of users.

The purpose for alpha and beta tests is to detect difficulties just before the release of a product. As a result, it may be too late in the development process to have much of an effect on product performance.

10.2.2.8 FOLLOW-UP STUDIES

Follow-up studies are surveys, interviews and observations that are used as means to collect information on usability while a product is in use.

Follow-up studies are used to collect information for a later generation of products. Those who have access to such information include members of the sales, customer service and maintenance staffs.

10.2.2.9 USABILITY AUDITS

A usability audit is the practice of comparing the design of a product with the items that are contained in human factor standards and checklists.

Checklists and standards are too general to provide specific guidance for the development of a product. Even so, performance of a usability audit is preferable to paying no attention to human factors issues.

10.2.2.10 USABILITY TESTS

Usability tests are an informal series of studies that can be performed early, midway through and at the final stages of development to discover product problems or opportunities for product improvements or new concepts. Usability tests can be used to explore concepts or to generate performance data on products that are being developed for release. Section 10.3 provides the information that is needed to develop and conduct usability tests.

10.2.2.11 RESEARCH STUDIES

A research study is a formal usability test that is conducted in order to determine whether a certain phenomenon exists. Such studies require close controls in order to ensure that the results fall within the standards of professional practice, as they are often intended to indicate how a product will perform in actual use among a very large population of users (e.g. nationwide).

As Chapter 5 explained, research studies do not get at the nature of a problem. Instead, they are useful when it is necessary to examine a specific aspect of product performance. Controlled studies (Chapter 11) describes how to design and conduct an experiment.

10.3 Usability testing

Can a user find, recognize, understand and use product features? Where does the user get stuck? Confused? Lost? Through usability testing, an investigator can learn the answers to these questions and more by watching actual users operate actual products.

10.3.1 Types of usability tests

Rubin (1994:30–41) sorts usability tests into three types (exploratory, assessment and validation) that can be performed on a single product or used to compare products. Table 10.2 summarizes their role in the development process.

10.3.1.1 EXPLORATORY

Exploratory usability tests are performed early in development in order to evaluate the effectiveness of a design concept in its early stages. The participant is given a prototype and asked to perform tasks using it. In contrast with other kinds of usability testing, exploratory tests evaluate the structure of a concept and its fit with the tasks that a user needs to perform.

Verbal protocol analysis (Section 6.5) is regularly used during exploratory tests in order to reveal insights into user reactions and thought processes. Interaction between participant and test monitor can be quite high as the monitor attempts to elicit information. Simulation (Section 8.1.5.2) describes the use of simple prototypes for exploratory usability assessment.

10.3.1.2 ASSESSMENT

Assessment usability testing is used in order to enrich the findings of exploratory testing. Performed early or midway through development, a participant is given an early functional prototype of a concept and asked to use it to perform tasks. The focus of attention is on more specific aspects of product operation (e.g. terms, control methods). Interaction with the test monitor is more limited in order to avoid conflict with the collection of quantitative measures (e.g. time to complete task).

Table 10.2 Types of usability tests—Rubin (1994:30–42) describes three types of usability tests that make it possible to assess products whether they are initial prototypes, preliminary products or pre-release production versions. Tests can be performed to assess one product, to compare a number of alternative concepts, or to compare a solution concept and competing products.

Type	Development stage	Purpose	Comparison
Exploratory	Early	Evaluate effectiveness of preliminary design concepts	Which alternative is a better conceptual fit with tasks
		Examine semantic, structural fit with required tasks	
Assessment	Intermediate	Evaluate usability of lower-level operations	Which alternative's features are best to perform tasks
		Examine grammatic, element fit with required tasks	
Validation	Late	Certify product usability by evaluating how it compares to a predetermined usability standard or benchmark	Which alternative compares most favorably with predetermined standard or benchmark data
		Examine utility, tool fit with required tasks	

10.3.1.3 VALIDATION

Validation usability testing is performed late in development in order to verify the product is usable by collecting performance data and comparing it with existing standards or benchmarks. Testing focuses on the most particular aspects of the product. Test objectives are stated in terms of quantifiable data (e.g. speed, accuracy). Because of the need to collect reliable data, there is little interaction with the test monitor.

Validation testing minimizes the risk of releasing a product with problems to customers. That can save the cost of loss of equity with customers, loss of market share to competitors, the cost of corrections to the product and costs of warranty replacement. The method can also be used to evaluate the cumulative performance of product components, to establish performance standards for future product development work and to provide valid data for marketing and customer information.

10.3.1.4 COMPARISON

Comparison testing is the evaluation of usability among two or more products at any of the three stages that were just described. Comparison is made on the basis of criteria that are of interest to the client (e.g. ease of learning and ease of use). The results of comparison testing tend to be used to select the best aspects of alternatives for combination into a hybrid.

10.3.2 Usability testing process

Table 10.3 summarizes the process of test development and execution that is similar regardless of the kind of usability test.

Step 1—Determine the need for conducting the test
Step 2—Write the test plan
Step 3—Recruit participants who will be test subjects

Step 4—Develop test materials

Step 5—Assemble and (if necessary) train the team that will run the tests

Step 6—Prepare the test facility and equipment

Step 7—Conduct a pilot test and adjust the test measures and materials based on what is learned

Step 8—Conduct the tests and (if appropriate) provide interim analyses to client

Step 9—Compile, summarize and analyze the data that were collected during the test

Step 10—Report and/or present test findings and recommendations for product improvement.

The test plan clearly relates the business goals of the client department or organization to the test process and results. The plan is used to develop a consensus among managers, development team members, testing staff and clients on the need for and nature of the test. Pilot testing makes it possible to discover flaws in the usability test before it is used to collect data. Interim reports on performance data can be used to focus development attention on products with a short development cycle. Final data analysis and summary information provide substantive guidance on improvements that are directly related to the client's business goals.

10.3.3 Usability testing traits

The term testing implies a set of conditions that must be met in order for an examination to be conducted according to predetermined standards. Usability tests must be reliable, valid, free from learning effect, adequate, consistent, objective and appropriate in duration.

Two other sections of this text provide guidance on testing: preparation (Section 5.3.2) and controlled studies (Chapter 11).

10.3.3.1 Reliable

To be reliable, the same result would be obtained if the test were to be repeated. Repeatability is a hallmark of inferential studies. However, usability studies are more difficult to repeat. That is due to variability in subject performance. Variability is expressed as the degree to which user performance deviates from a measure of central tendency (e.g. median).

Nielsen (1997a:1543–68) finds that individual differences between participants such as error rates and time to perform tasks can be quite large (e.g. the best user ten times faster

Table 10.3 Usability test process—regardless of the kind of usability test, the process of test development and execution is similar.

Determine the need

Write the plan

Recruit participants

Develop test materials

Assemble and, if necessary, train the team

Prepare the facility and equipment

Conduct pilot test and adjust the measures, materials

Conduct the tests and, if appropriate, provide interim analyses to client

Compile, summarize, and analyze data

Report and/or present findings and recommendations

than the slowest user). This high variability requires greater numbers of participants to be recruited into a sample in order to maintain same level of confidence.

10.3.3.2 VALID

To be valid a test must actually reflect the usability results that one wants to learn about. Validity refers to the relationship between the test and reality. Validity depends on test conditions, participant selection and tasks being as close as possible to actual conditions. The further away the test circumstances are from actual traits, the lower the validity of the test results (Nielsen 1997a:1543–68).

Confounding effects can erode test validity. Confounding is the condition in which two or more variables are combined in such a way that they cannot be separated. If complex task situations are used in a test and have not been carefully planned, variables can easily be confounded. A test plan that identifies clear objectives and measures to be collected can minimize the potential for confounded variables.

10.3.3.3 COUNTERBALANCED

The tasks that the test subject is asked to perform must be free from learning effect. To avoid this influence, tasks are arranged in random order, or balanced out. Without a random task arrangement, it would be possible for subjects to learn and improve performance over the span of the test. A counterbalanced test limits the effect of such learning transfer.

Another approach is to provide training for subjects until the rate of learning growth subsides. That can ensure learning is complete and effective before measuring performance.

10.3.3.4 ADEQUATE

The desirable goal (for true experimental design) is ten to twelve participants per condition. If resources do not allow, four to five can be sufficient to discover problems, but are not adequate in order to make high confidence statements regarding the general user population. The usability test plan (Section 10.3.6) provides guidance on how to ensure that tests satisfy these constraints.

10.3.3.5 CONSISTENT

Test sessions need to be conducted in the same manner, subject after subject. Changes to test session conditions can introduce variables that may affect test results. The use of written scripts, checklists and the same staff members will ensure test sessions are carried out consistently.

10.3.3.6 OBJECTIVE

Tests must be conducted in a manner that observes accepted professional practice in order to minimize the influence of subjectivity on results. To be objective, tests are qualified through a pilot test before they are administered to the user sample. In addition, participants who are recruited into the sample are representative of the actual user population according to essential traits (e.g. proportions of experience, gender, aptitude).

10.3.3.7 APPROPRIATE IN DURATION

Cost management and product release deadlines can exert considerable pressure to quickly obtain the answers that usability testing can provide. These pressures can compel a testing group to conduct tests in a length of time that is inappropriate for the information that is being sought. Tests can be configured to provide results in a short time frame. However, allowing less time to conduct the test changes what can be expected from test results.

Test duration can be planned according to formal, short-term and specific feature testing. Scale, cost, schedule and the type of product that is evaluated all affect the length of time to conduct a test. Much of the work in usability assessment is done in communications product development such as web sites and CD ROM instructional materials. Formal usability testing procedures for such communications products can be expected to last eight to twelve weeks. Short term testing requires strong collaboration between the usability team and the development team and can be completed in four to six weeks. Specific feature testing, in which an in-house team evaluates select portions of known product (e.g. a new web page on a site), can be accomplished in one week (Dumas and Redish 1994:100–5).

10.3.4 Usability testing roles and resources

A usability test can be conducted by an individual or by a team. The number of staff employed depends on the test budget and the purpose. Exploratory tests can be performed very simply using rough 'looks like' prototypes, a single video camera, basic log materials and a single investigator. Assessment tests require a more realistic 'acts like' prototype and an environment and equipment that will make it possible to operate the prototype authentically. In order to meet the criteria that are described in the previous section, validation tests require the accurate collection of a number of measures and would need to be performed by a team in a lab using a consistent approach.

Facilities do not need to be elaborate or fixed. An exploratory or assessment test that seeks to discover features or flaws of interest can be performed by an individual who uses a video recording camera and a note pad to document findings. Mobile 'fly-away' usability equipment can be transported to a site and set up to conduct a field study.

Whether performed by one or many individuals, the roles that are involved in usability testing include: monitor, technical staff (logger, videographer, timer), expert, observer and test participant (Dumas and Redish 1994:233–61).

10.3.4.1 MONITOR

Also referred to as the facilitator, the test monitor manages the administration of the test. The monitor is responsible for all test preparations and activities related to each session. This includes greeting the participant, collecting data, conducting the test and debriefing the test participant after the session. It also involves compiling the data that is collected during the day's sessions, debriefing the team members with regard to session results and ensuring that testing is producing results that follow objectives that were set forth in the test plan (Dumas and Redish 1994:30–2, Rubin 1994:213–24).

10.3.4.1.1 Approach—During the session, the monitor will employ one of four approaches to test subjects: co-discovery, active, neutral or retrospective review.

- *Co-discovery*—(Also referred to as constructive interaction, or paired user testing.) The monitor observes two test subjects who work together to perform tasks. The method can be useful when participants may be uncomfortable performing tasks alone (e.g. youngsters or workers who normally perform tasks collaboratively). Co-discovery can be used to collect qualitative comments from participants in exploratory and exploratory comparison tests.

- *Active*—The monitor asks questions of the test participant in order to learn the reasons for steps that the participant takes while performing tasks. Difficulty that the participant encounters and solution strategies that the user employs are what the testing seeks to discover. To optimize the potential for discovery, the monitor will typically be encouraging but will not reveal any thoughts about the problem or how to solve it. An active monitor role can be used in exploratory and assessment tests and comparison exploratory and assessment tests.
- *Neutral*—The monitor watches the test participant perform tasks in a location that is separate from the participant (e.g. from a control booth behind a two-way mirror). Contact with the subject is limited to instances when the participant needs assistance in order to continue. A neutral observer role is suited to validation and validation comparison studies. That is because their purpose is to collect accurate data that is related to benchmarks or standards.
- *Retrospective review*—User performs the task and watches a videotape of the session with the monitor, who asks questions about it.

10.3.4.1.2 Traits—The test monitor must possess a variety of personality characteristics and test and team management skills that are necessary for test success.

- *Personality traits*—Learns quickly, works well with people, is flexible and organized. Relates well with the participant (establishes rapport, listens well, is empathic and patient).
- *Test management*—Understands usability and its role in product development, can keep the big picture in mind, remembers what occurred during sessions and can tolerate ambiguity.
- *Team management*—Communicates well with team members and clients.

10.3.4.2 TECHNICAL SUPPORT

Three roles are necessary for the technical support that usability testing requires: logger, timer and videographer.

- *Logger*—Uses existing automated logging program or develops time-based event logs. Classifies critical activities and events into coded categories.
- *Timer*—Relies on an accurate, easy to use digital stop watch to record the start, end and elapsed time of test activities.
- *Videographer*—Prepares and operates video cameras in order to document each session. Collects and labels sessions according to test, session and participant. Nielsen (1997a: 1543–68) suggests the option of 'cameraless' recording by taping the computer video and microphone audio output to save the cost of operators and cameras.

10.3.4.3 EXPERT(S)

The test expert is an individual who is well versed in the product that is being tested. The expert ensures the product does not malfunction during the test and if it does is able to restore it to usable condition so that the test can continue.

10.3.4.4 OBSERVER

The observer role is one or more individuals who do not conduct the test, but who have a stake in the results (e.g. product manager, members of the research and development team).

Watching users experience product difficulties can have a much greater impact on product developers and management than any report or account from an evaluation team.

10.3.5 Usability subjects

The subject is the single most important participant in the usability process. It is subjects who will provide the information that is necessary to the insights that will drive usability knowledge and improvements. For that reason, care must be taken with subject sample development through attention to the characteristics, numbers and sources for test participants. The purpose of the usability assessment will determine how to proceed in each category.

10.3.5.1 PARTICIPANT PROFILES

The traits of subjects and the proportion of subjects with certain traits who participate in the test will influence their performance. Requirements for participants can be used for two purposes. The first is to develop the screening questionnaire that interviewers can use to qualify individuals to be test participants. The second is to develop the test design. Requirements can be obtained from product managers in marketing or research and development departments, marketing studies and functional specifications for products.

Individuals can be identified according to particular categories of characteristics that may pertain to the test group. They include:

- *Demographic information*—Demographic information is used to describe a sample according to descriptive traits such as age, gender, physical ability/disability.
- *Primary skills*—Primary skills are those that are employed in individuals' main work role, such as clerical assistant, web site developer, oncologist.
- *Secondary skills*—Secondary skills are employed in a role that is performed in addition to primary skills such as a clerical assistant who has web research ability.
- *Experience*—Degree of time spent in performing primary and/or secondary skills. Candidates can be classified as novice, intermediate, or experienced. The amount of time that is used for each category depends on the project (e.g. a novice clerical assistant can have under six months of experience, while a novice oncologist may have less than fifteen years). Rubin (1994:126–31) recommends including at least a few 'least competent users' in the test in order to ensure the test does not overlook those who are under-qualified yet may need to use the product.

Traits can be used to organize the test sample into a formatted table or matrix. Each cell of the matrix identifies a subgroup. Use of a matrix ensures that the sample is counterbalanced and exposes each subgroup to the conditions that are to be tested.

10.3.5.2 NUMBER OF PARTICIPANTS

Time and budget limitations always require a clear answer regarding how many participants should be included in the test sample. Rubin prefers a minimum of eight subjects in a test to rule out the effect of individual differences and no fewer than four subjects per subgroup. Dumas and Redish (1994:127–9) recommend no fewer that three to five subjects per subgroup. They describe a typical usability test as lasting half of a day, making it possible to test ten groups a week. The fastest rate is a test that lasts one to two hours, making it possible to test three to six subjects per day.

10.3.5.3 PARTICIPANT RECRUITMENT AND COMPENSATION

Participants can be recruited by the personnel department of one's own firm, by employment agencies and by market research firms. The sales and customer service departments can reach current customers. College students, members of user groups and the general public can also be recruited via handbills and advertising.

Compensation for participation in a usability test is based on the subject's qualifications and the length of the test. For example, college students who are asked to use a product for one hour might get a gift certificate to a retail store. Professionals should receive compensation that is commensurate with the hourly rate (or hourly equivalent) for their work.

See Chapter 5, Preparation (Section 5.3.2) for additional discussion of subject sample selection and development.

10.3.6 Usability test plan

The plan is used to relate testing activity to the business plan of the client by serving one of two purposes: to discover strengths, weaknesses and opportunities related to a product (a formative evaluation) or to compare and select among alternatives (a summative solution).
The plan is structured in order to relate test purpose to test conclusions. The problem statement identifies questions of interest. A user profile describes user traits.

The method section describes how the test will be conducted based on what needs to be learned about the product and/or participants. Within methods, one of two test designs is specified: independent groups (also known as 'between subjects') or within subjects. In independent groups tests, each individual participant uses one product in one session. While this approach is most valid and is simple, it is subject to large variations in individual performance and can require a large sample size in order to offset random differences between participants. In within subjects tests, all participants use all products that are being tested. While within subjects is more efficient, the test must be counterbalanced (divided into groups with each group having a different order of products to use). Counterbalancing will avoid participants learning skills and transferring them to the use of subsequent products.

A task list describes what the user is to do, equipment, successful completion criteria, as well as the expected and the maximum time to accomplish the work. Not all tasks are good candidates. Instead, choose tasks that are performed most often or that the investigator believes may have difficulties. Nielsen (1997a:1543–68) recommends that tasks should be kept small enough to complete within the allotted time, yet not made so small that they become trivial.

Additional test plan sections describe the intended physical environment for test, what the test monitor will do, what performance and/or preference measures will be collected and how results will be communicated (e.g. report, presentation).

Appendices to the test plan can include additional pertinent activities and materials. For example, pre-session training can be used to ensure that test participants are at the same level of experience/ability, or to move participant learning to a level that will focus on a certain aspect of product (e.g. advanced functions).

10.3.7 Usability test materials

Before test sessions begin, a selection of materials needs to be developed and produced. As Table 10.4 shows, materials include a screening questionnaire, orientation script, background questionnaire, data collection instruments, non-disclosure/tape consent form, pre-test questionnaire, task scenarios, prerequisite training materials and debriefing topics guide.

Table 10.4 Usability test materials—before test sessions begin, a selection of materials need to be developed and produced.

Item	Use
Screening questionnaire	Guide while interviewing candidates
Background questionnaire	Collect information on participant experience
Consent form	Protect sensitive information from disclosure and obtain permission to record on tape
Pre-test questionnaire	Capture first impressions of the product
Logs	Streamline data collection, reduction
Orientation script	Prepare the participant for the session
Task scenarios	Finished instructions for the participant
Training materials	Train participants before a usability session begins
Debriefing guide	Guide usability team review after session

10.3.7.1 SCREENING QUESTIONNAIRE

The screening questionnaire is a guide to use while interviewing candidates to participate as members of the test sample. The questionnaire prompts the interviewer to collect two kinds of information: to identify the candidate and to determine whether the candidate matches traits that are required for membership in the sample.

10.3.7.2 BACKGROUND QUESTIONNAIRE

The background questionnaire is used to collect information on participant experience after acceptance into the participant sample.

10.3.7.3 CONSENT FORM

A non-disclosure/tape consent form is used in order to protect sensitive information from disclosure and agree to permission to record them on tape. Participants are asked to sign the form prior to the session. The form also sets limits on how the recordings will be used (e.g. only for internal review related to the product's usability).

10.3.7.4 PRE-TEST/POST-TEST QUESTIONNAIRE

A questionnaire can be administered to the participant before the test session in order to capture first impressions of the product, its ease of use, utility, or value. The questionnaire can also be administered after the test session in order to determine changes that occurred to any of the participant's attitudes as a result of using the product. Post-test discussion can also be used to elicit further preference information (e.g. activities during the test session, information that is necessary for test results). Section 9.1 provides guidance on questionnaire development.

10.3.7.5 LOGS

Observation during sessions requires great concentration. Formatted preprinted sheets called logs can be used to streamline the process of data collection. Data files can be used if the process is computer supported. Categories of activities are listed, along with columns that can be used to note time of occurrence, duration and comments. Using prepared logs simplifies the investigator's task by making it easier to record observations. Logs also simplify data reduction after the session is completed.

Formatted spreadsheet software program files are an efficient way to organize data. Embedding formulae in the spreadsheet makes it possible to perform necessary computation automatically, that can streamline data reduction and summarization.

10.3.7.6 ORIENTATION SCRIPT

An orientation script is a brief description of what will happen during the session and is used to prepare the participant. It can be read aloud by the monitor or presented to the participant to read. The script describes the test facility, the test purpose (noting that the product, not the person, is being evaluated), test expectations, any unusual requirements (e.g. the use of verbal protocol analysis) and an invitation for the participant to ask questions.

10.3.7.7 TASK SCENARIOS

The task scenario converts the task list that was approved in the test plan into a finished instruction for the participant. The scenario describes the state of the product or system before starting to use it, describes why the work needs to be accomplished, describes the information that the user will experience and results the participant will pursue during the test. Consisting of one or a few instruction sheets, the scenario may be read aloud by the monitor (e.g. during an exploratory test) or given to the participant to read.

10.3.7.8 TRAINING MATERIALS

As mentioned in Section 10.3.2, materials may be developed and used to train participants before a usability session begins. The instructions can be in any convenient medium (e.g. CD ROM, web site, printed material) and should be evaluated before administering them to test participants.

10.3.8 Test data review

After each session and each day of data collection, compile, or organize, data so that it is easy to compare them according to measures and to detect patterns. The compiled data are summarized in order to convert them into meaningful categories. Taking note of how collected data compare with expected performance can be used to develop well-considered preliminary analyses. Use the post-session debrief to compile observations and comments which were not captured in the logs or other formal instruments.

Test results are summarized in order to allow the team to draw conclusions about findings. Data for assessment and validation tests are routinely described using statistics that are explained in the next section. Summarized data are then analyzed, or examined, in order to determine their characteristics.

Analysis forms the basis for recommendation to change the product by converting observations and measures collection into recommendations to improve the product that has been tested. It can be prudent to provide a preliminary analysis as soon as patterns become evident. Identifying key areas of concern can guide development team members onto areas most in need of attention. Providing a preliminary analysis can pre-empt glaring product errors weeks or months before a final report would be available, block presumptions about the testing or product and confirm the value of usability assessment. A comprehensive analysis provides all findings, including those in preliminary report.

Recommendations for improvement are then presented to members of the development team and, ultimately, the management or the client for consideration.

Table 10.5 Descriptive statistics—a descriptive statistic is used to account for the characteristics of data in ways that make it possible to see patterns.

Measures of central tendency

The measurement from a reference point of interest to the center of the distribution

Mode—The point at which the greatest frequency occurs.

Mean—Derived by adding all of the observations and dividing that sum by the number of participants.

Median—Represents the middle of a range of results. Derived by identifying which value has an equal number of observations above it and below it in the distribution.

Measures of variability (dispersion)

Describe characteristics of the data points in the population

Range—The highest and the lowest value along a dimension.

Standard deviation—Describes the pattern of scores in relation to the mean in terms of their typical departure from it

10.3.9 Descriptive statistics

Exploratory tests are conducted in order to discover insights into a product. Understanding how the user perceives a product and understands its use can produce new thinking about a concept or about user preferences and performance.

Assessment and validation tests, on the other hand, are used to collect objective data about how users perform while using a product. Patterns of superior and inferior performance can indicate features that are successful or that cause problems. Such insights can be valuable to an investigator. Those patterns are not immediately apparent, though, when one examines the raw data. Instead, patterns of performance need to be described collectively through the use of descriptive statistics.

Phillips (1982:17) defines a statistic as any measurement that is made of a sample. 'Every statistic obtained from a sample is an estimate of a particular population characteristic (parameter).' A descriptive statistic is used to account for the characteristics of data in ways that make it possible to see patterns, as Table 10.5 shows. Inferential statistics, which are covered in Chapter 11, are used to infer aspects of a larger population from the test participant sample.

Rather than explain descriptive statistics in the abstract, it will be more informative to show how they are used in the context of usability studies. Usability studies describe performance according to quantitative (e.g. time, accuracy, significant events) or qualitative measures. Many descriptive statistics can be figured using spreadsheet function features in personal computing software programs such as Microsoft Excel.

10.3.9.1 QUANTITATIVE MEASURE ANALYSIS

Basic descriptive statistical analysis is useful in order to represent particular aspects, or measures, of participant sample performance. Each measure can be used as a dimension along which performance data can be plotted. Two types of descriptive statistics are explained here: measures of central tendency and measures of variability. The measure that is used as an example is 'task completion time,' which is the elapsed time that a person takes to finished an assigned task.

10.3.9.1.1 Measures of central tendency—In a distribution of data along a dimension, measures of central tendency are the measurement from a reference point of interest to the center of the distribution. Measures are preferable if they are stable. A measure is considered to be stable if it is not affected by significant variations in a single data point. Three measures of central tendency can be used to describe usability performance data: mode, mean and median.

Mode—The mode is the point at which the greatest frequency occurs. For example, if most of the subjects take 35 seconds to complete a task, 35 is the mode. The mode can be useful when a quick estimate is needed such as doing an interim analysis while compiling usability test data.

Mean—This measure is what most people think of when they hear the word 'average.' To derive the mean, add all of the observations (e.g. all of the times that participants took to complete a task) and divide that sum by the number of participants. For example, if five participants each took 20, 35, 40, 45 and 70 seconds to complete a task, the mean would be 42 seconds (20+35+40+45+70=210/5=42). The mean gives an indication of the performance by the entire participant sample. Data points that differ significantly from the mean can have a big effect on it. The mean is considered to be the most stable measure of central tendency because it varies least from sample to sample of a given population.

Median—The median represents the middle of a range of results. Using the above example, the median among the set of five task completion times (20, 35, 40, 45 and 70) is 40. Among the five, two values are above 40 and two are below. Use of the median may be appropriate when the data for a few cases are significantly higher or lower compared to the population.

Which measure is the right one to choose? The investigator's attitude toward the data will influence the choice of using mean or median to measure central tendency. A brief example will help to explain why. While collecting data on task performance, the investigator may notice that a few of the participants were significantly delayed in finishing their assignment. The delays would cause their task completion times to be very long. Compared with the rest of the sample, their times would be considered outliers. If the cause was test-related such as a faulty piece of test equipment, the investigator may prefer to use the median as it is less influenced by outliers. As outliers can be caused by individual differences in performance, the reason for those differences may be informative in their own right. A very short time to complete a task might be the result of the participant using highly effective work procedures. Very long time to complete can indicate members of the participant sample who cannot understand the product yet still need to use it.

10.3.9.1.2 Measures of variability (dispersion)

—The mode, mean and median are used to describe entire populations. They do not, however, describe characteristics of the data points in the population. For that, it is necessary to use measures of variability or dispersion: the range and standard deviation.

Range—The range is the highest and the lowest value along a dimension. For example, among times to complete tasks of 20, 35, 40, 45 and 70 seconds, the range is 20 and 70 seconds. Outliers play a role when reviewing data using the range. Why would they occur? They may be an indication that the product needs to be changed, or that there are users who have characteristics that make them significantly different from the rest in the sample. Post-session interview or questionnaire content may provide qualitative explanations for the outliers.

Standard deviation—Recall that the mean is derived by adding all of the observations (e.g. all of the times that participants took to complete a task) and dividing by the number of participants. Standard deviation represents the typical departure from the mean. As a measure of variability in the data, standard deviation describes the pattern of scores in relation to the mean. Are they all close to the mean? If so, the standard deviation will be low. A standard deviation that is low indicates that the task performance time by individual participants in a sample is quite similar. Are some scores far from it while others are very close? A standard deviation that is high indicates that the task performance by individual participants is very dissimilar. If this is so, what causes it?

In addition to time, accuracy and significant events are also of interest in usability testing.

Accuracy—The percentage of participants who successfully performed a task within a time benchmark is an indication of correct performance. The percentage of participants who successfully performed a task is an indication of test accuracy. The percentage of participants who performed a task successfully (including those who needed assistance) is an indication of product utility.

Significant events—Occasions when participants request help, hesitate, or otherwise demonstrate extraordinary behavior are cues to underlying issues. The session log should note the time during the session when such events occur. Post-session debriefing can solicit more information about the event. Patterns of similar occurrences can indicate difficulties that may merit redesign.

Chapanis (1959:110) finds that 'just two numbers, the mean and standard deviation, usually do a pretty good job of summarizing an entire group of data.'

10.3.9.2 QUALITATIVE MEASURE ANALYSIS

The results of pre-session and post-session questionnaires can also be analyzed and represented quantitatively as frequencies. Sort responses to limited choice (e.g. 'multiple choice') questions and describe their distribution. Aggregate preferences according to categories that are meaningful to product features (e.g. product strengths and weaknesses, or product features).

10.3.9.3 BY GROUP, OR BY VERSION ANALYSIS

Different groups of test subjects may have different performance or preference patterns. For example, a delivery firm tracking program may find great favor with headquarters staff while delivery drivers in the field find it difficult to use.

Sorting results according to product version can make it easier to compare existing products with proposed products, or client products with competitor products.

10.3.10 Analysis

The analysis phase translates summarized data into insights about problems or opportunities that exist. If the tests uncovered problems with using the product, identify:

- Tasks that were difficult for participants to perform
- Errors that occurred during task performance
- Sources of errors.

Severity is the degree to which a problem impedes task completion. Problems that prevent task completion render the product unstable. Those that impede it are severe to moderate. Problems of a lesser degree can be considered irritants. Frequency of occurrence is the product of the percentage of users who will be affected by the problem and the probability that a user who might be affected will actually be affected. Attend to problems that are most critical (have the highest sum of severity and frequency ratings).

When considering multiple products, compare them according to quantitative measures, qualitative measures and problem criticality.

Lund (1998:688–91)contends that human factors practice needs to move beyond case-by-case analysis to a broader industry-wide set of standards for usability assessment. 'What is needed is a taxonomy of user interface methodologies,' he argues, 'a way of organizing them based on the aspects of usable design that they address most effectively.'

10.3.11 Recommend improvements

Recommendations put the insights from analysis into action by developing effective communications that describe directions the organization can take. The compiled, summarized and analyzed data are used to provide a compelling argument for product improvements. Changes can be proposed for the entire product, for portions of the product, or for specific details. Changes can also be identified according to opportunity ('low-hanging fruit') or time to implement ('quick fixes' versus major overhauls).

Usability tests regularly make discoveries that can be used to further improve current products or new offerings. Recommendations should also point to areas for further research.

The way that an organization views changes will have a significant effect on its likelihood of success. Rubin (1994:276–93) recommends that the investigator identify potential causes of resistance, organize information to engage potential resistance, focus on benefit to firm and describe the nature and scope of change in specific terms. Demonstrate the implications for recommendations. Tie recommendations to results, results to measures, measures to the problem, the problem to the business mission, the business mission to the user and the user to market share. See Chapter 14 for guidance on reports and presentations. See Chapter 13 for guidance on change management.

10.4 Usability benefits and issues

Openness to new thinking in an organization has much to do with the vision and agenda of its senior leadership. By the mid-1990s, Hamel and Prahalad (1994b:64–70) found that in U.S. firms '… on average, senior managers devote less than 3% of their time building a corporate perspective on the future. In some companies, the figure is less than 1%.' Such short term thinking can limit usability assessment's potential. Dumas and Redish (1994: 368–74) mention additional concerns that affect the organization's view of usability. Serial (versus concurrent) engineering and inter-departmental rivalry can prevent information from getting to those who need it. Internal competition can erode mutual respect among colleagues. Leaders can focus on cost management instead of market share. Usability and documentation are seen as ancillary support services rather than development methods. What can be done? Senior managers can press for their own organizations to change. Firms can implement concurrent engineering to draw departments into integrated teams. Firms can include usability assessment in the product development cycle.

10.4.1 Benefits

The benefits of performing usability assessment reflect many of Pew's objectives for human factors work that were covered in Section 1.6.2. Usability assessment increases the likelihood that a product will be right before it is released for production. It ensures products are developed according to actual (instead of perceived) user needs. Users are more likely to be productive, as they are able to perform tasks without errors being induced. Users can find product features on their own without the need to ask others (e.g. colleagues, customer service representatives) for assistance. Firms that employ usability assessment early in the development process can make significant changes when those changes are least expensive to correct. Improved features and ease of use are likely to result in improved product sales and reputation. Fewer problems with products means that less time is spent responding to customer inquiries and producing corrections and revisions. There is also less need to make corrections to products that are already in use and there are lower support and training costs (Dumas and Redish 1994:14–9).

10.4.2 Issues

A number of issues can impede the effectiveness of usability efforts.

Focus on thing, not user performance—Development team members can miss aspects of user performance that may influence product outcomes. Section 4.2.2.1 described how most human performance work is done outside of the commercial environment. Staff may try to apply human factors principles without benefit of human factors knowledge. Because it is harder to understand human behavior than it is to fix a finite object, those who are more facile with programming can favor what is more familiar and thereby miss insights into user behavior.

Process focus—Organizations can experience certain gaps that Bannon (2000) refers to when development team members translate work to and from the laboratory environment. There may be a mismatch between the artifact and the other applications and work it is intended to support. There may be confusion between usable (what works) and useful (what is valued). Emphasis on process of assessment may set expectations too low to realize a significant benefit to the product that is being assessed.

Effects—Differences can exist between between perceived and actual processes. Did the usability assessment identify and demonstrate the *actual benefits* of the product being tested? Did the development team *understand* the scope of what usability assessment discovered? Results are not necessarily captured and appreciated by development teams.

Peter Mitchell (1995:246–64) cautions that some issues make usability assessment difficult to perform, particularly in business sectors that have short development cycles such as consumer electronics. Teams may get only one chance to perform research. Testing can be considered too time-consuming or costly. Customers may not be willing to pay more for certain products such as electronics that they consider inherently easy to use. Geographic and organizational barriers can impede human factors and design team collaboration. Testing programs must account for the continual evolution of products. Human factors is often viewed as a service cost at the end of the development cycle. Managers and engineers who work in interface development may not see the value in usability testing. These issues challenge human factors staff to use resources wisely, identify tangible benefits and present them to management in a compelling manner.

10.5 Summary

Usability is the condition of a product, system or service being suited for human use.

Usability assessment can be performed either to determine whether there are difficulties with using a product or to uncover opportunities for improvements. Products that people operate and use are all candidates for usability assessment. Products that people maintain (but do not necessarily operate) also benefit from usability assessment when it is incorporated in design for maintenance programs.

Eleven methods have been developed to assess usability: participatory design, expert evaluations, walk-throughs, paper and pencil evaluations, focus groups, surveys, alpha and beta tests, follow-up studies, usability audits, usability tests and research studies.

Usability tests are an informal series of studies that can be performed early, midway through and at the final stages of development to discover product problems or opportunities for product improvements or new concepts. Exploratory, assessment and validation usability tests can be performed on a single product or used to compare products. Regardless of the kind of usability test, the process of test development and execution is similar. Determine the need. Write the plan. Recruit participants. Develop test materials. Assemble and (if necessary) train the team. Prepare the facility and equipment. Conduct the tests and, if appropriate, provide interim analyses. Compile, summarize and analyze data. Report and/or present findings and recommendations. The roles that are involved in usability testing include: monitor, technical staff (logger, videographer, timer), expert, observer and test participant.

The test plan relates testing activity to the client's business plan. Test materials include a screening questionnaire, orientation script, background questionnaire, data collection instruments, non-disclosure/tape consent form, pre-test questionnaire, task scenarios, pre-requisite training materials and debriefing topics guide.

Patterns of user traits and of user performance are described collectively through the use of descriptive statistics that include measures of central tendency and measures of variability (dispersion). Measures of central tendency include the mode, median and mean. Measures of variability include the range and standard deviation.

Usability affords a variety of benefits to the end user as well as the firms that practice it. In particular, usability assessment builds product benefits on the basis of actual (instead of perceived) needs. It also strengthens market share through sales that derive from superior product performance.

11 Controlled studies

When is a controlled study the right choice?
Basic controlled studies (11.1)

How do I run a controlled study?
Process (11.2)

How do I design an experiment that will produce valid, reliable, accurate data?
Experimental design (11.3)

How do I infer a population's performance from the study of a sample?
Inferential study (11.4)

Observation is methodical collection of information in the confusion and tumult of the operational environment. The observation methods (also called field studies) that are provided in Chapters 6 through 9 can be used to guide an observer through the operational environment as it is encountered in daily life. There are occasions, though, when it is preferable to take the user and product out of the hubbub of daily life and observe them in a controlled environment. These occasions are best handled through the use of controlled studies. Controlled studies are often referred to as laboratory studies, even though there is no need for them to be conducted in a laboratory.

As Table 5.1 showed, this text covers the four types of human factors research studies: observational, evaluative, descriptive and inferential. The observational studies in Chapters 6 and 7 account for the operational environment as it is encountered. They are of great benefit in getting at the nature of a problem or opportunity. Evaluative studies in Chapter 8 account for the effect of a product or products on human performance. The interviews and questionnaires in Chapter 9 can be used as evaluative methods to learn the attitudes of a subject regarding a situation or product. The usability assessment exploratory test in Chapter 10 is also an evaluative study method. Chapter 12 deals with cost-effective analysis, an evaluative approach that focuses on cost as a variable of interest.

Observational and evaluative studies can be used to develop a keen awareness of the nature of a problem. However, awareness of attitudes may not be enough. A more precise approach may be needed in order to obtain more precise data. Such research is best served by a study that is conducted under controlled conditions. Most of these controlled studies involve mathematical analysis; specifically, the use of statistics.

Human factors practitioners who work in industry tend to use statistical methods less often than the behavioral specialists who work in the laboratory. That is because statistical methods are primarily of value in experiments and experiments are rarely performed during system development (Meister 1985:458–96). Even so, a book on human factors research methods would be incomplete if experimentation and statistical methods were not included.

One caution: the development and conduct of controlled studies requires the use of judgment that is based on experience. Without it, an investigator can go through the process and produce an incorrect result without realizing it. For example, a professional at a U.S. Air

Force base who did not understand statistics once tried to 'plug and chug' his way through an analysis of altimeter data. He found an example in a statistics text and inserted his data. However, he missed one small point. As a result, rather than comparing the results of performance among altimeters he demonstrated that the human subjects were significantly different from each other in flying ability. After adjusting for the oversight, the results (correctly) showed that there was little difference among the altimeters (Bateman 2000b). To avoid a similar fate, the investigator who is new to experimentation should consult with a human factors professional who has experience with statistical analysis.

Many texts have been written on the nature and use of inferential statistics that will not be duplicated here. Instead, this section will explain the use of a few basic inferential statistics in order to show how they can be used as part of human factors research. This chapter provides the theory and fundamentals of controlled studies based on work by Robert Williges (1985), Alphonse Chapanis (1959:96–147, 148–207, 208–52) David Meister (1985:458–96, 1989:141–216), Robert Bailey (1989:525–56), Kenneth Bordens and Bruce Abbott (1998:196–202, 349–83) and Mark Sanders and Ernest McCormick (1993:23–43). It also describes the use of simple inferential statistics in order to analyze the data that these studies produce and provides the steps that are necessary to develop and conduct a simple experiment. The reader who is interested in a comprehensive discussion of statistics and complex experimental design is encouraged to consult Bordens and Abbott and work by Robert Williges.

> *Science just means being careful not to fool yourself.*
> (S. Carlson)

11.1 Basic controlled studies

As Table 11.1 shows, the process of planning and executing a controlled study involves a series of finite steps. In that way, it is straightforward and fairly simple. However, each step involves many decisions as to how to proceed. In that way, it is quite complex. Controlled studies that are covered in this text include frequency distributions and percentiles, descriptive studies, correlations and inferential studies. Of the four, only inferential studies are referred to as experiments because only experiments rely on the control and manipulation of variables. The steps that are explained in this chapter provide a structure of how to go about conducting a basic valid, reliable controlled human factors study.

11.1.1 Purpose

Controlled studies are used for two purposes: data analysis and significance evaluation.

Data analysis—Some types of controlled studies are used to condense data and thereby reveal information about it (such as patterns). Frequency distributions, percentiles and descriptive studies are used for data analysis.

Significance evaluation—Other kinds of controlled studies are used to evaluate the significance of the differences or the effects that are found through research (Williges 1985). Correlational studies and inferential studies are used to evaluate significance. Significance evaluation is the kind of study that Bailey refers to when he explains that human performance studies look for the existence of relations. Significance evaluation asks 'How much of the observed performance can be attributed to differences in the objects or methods being compared?' and 'How much of it can be attributed to other sources?' When things go together in a systematic manner, a relationship exists that can be explained through the use

Table 11.1 Controlled study process—the process of planning and executing a controlled study involves a series of finite steps. In that way, it is straightforward and fairly simple. However, each step involves many decisions as to how to proceed. In that way, it is quite complex. The steps that are shown below provide a structure of how to go about conducting a valid, reliable, controlled human factors study.

Plan the research

Define the problem

Define variables: independent, dependent, control

Select test conditions: location, equipment, apparatus

Prepare instructions and test protocols

Develop subject sample: determine composition, size, and treatment assignment

Choose an experimental design

Determine observational methods

Run the test

Conduct preliminary testing

Make test adjustments

Collect data

Analyze and interpret the data

Select pertinent statistics

Analyze the data

Interpret the data

Report the results

---

of statistics. Table 11.2 summarizes controlled studies and the statistics that are used to analyze their results.

Most relationships that are of interest to research and development teams concern the effect that design decisions have on the four standard issues: errors, processing time, training time and job satisfaction (Bailey 1989:525–6). Significance evaluation studies rely on inferential statistics to provide a level of precision that controlled circumstances make possible.

11.1.2 Issues

Regardless of the type of controlled study that is performed, there are significant issues that any investigator needs to be aware of before starting: control, repeatability and validity.

11.1.2.1 CONTROL

Controlled studies rely on the collection of information with regard to human performance in a laboratory, a prototype or actual operating conditions. Why does control matter? Control makes it possible to focus on one variable in order to observe how it changes when another variable is changed, without disturbance by external influences. Controlled studies follow a predetermined plan in order to ensure that the procedure is in keeping with professional standards of practice. The procedures are necessary in order to collect and interpret data and information accurately. In a controlled study, the investigator allows one and only one variable to change at a time. By exercising this control, the investigator is assured that

Table 11.2 Selecting controlled studies and their statistics—controlled studies make it possible to examine variables of interest by removing sample subjects from the hubbub of daily life. Knowing which type of study to perform matters, because it guides the series of decisions that will be made to recruit and schedule participants, collect data and analyze and interpret data. The table below lists questions that human factors researchers typically ask along with the studies that can be used to answer them. Statistics that pertain to descriptive study are covered in Section 10.3.9.

Question	Study/statistic
	Descriptive study
	Measures of central tendency
What is the middle of the range of values?	• Median
What is the average of values?	• Mean
What event happens most often?	• Mode
	Measure of variation
What are the lowest and highest data values?	• Range
How typical is the mean for all data values?	• Standard deviation
How many events occur and what is their value?	*Frequency distribution*
What are the proportions of the data set?	*Percentile*
	Correlational study
Does a relationship exist between 2 variables?	• Pearson *r*
Does a change in an independent variable cause a significant change in a dependent variable?	*Inferential study*
	• Kolmogorov-Smirnov one sample test
	• Wilcoxon matched pairs signed ranks test
	• Mann-Whitney *U* test
	• Friedman two-way ANOVA
	• Kruskal-Wallis one-way ANOVA

differences in the measured outcome (referred to as the dependent variable) can be explained only by changes that were allowed to happen.

Failure to create controlled conditions makes the procedures and data vulnerable to influences from outside of the bounded domain of the study. Such influences make it impossible to know what changes occur and why they occur. Control also makes it possible for studies to be repeatable and valid.

11.1.2.2 REPEATABILITY

The strength in controlled studies does not lie in one study's findings. Instead, it lies in the ability of many others who share an interest in the topic to follow the same approach as the first study and either get the same results (confirming the first study's findings) or extend the first study's work. This is the reason why strict methodology is followed and is explained in any article, paper or report of study results. Repeatability demonstrates that results were not obtained by chance or affected limited conditions (such as a small sample size).

11.1.2.3 VALIDITY

Two types of validity pertain to research design: internal and external. Internal validity refers to the relationship among variables that is related to a specific set of conditions and tends to be high. External validity refers to observed relationships among a range of variables for many different conditions and can range from high to very low. Cushman and Rosenberg (1991) offer three methods to improve external validity:

- Simulate the worst conditions of product use. In the study of fare card readers, this would require observation during the heaviest rush hour periods in the most densely traveled areas of the metropolitan area during adverse weather conditions.
- Create a laboratory environment that is similar in all significant ways to the actual product use environment.
- Weight tests according to the likelihood that test conditions will be encountered in actual use. For example, a run of abnormally good weather may occur during the test period with only two days of adverse conditions. Noting test conditions (such as variations in weather) would make it possible to adjust results to reflect what is normal for the area (Wickens 1998:18–22).

11.1.3 Example

The following example will show how controlled studies can be used to understand a simple data set. Stored value cards are being used increasingly instead of cash to conduct transactions from charging the cost of phone calls to the collection of public transportation fares. Consider a hypothetical sample of subjects who have been asked to use a stored value card to pay for their fare on a bus or train. The bus ticket processing unit that is shown in Figure 11.1 is an example of such a

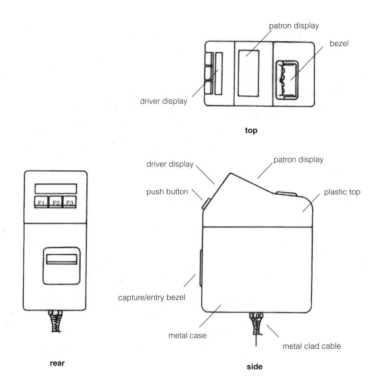

Figure 11.1 Bus ticket processing unit—designed to scan value-added fare cards, the proposed unit (254 mm high × 117 mm wide × 203 mm deep) is mounted on a pedestal on over 2,500 Chicago Transit Authority busses. Riders insert a fare card with coded magnetic stripe into the bezel. Messages (such as the amount debited from the card) appear on both the patron and the driver displays. How the driver and riders use the reader affects the time riders take to pay fares, errors that occur and, ultimately, rider satisfaction with the transportation system.

Source: Reproduced with the permission of the Chicago Transit Authority

fare card reader. The reader is installed near the door where riders enter. Typically, the rider will arrive at the reader, produce a card that has cash value stored on a magnetic strip and insert the card into the reader's port. The unit debits the correct amount of fare, shows the amount on a display and returns the card. The rider then retrieves the card from the reader and enters the bus.

The time that is required to complete the task of reading a fare card has an effect on fare collection, rider satisfaction and public transportation schedules. During peak use periods, many riders may need to board a bus. Each bus might be limited to spending no longer than 5 minutes at a stop in order to stay on schedule. While this time limit would not be an issue during off-peak period when a few riders would board or get off, rush hour can present a different set of conditions. Loading 20 passengers within 5 minutes would allow no longer than 15 seconds for each individual to arrive at the reader, insert and retrieve the card and leave the reader. Among the four variables that Bailey identified which were mentioned above, this 15 second cycle would be a processing time variable.

The card reader can function in two ways in this example: as equipment and as a data recording apparatus. The reader would serve as the equipment that riders would be asked to operate. The reader can also be used as an apparatus to capture information on data such as senior citizen discount rates and peak or off-peak use. That makes it possible to learn about card use patterns.

Two methods can be used to start understanding data that are collected about card reader performance: frequency distributions and percentiles.

11.1.4 Frequency distributions

Frequency distributions are the first step in data set description. Asking a hypothetical recruited sample of users to insert a value added card into a reader would result in a variety of times that the participants take to complete the task. Some subjects (such as adolescents) could be quite fast. Others (such as adults who are carrying parcels) might take longer. A few riders might fumble with the card or miss the reader port. Some riders might have difficulty understanding which end of the card to insert and have to try a number of times before succeeding. One rider might drop the card and need to pick it up and wipe mud off of it before inserting it into the card reader. An observer who collects this data would end up with elapsed times to complete the reader task: 16 seconds, 10 seconds, 18 seconds and so on. Assembling the times into a table such as Table 11.3 would make the data easier to review. Because the time that each rider took is arranged according to arrival on the bus, no pattern is apparent yet. Ordering the times from shortest to longest (in order of magnitude) as the lower half of Table 11.3 shows helps to better understand the data.

Creating a diagram would make it easier to detect any patterns in this data set. The time that each rider took to complete the task can be used to create a diagram to show how frequently certain completion times occurred. The frequency that results might look something like the histogram that is shown in Figure 11.2. The chart portrays time along the horizontal axis and the number of individuals who were observed along the vertical axis. Each dot in the distribution represents an individual (206 in all) and all are distributed along the single variable (univariate) dimension of time.

In order to learn more about the example, the data can be divided into subgroups. Separating the data makes it easier to see patterns, although how to divide the data is not a simple decision. The challenge is to make the right number of divisions that range on a scale from a few groups that contain many data points to many groups that each contain a few data points. Too few divisions would result in too great a compression of data. Too many groups would present too much detail, making patterns harder to detect. For our task time data,

Table 11.3 Bus rider data—the table shows the time it took for a hypothetical group of 206 bus riders to complete the task of scanning a value-added fare card as each boarded. The times were collected as riders stepped onto the bus. The lower data set shows the scan times that have been sorted in order from shortest to longest. The data can be used to create a graphical plot (Figure 11.2). That diagram can then be used to determine traits such as distribution.

Scan times in order of arrival

16, 17, 9, 20, 17, 17, 20, 20, 17, 17, 14, 17, 13, 17, 17, 16, 17, 12, 18, 12, 12, 25, 12, 12, 12, 12, 24, 23, 13, 11, 13, 13, 13, 13, 16, 16, 16, 18, 19, 19, 7, 19, 19, 19, 19, 11, 19, 16, 16, 19, 19, 14, 13, 13, 14, 17, 14, 15, 19, 15, 15, 8, 15, 19, 19, 19, 16, 19, 19, 10, 16, 16, 18, 16, 16, 13, 13, 18, 13, 10, 13, 13, 13, 18, 18, 12, 18, 18, 23, 13, 13, 13, 13, 14, 14, 14, 17, 14, 14, 9, 14, 21, 21, 21, 9, 21, 21, 19, 19, 21, 11, 21, 20, 20, 15, 15, 15, 15, 8, 14, 14, 14, 14, 11, 14, 14, 14, 20, 20, 14, 14, 10, 15, 15, 15, 7, 15, 15, 15, 14, 14, 11, 15, 15, 15, 15, 17, 15, 15, 25, 18, 18, 18, 19, 24, 15, 15, 15, 16, 16, 16, 16, 11, 16, 16, 16, 12, 16, 16, 16, 16, 17, 17, 11, 22, 22, 10, 22, 22, 17, 17, 17, 17, 17, 23, 17, 17, 17, 17, 16, 16, 18, 18, 18, 12, 18, 18, 18, 20, 20, 20, 18, 18, 18, 18, 15

Scan times in order of magnitude

7, 7, 8, 8, 9, 9, 9, 10, 10, 10, 10, 11, 11, 11, 11, 11, 11, 11, 12, 12, 12, 12, 12, 12, 12, 12, 12, 12, 13, 13, 13, 13, 13, 13, 13, 13, 13, 13, 13, 13, 13, 13, 13, 13, 13, 13, 14, 14, 14, 14, 14, 14, 14, 14, 14, 14, 14, 14, 14, 14, 14, 14, 14, 14, 14, 15, 15, 15, 15, 15, 15, 15, 15, 15, 15, 15, 15, 15, 15, 15, 15, 15, 15, 15, 15, 15, 15, 16, 16, 16, 16, 16, 16, 16, 16, 16, 16, 16, 16, 16, 16, 16, 16, 16, 16, 16, 16, 16, 16, 17, 17, 17, 17, 17, 17, 17, 17, 17, 17, 17, 17, 17, 17, 17, 17, 17, 17, 17, 17, 18, 18, 18, 18, 18, 18, 18, 18, 18, 18, 18, 18, 18, 18, 18, 18, 18, 18, 18, 19, 19, 19, 19, 19, 19, 19, 19, 19, 19, 19, 19, 19, 19, 19, 20, 20, 20, 20, 20, 20, 20, 20, 20, 20, 21, 21, 21, 21, 21, 21, 22, 22, 22, 22, 23, 23, 23, 24, 24, 25, 25

rounding to the nearest whole second is a reasonable grouping. Raw data are entered into the distribution tallies and the number of entries in each distribution category is totaled (Chapanis 1959:97–104). The horizontal axis in Figure 11.2 is divided into seconds and yields 19 groups.

Reviewing the data makes it possible to do two things at the same time. First, each category in the distribution can be identified individually. For example, 10 individuals took 12 seconds to finish the task. Second, patterns in data distribution can be detected by reviewing all of the distributions (e.g. whether the distribution is normal or skewed). The greatest number of subjects took 16 seconds to complete the task, making 16 the mode. Recall that the maximum cycle time that is preferred is 15 seconds.

Distribution becomes significant when choosing which statistics to use in the analysis of human factors study data. Large data sets that are distributed symmetrically on either side of the mean may be referred to as normal. Many biological and physiological measurements are distributed normally. Because human factors data are biological and physical, they often follow the normal in their distribution as long as the sample is large enough (Chapanis 1959: 97–104). The shape of a normally distributed data set produces a symmetrical curve with a shape that looks like a bell. Figure 2.2 illustrated a distribution according to male height and its symmetrical appearance indicated a normal distribution of data. The distribution in the bus rider example shown in Figure 11.2 appears symmetrical around the mean, or normal. Because the data in Figure 11.2 look like a normal distribution, we can use it to ask further questions. For example, how many riders in the city are likely to take over 20 seconds to scan their cards?

11.1.5 Percentiles

Percentiles give an indication of the variability of measurements and help to quickly understand proportions as they are related to the whole distribution. In day-to-day human factors

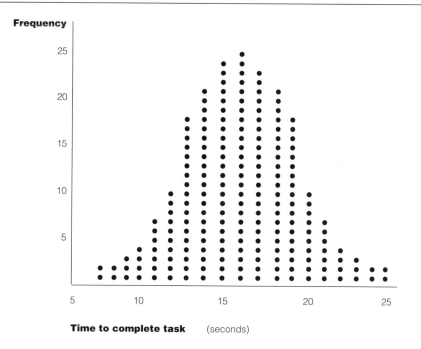

Figure 11.2 Frequency distribution—the hypothetical example shows data for 206 transit riders whose times to scan a value-added fare card are shown in Table 11.3. Each dot represents a member of the study sample and the time it took for that individual to complete the task. Task completion includes approaching the reader, producing, inserting and retrieving the card and leaving the reader. The data appear to be a good match for a normal (bell curve) distribution.

work, anthropometric data are most often discussed in terms of percentiles. See Section 2.4.2.2.1 for an example.

Percentiles are figured in the same way as the median that was described in measures of central tendency (Section 10.3.9.1.1). The middle value that divides the data set in the lower half of Table 11.3 into two equal parts is the median. Using the same approach, three values can be used to divide the data set into four equal parts by values that are usually referred to as Q_1 through Q_3. The value Q_2 corresponds with the median. Values that divide the data set into ten equal parts are called deciles (referred to as D_1 through D_9). The fifth decile corresponds with the median. Values that divide the data into 100 equal parts are called percentiles and referred to by P_1 through P_{99}). The 50th percentile corresponds to the median, while the 25th and 75th percentiles correspond to the first and the third quartiles (Spiegel 1988:63).

How would percentiles be applied to the fare card data set? It might be helpful to know two percentages: what percentage completed the scanning task at or within the target time, and what percentage of riders took longer. The lower half of Table 11.3 shows all of the times in order of magnitude (from shortest to longest). The P_{50} value in the data set (206 * 0.50) is 103. A review of those times shows that 115 of the 206 riders took longer than 15 seconds to scan their card. The P_{56} value (206 * 0.56) is 115. Over half of the sample took longer than the target time to complete the task. Who were the riders who took longer than 15 seconds to scan their cards? What was the reason for the longer time? Could something be done to shorten the time it takes them? Further methods can be used to answer these and other questions.

11.1.6 Descriptive studies

Descriptive studies are conducted in order to analyze the characteristics of a sample and its performance. The usability assessment and validation tests in Chapter 10 are descriptive studies. Descriptive statistics are used to understand and compare the traits of groups through data analysis. Section 10.3.9 explains the descriptive statistics that are used to describe sample traits and results.

11.1.7 Correlation

Correlation is used to describe how close the relationship between measurements of one kind is with those of another kind. Correlations are performed when the investigator cannot control variables of interest such as subject sex or age.

In order to determine whether the relationship between variables is linear, construct a rough scatterplot. As Figure 11.3 shows, a scatterplot is a two-axis graph that represents two linear variables: one on the vertical axis and one on the horizontal axis. Plotting each data point according to vertical and horizontal axis values produces a series of points that are scattered across the chart. When no correlation exists, data points appear in the chart area without any apparent pattern to them. Data points that appear to scatter around a straight line may indicate some degree of linear relationship between the two variables (Bordens and Abbott 1998:337–40). The +0.50 and –0.75 scatter plots in Figure 11.3 indicate such kinds of linear relationships.

The most widely used measure of correlation is the Pearson product-moment coefficient of correlation, or 'Pearson r.' Pearson r reflects the degree to which the relationship between two variables is linear. The method relies on the standard deviation for each set of scores in order to calculate the coefficient. Figure 11.4 depicts a formula that can be used to compute the Pearson r. Spreadsheet software programs such as Microsoft Excel provide functions to figure Pearson r correlations.

In the formula, N represents the number of observations (in this example N is 206). X and Y are the two variables to be compared. In this case, time and age might be informative. While we do not have age information for the subjects, general age data for a large population will serve as an example.

A few factors affect the Pearson r: outliers, range restrictions, and linearity. A fourth factor, sensitivity to score distribution, will not be covered here.

Outliers—Data points that vary widely from the majority of those in the data set can change the correlation coefficient. Outliers may also change the correlation magnitude, sign, or both. In the fare card example, if a few riders had dropped or misplaced their cards and required an extended time to complete the transaction this would cause outliers. A careful investigator would identify those outliers, exclude them, then re-calculate the correlation. A review of Figure 11.2 shows that there are no outliers in the transit rider example.

Range restrictions—Restricting the range over which data values can vary tends to weaken correlations. For example, the correlation of card reader task performance to age will tend to be lower among just adolescents than it will among all members of the general population.

Linearity—The Pearson r can underestimate the degree of relation if the relation is non-linear.

The term ΣX indicates the sum of all times. The term ΣY indicates the sum of all ages. In the fare card reader example, it might be informative to determine what correlation might exist between the time to complete the task and other available data. Because the reader can capture descriptive information, other data sets that would be available include riders who are

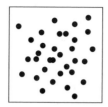

No correlation ($r = 0.00$)

There is no relationship between the two variables. Apparently age does not affect ability to perform the task of scanning a fare card in a reader.

Positive correlation ($r = +0.50$)

There appears to be some relationship between age and ability to scan a card (an increase in age yields an increase in task time).

Negative correlation ($r = -0.75$)

There appears to be a relationship between age and scanning a card (an increase in age yields a decrease in task time). Whether 0.75 is significant enough is up to the investigator to decide.

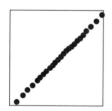

Perfect positive correlation ($r = +1.00$)

Strong evidence that as age increases, time to complete the task also increases. In practical studies, a perfect correlation is unlikely.

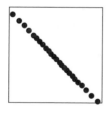

Perfect negative correlation ($r = -1.00$)

Strong evidence that as age increases, the time that is required to scan a value added card in a reader decreases.

Figure 11.3 Correlation plots—scatterplots are 2-axis charts that represent data according to two variables. Correlation represents the relationship between two variables. The degree of correlation can range from −1.00 to 1.00. The fare card reader example used in this chapter could compare age and time to complete task of reading a fare card, producing the correlations shown with these interpretations.

students, who are elderly, who ride at rush hour or who ride during off-peak hours. While age data is not available in this instance, age can serve as an example of how correlation can be used. A significant positive correlation might be found to exist between task time and age, indicating that age is related to increased time to complete the task. Such a correlation would invite attention to understand what causes that phenomenon. For all the investigator knows, it could be that managing grandchildren while boarding causes delay instead of mobility or visual acuity. Figure 11.3 shows a number of correlation plot examples along with possible interpretations that are related to the fare card reader example. The plus and minus signs in correlations indicate the direction of the relationship. A positive correlation indicates a direct relationship: as one variable goes up, the other goes up. A minus correlation indicates an inverse relationship: as one variable goes up the other goes down. The size of the correlation number indicates how strong that relationship is (Chapanis 1959:113–20). The farther the number is from zero, the stronger the relationship.

A correlation of the data from the bus rider example would plot scan times along one axis with the age of each rider on the other. Figure 11.5 shows the kind of plot that such a correlation might yield. The plot may show that very young riders would take longer. This could be due to such reasons as unfamiliarity with bus queues, scanners, or coordination in manipulating the card. Individuals in the middle of the distribution might be young professionals, motivated to get to work on time, who take the least time to scan their card. Individuals at the upper right who are older might take longer to complete their scan. This could be due to factors such as a decline in eyesight or physical agility. Only further investigation would reveal the actual reasons.

11.1.8 Combined use of studies

Frequency distributions and percentiles are helpful tools, yet they cannot reveal everything an investigator might want to know about rider fare card performance. If a development team was assigned to create a new fare card reader, observation, usability assessment (descriptive study) and controlled studies (correlation, inferential study) could all be employed.

$$r_{XY} = \frac{N\Sigma XY - (\Sigma X)(\Sigma Y)}{\sqrt{N\Sigma X^2 - (\Sigma X)^2}\sqrt{N\Sigma Y^2 - (\Sigma Y)^2}}$$

r_{XY} Degree of correlation between variables X and Y

N The number of cases observed

X, Y The two variables that are being compared

ΣX The sum of all values of X

ΣY The sum of all values of Y

Figure 11.4 Correlation study: Pearson r—the Pearson product-moment coefficient of correlation ('Pearson r') from Chapanis (1959:113–20) is a widely used measure of correlation to determine whether a relationship exists between two variables. While any of the three mathematically equivalent formulae can be used to figure the measure, the formula shown here is considered to be easiest when the data set is small and a calculator is available.

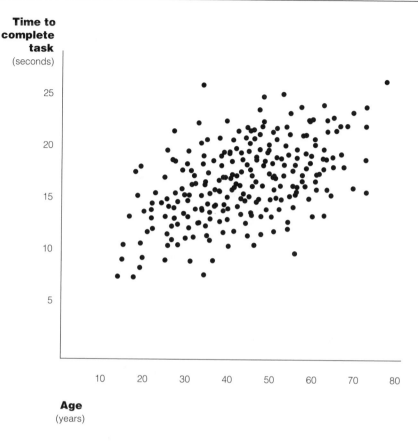

Figure 11.5 Time vs. age correlation—correlation represents the relationship between two variables. This case compares age and time to complete task of reading a fare card. From the plot shown, it appears that age does play a role (although not a strong one) in the time it takes to complete scanning. The shortest times appear among those who are 15–20 years of age. If there appears to be an inexact linear relationship, the Pearson *r* can be used to measure the correlation between variables more precisely.

Observation—Research about human performance in this case may need to focus on observational methods such as activity analysis (Section 6.4) to understand how people actually use value-added cards and readers. Observation would learn about how each passenger arrives at the reader, produces the card, finds the reader port, inserts the card, retrieves the card and leaves the reader. It would also be likely to discover a wide range of variations and complications such as riders who have no fare card, an expired card, mistake a credit card for a fare card, drop the card, insert it improperly or do not understand that a card is required for payment.

Descriptive study—A descriptive study could be used to learn about the types and numbers of users in the population (e.g. include children, riders who have disabilities and the elderly) who may have special needs for card use.

Correlation—While conducting activity analysis, an investigator could have also kept track of test subject errors. There may be an interest in understanding how subject task completion time and error rate are related. A correlation study could be used to check whether error rates and time to complete the scanning task are positively correlated. Such a result would indicate errors are a source of delay. If no correlation existed, it might be that some other cause such as visibility or difficulty with inserting the card in the reader port is causing delays.

There may be a need to know how different populations use fare cards. For example, transit riders with disabilities, children, the elderly and visitors from other cultures may all have different issues related to card use. A correlation study might be used to show which populations purchase value-added cards and when and where they use them.

Usability assessment—If activity analysis showed that users have frequent problems with card orientation and reading, the team may want to develop new card and reader interface design concepts. Such large-scale changes cannot be made without a basis in fact. The choice of a particular reader design to implement may require data that demonstrate the type and amount of improvement that use of that version causes in task performance. An assessment usability test could be used to compare two design alternatives. A validation usability assessment could be used to verify the time to perform a task in comparison to a benchmark. What benchmark would be appropriate? The 15-second cycle time per rider would leave roughly 5 seconds available for card reading (in addition to the time required to arrive at and leave the reader). The validation benchmark to compare the two proposed card reader solutions might be set at 5 seconds.

Controlled study—Usability tests might show that inserting the card in the reader port remains a problem. This could be due to issues that are related to visibility as light conditions vary widely, influencing perception. It might be due to physical configuration (the shape, position and form of the reader 'target') that requires greater dexterity and kinesthetic sense than most riders have. Should the team choose to examine the effect of reader port configuration on user performance, it would be necessary to set up a controlled inferential study to examine how reader port configuration affects rider performance.

11.2 Process

The four steps that are summarized in Table 11.1 are necessary to create and run an inferential study that discovers relations between variables: plan the research, run the test, analyze and interpret the data and report the results.

11.2.1 Plan the research

Preparation for the conduct of an experiment involves problem definition, variable definition, location and equipment selection, instruction and protocol preparation, sample subject selection, experiment design and the choice of observational methods.

11.2.1.1 DEFINE PROBLEM

Issues in human factors research are primarily concerned with the relationship between two items: human behavioral events that can be observed and the external and internal conditions that are responsible for these human behavioral events.

Definition is a clear statement of what the problem is and how it should be approached. The possible solution to the problem is termed the hypothesis. Data collection is intended to determine whether the hypothesis is plausible.

When possible, evaluate relatively small products or parts of products instead of the entire product (Williges 1985).

11.2.1.2 DEFINE VARIABLES

Identify and assign variables to the status of either independent, dependent or controlled. Section 11.3.1 explains how to assign variables.

11.2.1.3 SELECT TEST CONDITIONS

Test conditions include the experiment's location and equipment that will be used by subjects, including apparatus that may be used to collect data.

Location—For theoretical studies, the controlled environment of a laboratory makes it possible to isolate the effects of one or more variables. Exerting control over variables makes the laboratory by definition an artificial environment. Because it is one or more steps removed from actual conditions, it is less valid than an actual setting. The decrease in validity is traded off for an increase in precision that would not be possible to obtain in a real world setting.

Equipment and apparatus—Equipment is the device that will be operated by subjects in a study. Equipment may range from a current product to a proposed concept prototype.

The apparatus is equipment that will be used to both manipulate the independent variables and measure the dependent variables in the experiment. Recorders that collect data on such information as computer keyboard keystrokes or keying errors are experiment apparatus. Chapter 8 cautioned that the use of apparatus typically requires a good deal of expertise. The investigator who needs to use apparatus in an experiment would be prudent to retain a consulting firm or a university laboratory staff to collect data.

11.2.1.4 PREPARE INSTRUCTIONS AND TEST PROTOCOLS

A written set of instructions specifies what will occur during the test and how it is to occur. Instructions include the types of directions that are given to participants, the amount of practice that participants may be asked to perform and the number of rest periods that participants are allowed to take. Test protocols indicate the format that will be used to collect data and to account for results. The development of protocols well in advance of the test makes it possible to organize data collection. It also makes it easier to reduce the data after they have been collected.

Chapanis (1959:228–9) recommends five rules for good instructions:

- Decide what kind of instructions to give to subjects (e.g. strive for speed, or accuracy, or both)
- Tell subjects what is expected
- Write simple, clear and direct instructions
- Read instructions aloud to subjects so that every participant gets consistent direction
- Ask subjects to briefly repeat their directions in order to ensure that the instructions are understood.

Note that changes to directions can also cause changes in subject performance.

11.2.1.5 CHOOSE SAMPLE SUBJECTS

Three steps are necessary to develop a sample of test subjects.

- *Sample acquisition*—The recruitment of a sample of subjects who represent the population that is of interest to the investigator (see Section 11.3.2.1)
- *Sample size*—The determination of how many subjects will be needed (see Section 11.3.2.2)
- *Group assignment*—Determine how subjects will be assigned to treatment groups (see Section 11.3.2.3).

11.2.1.6 CHOOSE AN EXPERIMENTAL DESIGN

An experiment that uses statistical analysis studies the effect of two or more factors at the same time, detects interactions between the factors when they exist and guards against possible bias. It also evaluates the effect due to a certain factor by comparing variation that is due to that factor with the quantitative measure of experimental error. Three steps are involved in the choice of experiment design.

- Treatment conditions—A treatment is any diverse variety of conditions, ranging from training to task performance. To specify treatment conditions, the investigator chooses the condition (e.g. tasks that subjects will be expected to perform) and the measures that will be collected to evaluate the results. Section 11.3.1 addresses considerations that are related to the types and characteristics of measures.
- Subject assignment to conditions and subject test order—Section 11.3.2.3 explains the assignment of sample members to treatment groups.
- Statistical analyses that will be performed—Section 11.4 explains the selection and use of inferential statistics.

11.2.2 Run the tests

The performance of an experiment consists of two steps: preliminary testing and data collection.

11.2.2.1 CONDUCT PRELIMINARY TESTING.

Before proceeding with data collection, verify that the experiment design works as expected. Conducting a preliminary test will make it possible to determine the appropriate levels for independent variables, test the adequacy of experiment instructions, check the reliability of experiment and apparatus and review how appropriate repeated measures are on subjects and pre-test experiment procedures. If the preliminary testing indicates that any adjustments are necessary, they should be made before data collection.

11.2.2.2 COLLECT DATA.

Proceed according to the experiment plan.

11.2.3 Analyze and interpret results

Select pertinent statistics, analyze data and interpret analysis results. Section 11.4.5 provides a selection of inferential statistics.

11.2.4 Report the results

Chapter 14 (Communication) provides guidance on how to communicate the results of controlled studies.

> When in doubt, measure.
> (Anonymous)

11.3 Experimental design

The experimental design is the plan that is created to organize the type and order of observations that will be made. The experiment must be designed in advance of data collection and analysis. Without the deliberate identification and management of variables, it is possible to wind up with variables that are confounded and data that are incorrectly interpreted. A plan ensures that the investigator will be able to summarize and evaluate the data that are collected. It will also make it possible to detect weaknesses or fallacies before any data are collected.

The principal elements of an experimental design are to identify the variables that will be investigated, to determine the type and number of subjects and to plan how to manage the order in which experimental trials will be conducted.

11.3.1 Identify variables

The decision regarding exactly what is the problem poses the most difficult part of most experiments. What does the investigator want to find out by conducting an experiment? That decision influences the choice and assignment of variables (Chapanis 1959: 208).

Sanders and McCormick (1993: 23–43) describe three kinds of variables that are created for inferential study: independent, dependent and controlled.

11.3.1.1 INDEPENDENT VARIABLES (IV)

Independent variables are manipulated for their effect on human performance. Human factors research usually involves three types of independent variables that are related to the task, environment or subject. Variables that are task-related include equipment such as the type of visual display and procedures such as the manner in which a task is performed. Variables that are environmental include variations in conditions such as illumination and noise. Subject-related variables include traits that are related to individuals such as age, gender and experience.

Independent variables can be uni-dimensional:

- Variable along a single dimension, such as varying amounts of illumination
- Discrete (quantitatively different), such as different display designs.

11.3.1.2 DEPENDENT VARIABLES (CRITERION VARIABLES)

Dependent variables are measures of resultant performance. Dependent variables are observed in order to determine the effects that manipulation of the independent variable has on performance. For example, manipulating the independent variable of 'required task performance speed' may have an effect on the dependent variable of 'operator error rate.'

Dependent variables are also referred to as criterion variables or measures. Criterion measures can be used to describe the system, task performance or human performance. System criterion measures are typically used in evaluation research and include such aspects as equipment reliability, operation cost and engineering specifications. Task criteria and human criteria are so closely related that it can be difficult to distinguish them. Task criteria reflect the outcome of a task such as output quantity and quality or task duration. Human criteria are more specific than task criteria and address human behaviors and responses while performing a task. Human criteria may include workload, degree of fatigue, number and seriousness of injuries and human errors. Such criteria can be measured by performance measures, physiological indices and subjective responses.

11.3.1.2.1 PERFORMANCE MEASURES

Frequency, intensity, latency, duration and reliability are basic performance measures. Performance measures (and an example for each) include:

- *Frequency*—Number of queries for help
- *Intensity*—Degree of pressure applied to a joystick
- *Latency*—Reaction time or the delay in changing from one activity to another (see Section 2.4.2.1)
- *Duration*—Time that is taken to from initiation to operation
- *Reliability*—Probability of errorless performance (see Section 2.8.1).

Performance measures can also be combined. For example, a study could collect data on the number of help queries during a defined period of time (in this case, frequency and duration variables).

11.3.1.2.2 Physiological indices—Physiological indices are used to measure physical strain that results from mental or physical work. Physiological indices are based on the body's biological systems: cardiovascular heart rate or blood pressure, respiratory oxygen consumption, nervous muscle activity or electric brain potentials, sensory blink rate or hearing acuity and blood chemistry levels of catecholamines (adrenal hormones that increase heart rate and blood pressure).

11.3.1.2.3 Subjective responses—Subject opinions, ratings or judgments regarding criteria of interest such as workload, ease of use and comfort.

The time when measures are taken also plays a role in experiment design. Intermediate measures that are taken during a task can be used to diagnose or explain performance problems. Terminal measures that are taken at the end of a task describe the task's ultimate outcome.

11.3.1.3 CONTROL VARIABLES

Control variables are separated from independent and dependent variables to prevent them from influencing the dependent variable. For example, changes in task performance have been shown to be related to the day of the week. An experiment that fails to control for such an influence could wrongly attribute an improvement or decline in subject performance to product features.

Three steps are available to control a variable: keep it at a constant state, randomize it or block it.

- *Keep constant*—Performing tasks at the same time of the day for every test would maintain it at a constant rate. Holding every test session at 1:00 p.m. each day would control for the time variable by keeping it constant. Practical considerations such as scheduling, though, would make keeping time at a constant state impractical.
- *Randomize*—A better approach might be to randomize assignment and make it equally likely for each subject to be assigned to each group. Using a random number table to make schedule assignments would ensure that time of day does not influence results. This ensures that assignment order is random and assignment within groups is random. To illustrate, a subject could be randomly assigned to the session that starts at 1:00 p.m. and be randomly assigned as the sixth of ten subjects during that session. Because a random number table is used to make the assignment, it would be equally likely that the subject could be assigned as the second member of the 9:00 a.m. group.
- *Block*—Another method that is available to control a variable is blocking. Blocked variables are collected in different stages (blocks) such as different times of the day or different observers.

The effect of color as a warning and its effect on driving performance provides an example of how variables might be selected. Color can be examined in order to determine the effect on visibility under various environmental conditions. Color and environmental conditions are independent variables. Visibility is the dependent variable. Location, terrain and car model are control variables.

11.3.1.4 CHARACTERISTICS RELATED TO MEASURES

Measures are the means that are used to collect data on a variable. Four issues pertain to measures: reliability, validity, sensitivity and accuracy.

11.3.1.4.1 Reliability—A measure that has high reliability produces the same result when repeated measurements are taken under identical conditions. Section 5.3.2.7.1 described a reliable measure as one that is consistent and repeatable. Determining reliability needs to be completed before establishing validity. Procedures that are used to determine reliability vary according to the kind of measure that is being evaluated: physical measure, population estimates, judgments or ratings by several observers and psychological tests or measures.

- *Physical measure*—Physical measure (e.g. weight) reliability is determined by frequent comparisons to a gauge. For example, a weight that is known to be one kilogram would be placed on various scales that are under evaluation. After repeated measures, the scale that demonstrates the least amount of variation over or under one kilogram is most reliable.
- *Population estimates*—In population estimates, the margin of error (expressed as a range of variability above and below the reported values) is the range of variation that may be expected when the measure is repeated.
- *Judgments or ratings by several observers*—Reliability among individuals needs to be established when more than one individual makes observations. The methods in Part II of this text (particularly the analysis methods in Chapter 6) may require observations by multiple individuals.
- *Psychological tests or measures*—The reliability of tests on psychological variables such as intelligence are difficult to determine. One accepted practice is to administer the test to two large groups and determine the correlation between the scores using Pearson r (described in Section 11.1.7). Another is to compare the results of the odd-numbered items with the results that are produced by the even-numbered items.

The higher the correlation, the higher the reliability. Three approaches have been developed to determine the reliability of psychological tests (Bordens and Abbott 1998:79–81).

1.3.1.4.2 Validity—To be valid, a measure should reflect the aspects of the problem that will be relevant to how the system will perform when it is in actual use. Section 5.3.2.7.2 explained that a valid measure reflects the ultimate performance of a system in actual use.

Four types of validity pertain to measures: face, content, criterion-related (concurrent and predictive) and construct.

- *Face*—The test appears to measure what it was intended to measure (e.g. drivers are asked to operate a steering wheel to evaluate driving ability). Face validity can inspire confidence in the test among those who take it.
- *Content*—The test samples the behavior that is representative of the domain of behavior the test was created to sample. For example, a test of ability in math should contain math problems and should not contain problems that depend on logic or reading comprehension.
- *Criterion-related*—How adequately a test score can be used to infer an individual's value on another measure (termed the criterion). Subjects who do well on a test of flying ability should later be successful in a flying career.
  There are two types of criterion validity based on when tests are administered: concurrent (when scores of the test and criterion are collected at about the same time) and predictive (the score of a test is compared with the criterion at a later time).

- *Construct*—The test is designed to measure a variable that cannot be observed (termed a construct) that was developed to explain behavior on the basis of a theory. A lifeguard's actual ability to save a drowning swimmer at a waterfront is a construct. Accuracy in hitting a target by tossing a buoyant life ring on a long lanyard can be measured. However, the measure has only partial construct validity for life saving ability (Bordens and Abbott, 1998:82–4).

11.3.1.4.3 Sensitivity—A sensitive measure collects data in units that complement the differences that an investigator would expect to find among subjects. Investigators can err on the side of under- or over-sensitivity. An under-sensitive measure fails to capture subtle differences that might be detected using a more sensitive measure. For example, asking about preferences using a binary (yes/no) scale might overlook variety that respondents might offer if they were given the opportunity. Overly sensitive measures invite more refinement than is appropriate. For example, asking respondents questions about their preferences regarding the disposable paper cups at retail coffee stores (such as Starbucks or Brothers) using a 100-point scale invites differences that may not exist. Respondents may only have preferences for coffee cups that can be accurately allocated along a 5-point or a 10-point scale. Using an overly sensitive scale (a 100-point scale in this instance) would decrease the survey's reliability (Sanders and McCormick 1993:39).

11.3.1.4.4 Accuracy—An accurate measure produces results that agree with an acknowledged standard. Measure accuracy is gauged by making the measurement many times and computing the average. The measure is considered to be accurate if the acknowledged standard and the average of the measure agree (Bordens and Abbott 1998:81–2).

11.3.2 Determine type and number of subjects

A population is all of the individuals in a definable group. Most populations are too large to study as a whole. Instead, representative members of the population are recruited to comprise a subset of the population that is called a sample. Sample members are retained to participate as subjects in a descriptive or experimental study. Figure 11.6 illustrates the relationships among population, sample, groups and the assignment of groups to one or more treatment conditions.

Section 10.3.5 described the recruitment of subjects for a descriptive study such as a usability test. In experimental study, the issue is whether test subjects will be affected by variations in the independent variable in the same way as the target population.

Subjects are chosen on the basis of relevant characteristics. The sample need not be identical in all ways to the population. It does, however, need to be matched according to traits that matter, or are relevant, to the study.

A sample of college students may be convenient to recruit for a study but that group will not be representative of any population (other than, possibly, college students). That is why it is important for members of an experimental sample selection to be representative of the population to whom the results will be generalized (Sanders and McCormick 1993:29–31).

11.3.2.1 ACQUIRING A SAMPLE

Five methods of recruitment are available to develop a representative unbiased sample: simple random, stratified, proportionate, systematic and cluster sampling.

- *Simple random*—Select a certain number of individuals by consulting a table of random numbers.
- *Stratified*—Divide a population into segments (strata) using meaningful criteria such as age or number of years of experience. Choose subjects at random in equal numbers from each segment.

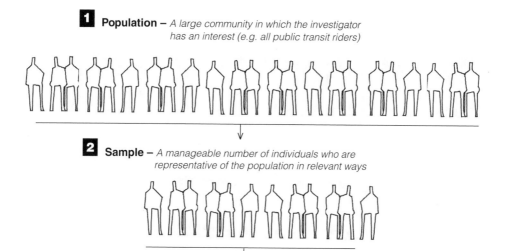

1 **Population** – *A large community in which the investigator has an interest (e.g. all public transit riders)*

2 **Sample** – *A manageable number of individuals who are representative of the population in relevant ways*

3 **Groups** – *Sample members who experience treatment conditions in one of two ways*

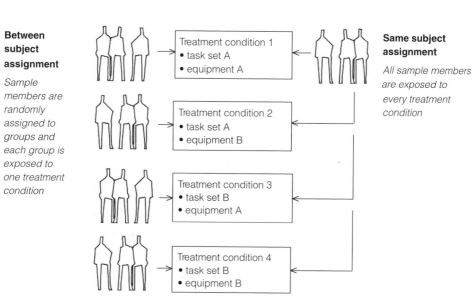

Between subject assignment

Sample members are randomly assigned to groups and each group is exposed to one treatment condition

Treatment condition 1
• task set A
• equipment A

Treatment condition 2
• task set A
• equipment B

Treatment condition 3
• task set B
• equipment A

Treatment condition 4
• task set B
• equipment B

Same subject assignment

All sample members are exposed to every treatment condition

Figure 11.6 Sample development and experiment design—controlled studies rely on representative samples of individuals to provide data on characteristics and performance that can be inferred about a population. A treatment condition is a unique set of tasks and equipment that an individual is asked to use. Exposure to a treatment condition creates quantifiable results that can be analyzed using one or more statistics. Sample members may be broken down into random groups and exposed to one treatment per group. If limitations such as cost prevent the recruitment of a large sample, investigators may use 'same subject' assignments to expose each subject to every treatment.

- *Proportionate*—The subjects in the sample are selected in proportion to the members of the population.
- *Systematic*—Start randomly such as making a random choice in a phone book. Then, select an interval and use it to make systematic selections (e.g. choose every fifth listing on the page). Systematic sampling makes better use of time and costs than random sampling.
- *Cluster*—Identify naturally occurring groups (e.g. all local chapters of a particular union) then randomly select certain groups (chapters) to comprise the sample. Cluster sampling is useful in circumstances such as recruiting a sample from a large population, where random and cluster sampling is too difficult. Under certain conditions, the clusters that are chosen may limit the variety of characteristics among sample members. In that case, multistage sampling can be used by randomly selecting among clusters then randomly selecting subjects from each cluster (Bordens and Abbott 1998:196–202).

11.3.2.2 SAMPLE SIZE

Stipends, staff time to make observations and process results and many other expenses contribute to the costs that always constrain sample size. Enough participants need to be chosen to ensure that a sample is valid. Yet, more participants result in higher costs. Chapanis (1959:236) cautions that one of the most serious mistakes that physical scientists, engineers and psychologists make is to recruit too few subjects.

How many subjects need to be included in a study? To be sufficient the number of subjects in a sample needs to satisfy two requirements. The sample will need to provide a dependable measure of average performance. The sample will also need to be large enough to provide an estimate of the amount of expected variability that can be obtained. Experiments that involve such aspects as learning and motor function will yield erratic results unless twenty to thirty subjects are used. Simple sensory tests (such as the ability to detect signals over the masking effects of noise) require fewer subjects, although those who are accepted into the sample must have passed a test to ensure they have normal sensory ability (such as vision, hearing and color perception) (Chapanis 1959:236–8).

Cushman and Rosenberg (1991:59) recommend (if time and resources permit) a minimum of ten subjects for informal tests and fifteen to twenty subjects for formal tests. The numbers should be increased when variability among subjects is expected to be large. Larger size samples provide more stable estimates of the parameters of the population they represent. How many is enough? How many is too many? The sample should be large enough to be sensitive to the differences that may exist between treatments. It should not be so large that the analysis produces results that are statistically significant yet practically trivial (Bordens and Abbott 1998:195–205). Would the 206 observations shown in Figure 11.2 be adequate for a large metropolitan transit authority? Not at all. One reason is that a transit organization such as Chicago's carries over 1.2 million patrons through the city and surrounding suburbs on a typical weekday. Another reason is that sample must be representative. To be representative, the sample must include not only members of the general population but also groups of riders who have special needs.

Sample size is one of four elements that affect the power of a statistical test, which is covered in Section 11.4.3.

11.3.2.3 DETERMINING GROUPS

Figure 11.6 illustrates the relationship among a population, a sample drawn from that population, groups that are developed from the sample and the assignment of groups to treatments. The choice of sample subjects needs to be made carefully in order to avoid bias or transfer effect. Transfer effect is the learning that can occur during a session. For example, a subject who operates a software program that is under evaluation will acquire skills. This improvement in ability can then be transferred to the subject's use of another software program that is also to be evaluated later in the same test session. The improvement in performance that occurs can be incorrectly attributed to the second program being better than the first. In reality, the performance improvement may instead be due to transfer effect. E.C. Poulton's work in 1974 demonstrated that asymmetrical transfer of learning can make the problem of transfer effect even more difficult. The only remedy to prevent asymmetrical transfer effect is to design a between subject experiment that prevents any subject from being exposed to a second treatment.

There will be variations in performance ability among members of any group. For that reason, subjects who are members of each group should have roughly the same traits. Minimizing differences will ensure that any differences that do occur in performance are the result of changes in the independent variable. Either random (between subjects) or same subject (within subject) assignment methods are used to ensure that groups share roughly the same traits (Bailey 1989:525–56). Mixed factor assignment is a combination of random and same subject assignment and is used in more complex multivariate studies.

11.3.2.3.1 Random (between subjects)—Each sample member is assigned to one group randomly. Random assignment avoids bias in assignment to groups and also avoids transfer effects. Random assignment can be performed in a number of different ways. The organizer can write names on pieces of paper and pick them out of a container. Another approach is to assign each individual a number and consult a random number table to make assignment choices.

Error estimate is largely affected by individual differences of subjects within groups. To solve this, match subjects in each treatment condition as closely as possible on criteria that will affect dependent variable variability (Williges 1985). Three approaches that are available to match subjects are: same subject (also called within subject), mixed factor and group assignment.

11.3.2.3.2 Same subject (within subject)—Each individual who is recruited for the study sample participates in all groups and is observed more than once.

The cost is lower, as fewer individuals are needed by comparison with random assignment. Same subject assignment allows for the possibility of transfer effects. For that reason, same subject assignment cannot be used to study training (what is learned during the first session will be carried forward into subsequent sessions). It can be used to study the effects of learning and of fatigue (Chapanis 1959:151–6). While fewer subjects are necessary, it is necessary to counterbalance the order of conditions to avoid order effects biasing results. That is, each subject or group experiences the conditions in a different order. This prevents order effects from influencing the results.

11.3.2.3.3 Mixed factor—Some independent variables (factors) are within subject, while other independent variables are between subject.

Mixed factor is most common in large multi-factor analysis of variance designs, particularly in behavioral studies. The mixed factor approach is used when the between-subjects are investigated in terms of some time-dependent factor that requires observations to be made more than once (repeated measures) (Williges 1985).

11.3.2.3.4 Group assignment—Chapanis (1959:157–65) describes four ways to design experiments in order to make simple comparisons that are summarized in Table 11.4: independent groups, matched groups, groups with matched subjects and each subject serves as its own control.

- *Independent groups*—Two separate groups of subjects are tested, one under one condition and the other under another condition. Random number tables are used to assign subjects in order to protect against bias.

Table 11.4 Experiment design: treatment plan—Chapanis (1959) describes four methods that can be used to make simple comparisons.

Independent groups

Two separate groups of subjects are tested

Matched groups

Two separate groups are tested. Both groups are matched according to relevant traits (e.g. experience)

Groups with matched subjects

Two separate groups are tested and pairs of individuals in each group are matched according to relevant traits

Each subject serves as own control

Each subject is tested under both experimental conditions

- *Matched groups*—Two groups of subjects are tested and both groups are matched according to a relevant trait (such as experience on the job or a measure of performance or skill). Matched groups usually requires a large initial pool of subjects because many candidates fail to meet matching characteristic requirements. Matched groups is not a foolproof approach. Matches are not exact. In addition, mismatches may occur among traits that the investigator did not deem to be relevant.
- *Groups with matched subjects*—This approach is the same as matched groups, except that individual subjects in either group are matched according to a trait instead of the two groups being matched. Groups with matched subjects is vulnerable to the same hazards as matched subjects.
- *Each subject serves as own control*—Each subject is tested under both experimental conditions. While this is an efficient type of design, it is vulnerable to transfer effect. For example, a subject learns how to drive an auto once. Studying how the subject learns how to drive in a second treatment will inevitably result in the subject transferring skills from the first treatment to the second.

11.3.3 Plan to manage the order of trials

A plan is an agreed-upon set of intentions on how to use resources in order to meet objectives. After variables have been identified, subjects recruited and organized into groups and procedures for conducting the study determined, all should be described in a clearly worded plan. The benefit of drafting such a plan is that those who are involved in conducting research understand what is expected and how to proceed. A clear plan also protects the integrity of the experimental design. Without the consistency that a plan provides, it is possible that the data that is collected may not be valid or reliable.

> *Inferential statistics are simply a guide for decision making and are not the goal of a research project.*
>
> *(Kenneth Bordens and Bruce Abbott)*

11.4 Inferential study

The variables have been identified and assigned. The subjects have been recruited and assigned to groups. Tasks have been developed and subject products organized in a setting for use. The plan was carried out and data have been collected. The next step is for the researcher to determine whether manipulation of the independent variable caused a change in the performance of the human subjects.

Traditional practice calls for the investigator to first assume (hypothesize) that there was no (null) change caused. If the independent variable has no effect, this is described as the 'null hypothesis' being true. If the independent variable does have an effect, the null hypothesis is false. Did the effect result from how the data were collected (sampling error)? Was the effect due to an actual influence that the independent variable exerted that was meaningful (significant)? Will the investigator choose to accept the null hypothesis (decide that there was no significant effect) or reject it (decide that there was a significant effect)? Inferential statistics are the tools that the researcher uses in order to make these decisions. Consider the transit rider example. The data set might result in ten groups of three to five hundred riders each. Half of the groups used the original fare collection equipment in order to collect baseline performance information. The other half used the proposed new generation equipment and their time to complete the fare collection task was shorter. Does that mean that the new generation equipment is better? Or, was this the result of other influences? Inferential statistics will make it possible to determine the correct conclusion.

The reader who is considering inferential study is encouraged to become familiar with relatively sophisticated popular software programs such as SPSS, SAS or BMDP. Each of the programs is available for use on personal computers and each can be used to perform statistical data analysis. The reader who is interested in the actual computation of statistics is encouraged to consult Bailey (1989) for demonstrations of how to calculate certain basic statistics that pertain to human factors experimentation.

Are the small differences in experimental treatments due to the treatments that the investigator has created or just due to chance? Tests of statistical significance have been created to make that determination. The process of testing for statistical significance is similar to detecting weak signals against a noisy background (Wickens 1998:18–22). There are many considerations to account for to employ statistics correctly. Three that are explained here include statistical error, significance and power.

11.4.1 Statistical error

Is the null hypothesis true or false? Do I accept or reject it? Four combinations are possible among truth and falsehood and acceptance and rejection, as Figure 11.7 illustrates. Two of the combinations are correct:

- The null hypothesis (manipulation of the independent variable does not affect dependent variable of subject performance) is true and the investigator does not reject it.
- The null hypothesis (changes in the independent variable have no effect on subject performance) is false and the investigator rejects it. This means that the investigator correctly infers that there is an effect due to the independent variable.

The two other possible options are erroneous:

- The investigator infers there is a result from changing the independent variable when, in fact, there is none (referred to as Type I error).
- The investigator infers that there is no result from changing the independent variable when, in fact, there is one (referred to as Type II error).

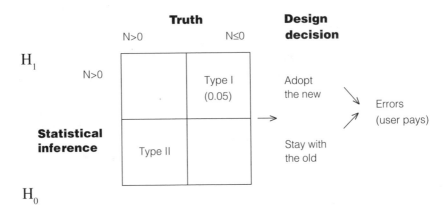

Figure 11.7 Statistical error decision matrix—during data analysis, there are two possible conditions for the data: the null hypothesis is false (H_0) or it is true (H_1). The researcher can either accept the null hypothesis or reject it. The table describes the possible options and outcomes in the comparison of an old and new system. Two of the conditions are considered correct: acceptance of a true null hypothesis and rejection of a false null hypothesis. The other two conditions can lead to incorrect conclusions and possible failure to adopt solutions that are actually an improvement.

Source: Reprinted with permission from *Ergonomics in Design*, Vol. 6, No. 4, 1998. Copyright 1998 by the Human Factors and Ergonomics Society. All rights reserved.

The relationship between the probabilities of two types of errors is inverse. Efforts to minimize Type I errors tend to increase the probability of Type II errors and vice versa. Wickens illustrates the implications of both error types in the decision to retain an old system or to replace it with a new system. The decision of whether to replace a paper map with an electronic map might lead to an experiment that compares these two types of decision aids. If a Type I error is made, the investigator recommends the new system when it is no better than the paper map. In this instance, the firm manufactures, sells and maintains (and the customer buys and uses) a product that provides no improvement. If a Type II error is made, the investigator recommends retaining the paper map when the electronic map is actually a better solution. In this instance, the user suffers by not gaining access to a system that was superior and could even enhance safety. For example, heavy equipment operators may work better wearing gloves rather then mittens. In a comparative study, the null hypothesis would be that performance is similar wearing either mittens or gloves. The researcher observes subjects performing work. The sample data indicate that those workers who wear gloves are able to perform better but only by a small margin. The researcher can accept the null hypothesis or reject it, as it might be true or false.

In light of costs due to either type of error, Wickens contends that the traditional practice of weighting heavily to avoid Type I errors is inappropriate. 'Instead, designers should be at liberty to adjust their own decision criteria (trading off between the two types of statistical errors) based on the consequences of the errors to user performance' (Wickens, 1998:18–22).

Designers are typically more interested in the size of an effect (its significance) than its statistical significance …

(Christopher Wickens)

11.4.2 Statistical significance

Both statistical and practical significance are important issues to experimentation. The observation that an independent variable has a significant effect on the dependent variable means that there is a low probability that the observed effect was due to chance. If chance could not have caused change in the dependent variable the inference is that the independent variable caused it.

For most experimental work, differences are considered to be significant if they could occur by chance less than 5 percent of the time ($p = 0.05$) and highly significant if it could occur by chance 1 percent of the time ($p = 0.01$). If the criteria (termed 'alpha level') are always set at the 0.05 level, the investigator runs the risk of saying that an experimental result is significant when it really is not. This Type I error can occur by chance 1 in 20 (5 out of 100) times when sampling error alone could have produced an apparently significant difference.

Traditional practice uses 0.05 as a threshold and results below it may be termed 'not significant.' In applied studies, low power and large effect size can weaken the ability of statistics to discern between significant and insignificant effects. Rather than use 0.05 as an absolute threshold, Wickens (1998:18–22) recommends the inclusion of further information (such as confidence intervals and standard error bars) that the investigator can use to evaluate the null hypothesis. The additional information would make it easier for an investigator to make a more informed choice among the options that were described in the previous section.

Sanders and McCormick (1993:23–43) explain four points about statistical significance:

1   Results may be due only to chance.
2   Other insignificant variables (other than the independent variable of interest) may also be influencing the dependent variable.
3   There is no way to know whether the experimental design was faulty or whether uncontrolled variables confounded the results. The best defense against confounded variables is to use randomization and large sample sizes.)
4   Results may be statistically significant but unimportant. That is because it is possible that a variable that shows an effect in the highly controlled laboratory becomes relatively unimportant (washes out) when it is compared with all other variables that affect performance in the real world.

This last point is a particular issue in product and service development which compels research to deliver useful results. The investigator who devotes time and resources to a study will be expected to produce results that have practical value that can be converted into revenue.

11.4.3 Statistical power

The weaker (less powerful) a statistical test is, the more difficult it is to detect differences in data that may show the independent variable did have an effect, resulting in a rejection of the null hypothesis. Dunlap and Kennedy (1995:6–7, 31) define statistical power as 'the probability of detecting a difference among groups by means of a statistical test when such a difference exists.' They contend that researchers avoid power analysis because this aspect of experiment design is considered to be too complicated. Wickens (1998:18–22) contends that applied studies have low power for three reasons: variability, rarity and limited resources. Applied studies deal with highly variable elements of human performance (including trade-off strategies and differences) that produce highly variable performance measures. At times it is necessary to investigate each subject's response to a rare event and the collection of only one data point per participant lowers the power. Investigators are often forced to keep sample populations low in order to fit within cost or time limitations.

Power is affected by four elements: sample size (addressed in Section 11.3.2.2), alpha level (discussed in Section 11.4.2), use of a one- or two-tailed test and the size of the effect that is produced by the independent variable. 'One- or two-tailed test' refers to whether one or both extremes of a normally distributed data set are used in order to evaluate statistical significance. A one-tailed test is conducted (using either the upper or the lower extreme of the data set distribution) if the investigator seeks to determine whether one alternative is better than another. The one-tailed test is more likely to detect a real difference if a difference exists and is therefore considered to be more powerful. But which direction (between better or worse) does the investigator choose? The answer to that question cannot be known without conducting a test.

11.4.4 Inferential statistic considerations

The choice of statistics incorporates a variety of considerations. Four are discussed here: parametric versus non-parametric, selection of statistic based on sample type, level of measurement and process to follow.

11.4.4.1 PARAMETRIC VERSUS NON-PARAMETRIC

As Figure 11.6 shows, a sample is a manageable number of individuals that are used to represent a population. The sample members must have certain traits that will matter in their effect on the experiment. Such traits are termed relevant. In a test of visual ability, visual acuity is relevant but eye color is not.

Parametric statistics are used to estimate the value of a characteristic (parameter) of a statistical distribution that takes a particular form such as normal. This requires the researcher to make certain assumptions about the population. For example, assumptions that scores have been sampled randomly from the population, or that the variances of the different groups in the sample are highly similar (homogenous within-groups variance) (Bordens and Abbott 1998:351, 375). Parametric statistics that find use in human factors studies include: t test, z test, f test, chi square and analysis of variance (ANOVA).

Non-parametric statistics make no assumptions about the distribution of the population from which the sample was drawn. For that reason, they are described as 'distribution free.' Non-parametric statistics are often used when data come from discrete (compared with continuous) categories (Phillips 1982:139).

Non-parametric statistics have very nearly the same power as parametric tests yet they do not require as many assumptions as parametric statistics. These assumptions require justification, which is open to scrutiny by those who review the research. Because they do not require as many assumptions, non-parametrics make the researcher less vulnerable to such criticism. Siegel (1956:20–1) finds that the power of non-parametrics can be increased by increasing the size of the subject sample.

11.4.4.2 SAMPLE TYPE

Statistics are selected according to the experimental design, as Section 11.2.1.6 described. The statistics that will be explained in Section 11.4.5 and are portrayed in Table 11.5 are organized according to sample type. In designs of two or more than two ('k'), samples, samples may be either related or independent.

- *Related*—Related samples are groups that are matched according to relevant characteristics. Relevant characteristics are those that have the prospect of influencing the experiment's results (such as experience). Matching is a method that is used to overcome the difficulty that can be imposed by extraneous differences between two groups. Differences are extraneous when they are significant but are not the results of the treatment. As an example, an investigator might try to compare two training methods by having one group of study participants taught by one method and a separate group taught using another method. If one of the groups has more motivated or more capable students, there will be differences in performance but they may as easily be caused by these extraneous variables of motivation and ability as by a better training method.
- *Independent*—Circumstances may make it impractical to use related samples. The nature of the dependent variable may make it difficult to use subjects as their own controls, or it may be impossible to design a study that uses matched pairs. Two samples may be drawn at random from two populations, such as assigning members from two classes using random selection. Alternately, either of two treatments may be randomly assigned to members of either one or another of two samples. For example, registration records of all of those who are in training might be collected and half of the records would be assigned at random to one instructor and half to the other (Siegel 1956:159–61, 174–5).

Three classes of samples are used in experimental design: one sample, two samples and more than two ('k') samples.

- One sample—One sample is used to expose a single aggregation of individuals to a single treatment. For example, forty bus riders in one location who share the same relevant traits and are exposed to one treatment condition (e.g. use of one fare card reader) is a one sample condition.

Table 11.5 Examples of non-parametric statistics—statistics are used to describe the characteristics of subject samples. Non-parametric statistics can be used to analyze the results of studies in order to determine whether the effect on subject performance was due to chance alone or due to changes that were made to the independent variable. Measurement type (nominal, ordinal, interval) and sample type (one, two, or multiple samples) can be used to select an appropriate statistic. Because most behavioral data cumulates in ordinal scales, tests that are used to analyze ordinal data are explained in this chapter.

	Nominal	*Ordinal*	*Interval*
One sample case	Binomial test	Kolmogorov-Smirnov 1-sample test	
Two sample case			
Related samples	McNemar test for the significance of changes	Wilcoxon matched pairs signed ranks test	Walsh test
Independent samples	Fisher exact probability test	Mann-Whitney U test	Randomization test for two independent samples
k-sample case			
Related samples	Cochran Q test	Friedman two-way analysis of variance	
Independent samples	x^2 test for k independent samples	Kruskal-Wallis one-way analysis of variance	

- Two samples—Two samples are used when the investigator wants to determine whether two treatments are different or whether one treatment is 'better than the other.' Tests are conducted to compare the group that has been exposed to the treatment with one that has not or that has undergone a different treatment.
- 'k' samples—More than two samples may be needed to account for all of the traits that are essential to the experiment. When using multiple samples, sample values nearly always differ to some degree. The challenge is to determine whether differences that are observed among samples signify differences among populations or whether they are due to chance variations that would be expected to occur among random samples that are drawn from the same population. Tests that are conducted on multiple samples are used to determine whether there is an overall difference among the samples before the selection of any pair of samples in order to test the significance of differences between them.

11.4.4.3 LEVEL OF MEASUREMENT

Four types of measurement are used in the evaluation of data for inferential statistics: nominal, ordinal, interval and ratio. Of the four, nominal and ordinal scales are the most common in the behavioral sciences and should be analyzed using non-parametric statistics, as Section 11.4.5.4 will explain.

Nominal—Also referred to as the classificatory scale, nominal measurement uses numbers, names or symbols in order to classify an object, person or trait. Nominal measurement is considered to be the least powerful form, as it simply names data elements and provides no further information. Nominal measurements can be subdivided into mutually exclusive subclasses. However, all of the subclasses have to be equal in the property that is being scaled. For example, license numbers that are assigned to all of the automobiles in a state are nominal in character.

Ordinal—Also referred to as the ranking scale, the ordinal scale is used to organize elements that are described according to relations between them. Size, quality and difficulty can be used as the means to identify the relationship between elements that are larger, better, or more difficult. Comparison among all elements in a data set will eventually result in a set that is sorted according to order. However, the distance between the elements is not known. For that reason, one item in the set may be significantly better than another and the ordinal scale will not reflect the difference. For example, the finishing sequence of runners in a race is ordinal (e.g. first, second, third). Ordinal scaling gives no indication of whether the finishers were separated by six inches or by six meters. According to statistics author Sidney Siegel, most of the measurements that behavioral scientists make culminate in ordinal scales.

Interval—The interval scale is rank-ordered, similar to the ordinal scale. However, the distance between elements is known. A common, constant unit of measurement is employed to assign a real number to all pairs of objects in the ordered set, although there is no set point that represents zero. While an ordinal measure may indicate one element is better than another, the interval scale can represent how much better it is. Because interval measures are constant, it is possible to relate one data set to another (such as the translation between fahrenheit and centigrade temperatures). Any statistical test can be performed with data that is represented by interval measures.

Ratio—The ratio scale is similar to the interval scale and has a true zero point as its origin. Only the unit of measurement is arbitrary. As a result, operations that are performed on the scale will not affect any of the information that is contained in it. Any statistical test can be used with data represented by ratio measures (Siegel 1956:21–9).

11.4.4.4 PROCESS

The use of inferential statistics typically follows a six-step process:

1   Define the null (H_0) and alternative (H_1) hypotheses—For the null hypothesis, state the condi-
     tions in which there would be no difference between observations (e.g. no difference exists in
     performance between Group A and Group B).
         The alternative hypothesis is the operational statement of the investigator's research. To
     formulate it, state whether a difference is expected based on the nature of the research hypoth-
     esis (the prediction of what the investigator expects). Include the direction of the predicted
     difference if it is known (e.g. Group A will perform better than Group B).
2   Choose a statistical test based on the manner in which the sample of scores was drawn, the
     nature of the population from which the sample of scores was drawn and the kind of measure-
     ment that was used to define the scores.
3   Specify the significance level and the sample size—Choose the level of significance (alpha)
     that will be high enough to avoid rejecting a true null hypothesis (Type I error) and low enough
     to avoid accepting a false null hypothesis (Type II error). The power of a statistical test
     describes the probability of (correctly) rejecting a false null hypothesis. Power is related to the
     nature of the statistical test that is chosen. Increases in sample size generally increase the
     power of a statistical test.
4   Determine the sampling distribution—Each statistic has its own sampling distribution value that
     is provided by statistical software programs as well as tables that are found in statistical texts.
     The value accounts for the mean of all of the possible samples, as compared with the mean of
     the population from which the samples were drawn. Using these sampling distributions makes
     it possible to determine the likelihood that a value of a statistic as large as or larger than the
     observed value could have occurred by chance (Bordens and Abbott 1998:349–50).
5   Specify the critical (rejection) region—On the basis of steps 2, 3 and 4, define the value range
     in the score distribution using a two- or one-tailed test that will enable the investigator to com-
     pare observed values with the alpha value (level of significance).
6   Decide whether to accept or reject the null hypothesis—Compare the alpha level chosen in
     step 3 with the obtained p ('the probability with which a value equal to or larger than the
     obtained value of the test statistic would occur if the null hypothesis were true') (Bordens and
     Abbott 1998:351–6). If the obtained p is less than or equal to alpha, the comparison is statisti-
     cally significant.

11.4.5 Examples of non-parametric statistics

Sample design and level of measurement can be used to create a decision matrix for the selection
of non-parametric statistics as Table 11.5 shows. Most non-parametric tests pertain to data in an
ordinal scale. For that reason the tests that are explained below are used with ordinal data.

11.4.5.1 KOLMOGOROV-SMIRNOV—ONE SAMPLE

Kolmogorov-Smirnov compares an observed frequency distribution with a theoretical distribution for the same data. The test compares cumulative distribution of observed scores and
the theoretical cumulative distribution of scores that would be expected under the null
hypothesis. Both distributions are compared, the point of largest divergence is identified and
evaluated according to whether such a large divergence could be caused by chance.

   To illustrate, an investigator might want to determine whether subjects have a hierarchy of
preference among shades (from light to dark) of a particular color (hue) that will be used in a

worldwide corporate safety program. The implications might indicate that certain shades result in a better response to warnings and recommendations to keep facilities safe. Subjects might be asked to review samples of ten shades (saturation levels) of each color from light to dark. The firm's preferred choice might be included (unknown to the subjects) as the middle example in each case. Subjects would be asked to select one preferred value from the examples presented. If shade preferences are unimportant to the subjects, ranks should be chosen equally often except for random differences. If there is a preference, subjects would consistently favor a particular portion of the shade selection (e.g. lighter, darker). Kolmogorov-Smirnov can be used to determine whether the preferences are statistically significant.

Phillips (1982:139) recommends Kolmogorov-Smirnov as an alternative to chi square when sample size (N) is small.

11.4.5.2 WILCOXON MATCHED PAIRS—TWO RELATED SAMPLES

Wilcoxon matched pairs is used to assess positive versus negative changes when subjects who are exposed to an experiment are compared to matched control subjects. When analyzing behavioral data, an investigator can often tell which member of a pair is 'greater than' and rank all members of the data set in order. Wilcoxon matched pairs gives more weight to a pair that shows a large difference between two conditions than it does to a pair that shows a small difference. The equivalent parametric statistic is the t test.

To illustrate, an investigator might want to determine the effect that safety training has had on participant abilities to perceive a hazardous condition. Showing participants a series of pictures that depict a variety of circumstances that include unsafe conditions would be used to score perceptiveness. Asking a standard group of questions about each obtains scores between 0 and 100 for each subject. While the investigator can be confident that a higher score represents better perceptiveness than a lower score, it is not clear whether the scores are exact enough to be treated numerically (e.g. whether a score of 60 indicates twice the perceptiveness of a score of 30). To test the effect of training on perceptiveness, the investigator obtains eight pairs of matched subjects. One from each pair is selected at random to participate in safety training and the remaining eight do not attend. At the end of the training, all sixteen individuals are given the test and scores for each individual are generated. Each pair (of one trained and one untrained individual) would show a difference in scores. The use of Wilcoxon matched pairs would make it possible to determine whether differences in perception were significant and, thereby, whether training improved perceptiveness.

11.4.5.3 MANN-WHITNEY U TEST—TWO INDEPENDENT SAMPLES

One of the most powerful non-parametric tests, Mann-Whitney is useful as an alternative to the parametric t test when populations are not normally distributed and/or are not homogenous in their variance (Phillips 1982:139). Mann-Whitney uses small samples and requires measurement that need only be in an ordinal scale. The equivalent parametric statistic is the t test.

An investigator may be interested in determining the effect of instructions on subject performance while operating equipment. The hypothesis might be that operating instruction learning is generalized beyond specific instances. The results could have implications for software and intelligent interfaces that would be sold in diverse cultures. The investigator might expose five subjects (Group A) to a treatment in which displayed instructions lead individuals through the operation of a piece of equipment. The five subjects might then be given a task to perform using a second piece of equipment. Four subjects (Group B) serving as controls would not receive the instructions, but would be required to perform the same task on a second piece of equipment. The comparison would determine how many trials it would take for members of both groups to perform the tasks correctly.

Mann-Whitney can be used to determine whether the scores of either group were significantly higher than the other. In this instance, the prediction would be that Group A would have a better performance than Group B, leading to a two-tailed test. If the difference is found to be significant, that could lead to further testing among other groups in order to determine how pervasive the pattern might be.

11.4.5.4 ANALYSIS OF VARIANCE

Analysis of variance is the statistical analysis method that is necessary to interpret data that contain much unexplained variation such as human performance data. Williges points out a number of advantages in the use of analysis of variance. ANOVA makes it easier to use the research hypothesis structure to organize data collection. ANOVA makes it possible to conduct a composite test of statistical significance on all variables simultaneously. ANOVA can be used to examine all possible interactions among independent variables and to generalize research results by providing a broader baseline of variables which is more representative of the complex nature of human behavior (Williges 1985).

Much human performance data consists of two variables: time to perform (speed) and errors. As a result, the analysis of such data can require the use of MANOVA. Multiple analysis of variance (MANOVA) is the statistical analysis method that is used to analyze how a set of independent variables affects two dependent measured variables (e.g. time to perform and errors) variables.

Both ANOVA and MANOVA can be performed using statistical analysis programs such as SPSS, SAS or BMDP. Interpretation of the results of these analyses can be quite complex. The investigator who needs to use these methods should consult an experienced statistician.

11.4.5.4.1 Friedman 2-way analysis of variance (ANOVA) by ranks—more than two related samples—Friedman 2-way ANOVA by ranks is useful for testing whether multiple matched samples have been drawn from the same population. Because they are matched according to relevant characteristics, the number of individuals in each sample is the same. The equivalent parametric statistic is the *F* test.

As an example, an investigator might have a need to evaluate the effectiveness of four separate approaches (e.g. self-paced interactive software, CD-ROM, hard copy manual and on-the-job training) to train individuals to repair a particular item of electronics equipment. The test can be set up according to between subjects or same subjects design, as Figure 11.6 illustrates. A between subjects design would organize test subjects into several groups. Each group would include the same number of individuals matched according to relevant traits (such as experience, intelligence, motivation). In this case, four groups each containing three individuals would be randomly assigned to each of the four training methods. Test criteria might include 'time to complete repair' and 'equipment deficiencies detected and corrected.'

Data are collected by evaluating the equipment repair performance by members of each group after being exposed to the particular training condition they were assigned to. Data are organized according to rows for each group labeled 1, 2 and 3. Each treatment condition (training method) is labeled A, B, C and D.

	A	B	C	D
Group 1	9	4	1	7
Group 2	6	5	2	8
Group 3	9	1	2	6

Scores across each row are ranked (e.g. 1 for the lowest raw score to 4 for the highest raw score). Ranks for each column are added together to produce a total rank score for each treatment (i.e. training method).

	A	B	C	D
Group 1	4	2	1	3
Group 2	3	2	1	4
Group 3	4	1	2	3
R_j	11	5	4	10

Is the training method that receives the highest score the best? Not necessarily. The question is whether the rank totals differ significantly. If the null hypothesis is true (i.e. the independent variable of training mode has no effect on subject performance) the distribution of ranks in each row would be a matter of chance. The Friedman test determines whether the rank totals differ significantly (Siegel 1956:166–72).

11.4.5.4.2 Kruskal-Wallis 1-way analysis of variance (ANOVA)—more than two independent samples—Because the values of multiple samples vary to some degree, the question is what caused the differences among the samples. Do the differences signify genuine differences or do they represent chance variations that could be expected among several random samples?

Like the Friedman 2-way ANOVA, Kruskal-Wallis 1-way analysis of variance (ANOVA) replaces each of the observations with ranks. Ranking begins with '1' for the lowest score continuing consistently through to the highest score. The sum of the ranks is determined for each treatment condition (column). The test determines whether these rank sums are so distinct that they are not likely to have come from samples that were all taken from the same population. Kruskal-Wallis requires at least ordinal measurement of the variable. The equivalent parametric test is the *F* test.

A study of the effect of authoritarian personality provides an example of the use of Kruskal-Wallis. An investigator might want to test the hypothesis that school administrators are typically more authoritarian than classroom instructors. The potential exists for possible contamination of data in this design. The reason is that some instructors may aspire to be eventually selected for a role in the administration (i.e. instructors take administrators as a reference group). In order to avoid possible contamination, fourteen subjects are divided into three groups: instructors who have no aspirations to an administrative position, instructors who do aspire to such a role and administrators. The investigator administers a standardized test of authoritarianism, the *F* scale, to the subjects. Kruskal-Wallis can be used to determine whether *F* scale score differences among the three groups are significant (Siegel 1956:184–93).

11.5 Summary

Controlled studies are used for two purposes. Frequency distributions, percentiles and descriptive studies are used to perform data analysis. Correlations and inferential study are used for significance evaluation. Frequency distributions portray collected data along a single measure. Percentiles give an indication of the variability of measurements and help to quickly understand proportions as they are related to the whole distribution. Descriptive studies are conducted in order to analyze the characteristics of a sample and its performance. Correlation is used to describe how close the relationship between measurements of one kind is with those of another kind and is performed when variables of interest (such as subject sex or age) cannot be controlled by the investigator.

Frequency distributions and percentiles are helpful tools yet they cannot reveal everything an investigator might want to know about subject performance. Typically, some combination of methods is employed. This may include observation, usability assessment (descriptive study) and controlled studies (correlation, inferential study).

Four steps are necessary to create and run an inferential study that discovers relations between variables: plan the research, run the test, analyze and interpret the data and report the results. The experimental design is the plan that is created to organize the type and order of observations that will be made. The principal elements of an experimental design are to identify variables that will be investigated, to determine the type and number of subjects and to plan how to manage the order in which experimental trials will be conducted.

Inferential statistics are used to determine whether the manipulation of the independent variable caused a change in the performance of the human subjects. Parametric statistics are used to estimate the value of a characteristic (parameter) of a population from the characteristics of a sample and require the researcher to make certain assumptions about the population. Non-parametric statistics are used when data come from discrete (compared with continuous) categories and when quick estimates are needed. Because they make no assumptions about the population (as parametric statistics must) they can be used more easily when conducting basic analyses.

Data interpretation follows data analysis. Interpretation can be used to effect change, as Chapter 13 describes. Reporting the results of analysis and interpretation relies on the communication tools that are described in Chapter 14.

Part III

Application

12 Cost-effectiveness

What is value in a product? How does value affect decisions?
Value (12.1)

What approach has business used to determine value? Does it apply to human factors?
Cost-benefit analysis (12.2)

How does cost-effectiveness affect research and development?
Cost-effectiveness (12.3)

How can cost-effectiveness be determined?
Determining cost-effectiveness (12.4)

To what degree is a condition made safer or more productive by investment in human factors?
Application to practice (12.5)

Economics always enters into decisions about product development. Companies, organizations and individuals all take cost into account when making decisions about products. For example, commercial manufacturers must optimize profits. Organizations, such as the government, must use budgets prudently. Users consider price and operating costs when making a purchase.

So far in this work, cost has been a minor consideration. In Chapters 6 through 11, issues related to costs have been considered as one of many variables in methods such as Pareto analysis (Section 6.1) and decision analysis (Section 7.10). Questions that are based on economics have a substantial influence on decisions about user-centered products and deserve substantial examination. Research and development teams need to understand how economic issues such as life cycle costs and perceptions of concept value influence choices among alternatives. For these reasons, this chapter discusses value, how value is created and how value is shared among those who make and use product. Cost-benefit analysis is compared with cost-effective analysis. Approaches to cost-effectiveness analysis are reviewed. A method to account for cost-effectiveness is presented, then demonstrated in two examples.

12.1 Value

The value of something is what a person is willing to give up for it, which can be illustrated through a brief story from Friedman (1996:17).

Two economists walked past a Porsche showroom. One of them pointed at a shiny car in the window and said, 'I want that.' 'Obviously not,' the other replied.

Because economics exists to explain and predict behavior, it is in taking action, making the choice to give up something, that value is realized. If the economist *actually* wanted the Porsche (the shiny sports car *and* its price tag), he would have already acted to buy it.

The same lesson holds true in research and development work (although the trade-off can be less direct than buying a sports car). Value is demonstrated by the actions that are taken by members of an organization. Typically, that action is the choice to make an investment in order to derive a benefit.

12.1.1 Value creation and transfer

Each person, department, or firm that participates in the development of a product determines its value. How many participants there are, which participants are most influential and how important aspects are to each participant creates a fabric of decisions that describes a product's value. Conditions, participants, their decisions and the type and amount of resources invested continue to change throughout the life cycle of the product.

Rouse and Boff (1997:1617–32) use the value chain shown in Table 12.1 to describe the creation and transfer of value among those who participate in the research and development process. This is not a decision-making scheme. It is, rather, a way to describe what goes on among those whose support is necessary to each project. In product development, value is made possible by new thinking that is used to create solutions to fulfill user needs. At the outset of a development project, resources (e.g. including funds, talent, time and effort) are given over to ('invested in') the project. Products are produced over a period of time. As products are produced over time, each of the stakeholders also receives benefits over time such as revenue and the use of the product. As stakeholders vary widely, the way that each perceives benefits and the way each understands the utility of those benefits can vary significantly. Participants determine whether they got less than, the same amount as, or more than they originally contributed. The results of that determination affect the willingness of participants to invest further.

The development of a software program can be used to illustrate the process. Marketing surveys might indicate a program to manage personal investments has good sales potential. The development team creates a software program concept that provides a new approach to graphically displaying the current state of a user's investments in real time. It might also make particularly insightful use of human factors methods such as cognitive task analysis (Section 7.7). The team proposes the initial concept to their management. Management agrees to invest a budget, staff and resources for production, promotion, distribution and sales in hopes of a payback. Upon completion and release, the product is sold and used by customers. The company receives the benefit of revenue from sales. The customer receives the benefit of improved personal investment analysis from using the software. If enough customers are satisfied, word of mouth and product reviews will lead to increased sales, increased revenues and more customers deriving benefit from using the program. Sales that pay back the company's investment and that meet profit requirements disposes management to be positive about future investment in later versions and related products. Customers who find that the product helps with investment tracking, analysis and management will be favorably inclined to buy updates or additional products from the same company.

Table 12.1 Value chain from investments to returns—Rouse and Boff (1997) propose a path from investment to returns, in the interest of explaining how benefits relate to costs. Satisfaction (or lack of it) among stakeholders with the utility that each derives from products influences their inclination to pay for it and, by extension, affects returns to investors.

Investments (costs)	to	Resulting products over time
Resulting products over time	to	Benefits of products over time
Benefits over time	to	Range of stakeholders in benefits
Range of stakeholders	to	Utility of benefits to each stakeholder
Utility to stakeholder	to	Willingness to pay for utility gained
Willingness to pay	to	Returns to investors

12.1.2 Value in research and development

A variety of participants need to support research and development efforts in order to succeed. Participants include management, finance, design, engineering, marketing, operations, sales and customer service. Each has a role in translating a concept from a notion into the reality of a product. Value is realized when that support is given.

Research and development has a number of characteristics that make the creation, transfer and realization of value difficult to track. Projects typically involve many participants who have an interest in the outcome. Projects can span long periods of time. Project complexity can make it difficult to isolate specific features that yield identifiable results. It is even more difficult to pin down specific results from investment in human factors. For example, human factors can produce value through the absence of a loss (e.g. accident prevention) rather than the presence of a gain. The value of human factors investment may be diffused among many system components rather than realized in a few discrete features. A component such as a web interface that is produced for one application may be used in many later versions of the product. Investment in human factors may produce results such as training and performance standards that are more abstract than physical things and may have value that is deferred. The value of human factors investment often accrues over the entire life cycle of a product (e.g. improved operation or maintenance). Human factors activity may reduce costs, but the reductions can be difficult to identify (Rouse and Boff 1997:1617–32).

12.2 Cost-benefit analysis

Cost-benefit analysis is widely known as a method that is used to compare the cost of an investment (e.g. money) with the benefit, expressed in dollars, that results. Two issues make cost-benefit analysis problematic when it comes to research and development: quantification and short-term focus.

Quantification—Cost-benefit analysis relies on the ability to quantify all costs that are involved in a given project. However, research and development and human factors benefits are difficult to express in purely economic terms. As the previous section explained, the specific benefits of R&D can be difficult to quantify and to tie directly to costs. Human factors benefits such as ease of use can even more difficult to quantify.

Short-term focus—Cost-benefit analysis focuses on the immediate transfer of value. By contrast, research and development typically considers costs and benefits over the life span of a product. As a result, cost-benefit analysis can miss some aspects of risks, costs and benefits. Orasanu and Shafto (1999:629–58) use software development team member performance as an example to demonstrate the difference between cost-benefit analyses and cost-effectiveness. The authors note that the use of cognitive task analysis (Section 7.7) is widely known to improve software design. However, software developers are known to resist using it. Why? Software developers are disinclined to follow good research practice due to rapid changes in technology, high product turnover and little personal investment in the long-term growth of a product line. The authors conclude that 'The true life-cycle risks and costs of bad design are rarely considered in shortsighted "cost benefit analyses."'

Without sufficient human factors research, the likelihood of loss due to product failure is high. In the case of mass produced products (e.g. software that is used in home and office, consumer electronics) the cost of loss remains low and can be spread over a long span of time and large number of customers. Cost-benefit analyses miss such conditions and, as a result, allow suboptimal product performance to survive.

For these reasons, cost-benefit analysis is a poor match for human-centered research and development. Instead, another approach has been developed to describe the benefits of activities such as human factors research: cost-effectiveness.

12.3 Cost-effectiveness

When benefits are not expressed in purely economic terms, the analysis is referred to as 'cost-effectiveness' instead of 'cost-benefit.' In cost-effectiveness analysis, attention is focused on how well objectives are achieved over the life of the product. When considering one or more concept alternatives, an organization will prefer the approach that satisfies objectives by using the most cost-effective approach. A number of authors have put forth proposals to determine cost-effectiveness. Three are reviewed here: Rouse and Boff, Chapanis and Van Cott, and Daalen, Thissen and Verbraeck.

12.3.1 Rouse and Boff

Rouse and Boff (1997:1617–32) consider cost-effective analysis as a means to formulate and resolve specific decisions, such as how to allocate resources (e.g. funding, staff, time). Their seven-step approach, shown in Table 12.2, is based on three items: which traits will be used in an assessment, how those traits will be mapped onto some means of measurement and whose preferences will be used to determine success.

The method begins by identifying each individual who has an interest in the project outcome, including those who authorize capital funds, create the concept, produce it and benefit from it. Benefits and costs are defined from the viewpoint of each stakeholder. Utility functions are determined for attributes. That is, each individual is expected to assign a value to each benefit according to how important it is to them. A decision is then made as to how utility functions should be combined across stakeholders and how the utility functions are perceived and valued by the stakeholders. Parameters within utility models are assessed. Levels of attributes (benefits and costs) are forecast and the expected utility of alternative investments is calculated.

Rouse and Boff identify difficulties in the process that they describe. The results of their method depend on the correct identification of stakeholders and attributes. The quality of inputs to calculations (i.e. estimates of model parameters, attribute level forecasts) can be problematic. Data are often unavailable.

Beyond their own observations, Rouse and Boff's model is difficult to apply for three additional reasons. First, neither benefits nor costs are entirely known by each stakeholder. As a

Table 12.2 Cost-effectiveness method: Rouse and Boff—various approaches have been proposed to determine cost-effectiveness in human factors contributions to research and development activity. Rouse and Boff (1997) propose a seven-step model.

Identify stakeholders in alternative investments

Define benefits and costs of alternatives in terms of attributes

Determine utility functions for attributes

Decide how utility functions should be combined across stakeholders

Assess parameters within utility models

Forecast levels of attributes

Calculate expected utility of alternative investments

result, both can be difficult to determine. Second, combining utility functions across all stakeholders would involve weighting each in accordance with its importance. That can also be difficult to determine because (in order to advance their interests) stakeholders are very likely to understate utility (in the interest of bargaining for more) and are likely to overstate potential loss (as a hedge against risk). Third, forecasting levels of attributes is also a challenge, as future costs can be difficult to know.

12.3.2 Chapanis and Van Cott

Chapanis and Van Cott (1972:701–28) take a more direct approach with a simplicity that makes it easier to use and is more reliable. Abstractions such as benefit are difficult to determine. Instead, it is easier to focus on improvements to performance. In new system applications, Chapanis and Van Cott find 'it is usually better to describe performance in terms of one or more functional relationships.' They reason that 'as a general rule, it is simpler to devise human engineering tests to yield comparative statements of relative value between equipments than to devise tests to give estimates of an absolute quantitative difference.'

Their approach uses performance and cost to make a comparison of value. As an example, they propose a cost-effectiveness ratio, shown in Figure 12.1, for new equipment (e.g. acoustic magnifier) that is used to augment human senses. The formula represents gain per dollar as the difference between targets that are detected with the device, less the targets detected using no performance aid, divided by the cost of the device in dollars. The performance improvement is expressed by the expression 'Targets that are detected with the device.' 'Targets detected using no performance aid' removes results that can be obtained without using the device. The 'cost of the device in dollars' accounts for the development investment. 'Gain' in detection ability is the objective and represents cost-effectiveness.

12.3.3 Van Daalen, Thissen, and Verbraeck

While Van Daalen, Thissen, and Verbraeck (1999:1063–76) do not address human factors directly, their analysis bears directly on decision making regarding cost-effectiveness. Like Chapanis and Van Cott, they contend that 'the choice among alternative designs or systems involves trade-offs between functional performances and costs.' Five considerations enter into their analysis: time preference, costs, market effects, indirect effects and external effects. Among them, time preference, costs and external effects have a bearing on research and development and are discussed here.

$$\text{Gain per dollar} = \frac{\left[\begin{array}{c}\text{Targets detected} \\ \text{with device}\end{array}\right] - \left[\begin{array}{c}\text{Targets detected with} \\ \text{no performance aid}\end{array}\right]}{\text{Cost of device in dollars}}$$

Figure 12.1 Cost-effectiveness method—Chapanis and Van Cott (1972:710) recommend the comparison of new concept performance to either current conditions or performance standards. Their example in this instance considers a new surveillance system as a function of the number of targets detected compared with the number that can be detected without the use of such a device.

12.3.3.1 TIME

Costs are generated during development and production. Benefits are realized at a later time, during use. As a result, costs are compared with benefits at different times. Equivalent present values of benefits and costs that occur some time in the future are figured using the principle of discounting. If net present value of a project is positive, it is considered economically reasonable to consider. Kantowitz and Sorkin (1983:17–21) also point out that present worth considerations (as well as inflation, tax laws and depreciation) influence the lifetime cost of a product by affecting the cost of money that is used to invest in production.

12.3.3.2 COSTS

The costs of materials, labor and equipment that are allocated for a particular project can be directly assigned to a development budget. Other costs are more difficult to assign. The cost of facilities, for example, may need to be handled as marginal costs (extra costs incurred by the project) or opportunity costs (what is foregone if the project is realized).

12.3.3.3 EXTERNAL EFFECTS

External effects are costs that can affect valuation. However, they are not bought and sold in an open market. As a result, they are not accorded a recognized value and because of this they are traditionally ignored. A firm can choose to accept the burden of an external effect. When it does, the external effect becomes an internal effect.

External effects include mitigation (e.g. cost of waste dump clean-up), prevention (e.g. changes to production facilities to prevent water pollution), compensation (e.g. a stipend paid to residents in exchange for their acceptance of excessive noise from a local factory) and curation (e.g. health care, lost labor costs due to a traffic accident). Human factors issues (e.g. the prevention cost to improve traffic safety) may also be included in discussions of external effects.

Which external effects will be taken into account and become internal effects, or ignored and kept as external effects? In the case of an auto manufacturer, one might presume that external effects such as prevention (action that is taken in advance in order to avoid costs of litigation over foreseeable hazards) and curation (costs that are incurred as a result of losses due to accidents that are experienced) would be internalized. In the case of U.S. auto manufacturer Ford Motor Company, it appears that they are not. In May 2000, Ford acknowledged that the sports utility vehicles (SUV) that they manufacture and sell contributed more than cars to the threat of global warming, emitted more smog causing pollution and endangered other motorists.

> Sport utilities are three times as likely to kill other drivers in a crash, but the death rate for the sport utility occupants is just as high as for car occupants because of sport utilities' tendency to roll over and their lack of crumple zones.
>
> (Bradsher, 2000:C2)

Unsafe? If hazards are known and the means to meet them are available, how can a firm ignore them? By Ford's own admission, the profits from these vehicles are substantial: US$10–15,000 for each Ford Expedition and Navigator and up to US$18,000 for a Ford Excursion (Bradsher 2000: A1–C2). The cost of changes that would avoid effects such as prevention and curation appear to be secondary (and external) compared with the dramatic profits that the SUVs generate.

12.3.4 Conclusion

Each of the above authors agrees that, as attractive as it is to reduce the decision to a single measure of performance, it cannot be done. Instead, a better approach would be to outline the steps that can be used to determine cost-effectiveness.

12.4 Determining cost-effectiveness

Table 12.3 summarizes an approach to cost-effectiveness that incorporates elements of Rouse and Boff's approach as well as Chapanis and Van Cott's model. Cost-effectiveness requires the ability to identify the aspect of performance that is intended to be improved, to collect performance data through simulation and/or operation and to evaluate performance for either current conditions, a performance standard, or a new solution.

12.4.1 Preparation

12.4.1.1 CREATE SOLUTION CONCEPT FUNCTIONAL PROTOTYPE(S)

The prototype may be entirely computer-based. The material configuration may be evaluated according to engineering software programs such as finite element analysis (FEA). Physical functional prototypes may also be developed to verify one or more aspects of the computer model, particularly if task performance involves the manipulation of hardware. See Simulation (Section 8.1). If a functional prototype cannot be developed, it will be necessary to estimate new concept performance.

Table 12.3 Method to determine cost-effectiveness—cost-effectiveness can be determined by comparing data related to the costs and the performance of a current set of conditions and those that are related to the development of new solution. The approach is based in part on Chapanis and Van Cott (1972).

Determine the purpose for performing the cost-effectiveness study

Determine costs
Account for costs that have been incurred in order to develop the solution alternative(s)

Identify variables of interest
Establish baseline for the variable of interest in the current stuation

Determine the new solution benefit
To estimate
    Review reliable literature
    Seek out examples of comparable improvements
    Propose the estimated benefit in terms of the variable of interest
To test
    Create a test plan
    Conduct the test
    Compile and review performance data

Determine cost-effectiveness
Compare the benefit realized with the new solution, less the benefits under the current conditions, divided by the cost of the solution in dollars

Review results of analysis with stakeholders
Determine whether performance improvement is sufficient to warrant development cost

12.4.1.2 COLLECT CURRENT PERFORMANCE DATA

Review records that account for current product performance during specified periods (e.g. by shift, hourly, daily, weekly, monthly, yearly). For example, work center records would document the number of units that are currently being produced. Review any applicable standard that specifies performance requirements.

12.4.2 Materials and equipment

Arrange to have a prototype (simulated computer-based model, physical functional prototype, or combination) of the solution for performance evaluation and software programs for simulation (e.g. solids modeling) and evaluation (e.g. finite element analysis or similar program).

12.4.3 Procedure
12.4.3.1 DETERMINE THE PURPOSE

Determine the purpose of performing the cost-effectiveness study. Purposes can include the comparison of:
- A solution under development with existing conditions. The example in Section 12.5.2 demonstrates the use of existing conditions as the basis for cost-effective analysis.
- More than one solution alternative
- A solution against a predetermined performance standard. The example in Section 12.5.1 demonstrates the use of a standard as the basis for cost-effective analysis.

12.4.3.2 DETERMINE COSTS

Account for costs to develop the solution alternative(s). If use cycle operating costs are an issue, collect information on current operating costs and expected operating costs for the solution alternative(s).

12.4.3.3 SPECIFY VARIABLE

Identify variables of interest. For example, if the objective is to improve safety, what variable is used to determine whether a condition is safe? The prevention of an injury?

12.4.3.4 ESTABLISH BASELINE

What is the value for the variable of interest in the current situation (e.g. what losses have been incurred as a result of exposure to hazard?) or that is directed by performance standards?

Use the time period for amortizing development costs as the basis for comparing performance that is realized either with or without the new solution.

12.4.3.5 DETERMINE THE NEW SOLUTION BENEFIT

Performance can be either estimated or proven through data collection. Estimation is faster, but is more vulnerable to challenge. The development and assessment of a prototype makes a much stronger case, yet involves time and expense. It can be effective to estimate early in the process (e.g. during simulation), then develop and assess as a project evolves.

12.4.3.5.1—By estimation

- Review published professional literature (e.g. journals) in order to substantiate the type and degree of performance that is being estimated.

- Seek out examples of comparable products or improvements to determine the amount of impact such a change may have.
- Propose the estimated performance improvement in terms of the variable of interest that was specified above in Sections 12.4.3.3 and 12.4.3.4.

12.4.3.5.2—By testing

- Create a test plan—If a solution will be tested to collect data, follow procedures described in Chapters 5, 10 and 11 to ensure a representative sample is recruited, conditions are as close to reality as possible and accurate data are collected on the variable of interest.
- Conduct the test—Operate functional prototype(s) and collect performance data according to measures.
- Compile and review performance data—Compare performance data with data that has been collected on current conditions (or performance standards if this is the first time such a concept has existed). Propose the estimated performance improvement in terms of the variable of interest that was specified above in Sections 12.4.3.3 and 12.4.3.4.

12.4.3.6 DETERMINE COST-EFFECTIVENESS

As Section 12.3.2 described, compare the performance improvement that is realized with the new solution, less the benefits under the current conditions, divided by the cost of the solution in dollars. See Figure 12.1.

12.4.3.7 REVIEW RESULTS OF THE ANALYSIS WITH STAKEHOLDERS.

Substantial discussion can occur over the type and amount of performance improvement and whether the improvement is sufficient to warrant development cost. Many options are available at this stage. The project can proceed intact. Changes can be made to the concept to shift costs among components (e.g. making standard items optional). The project can be extended to produce new approaches or even discontinued.

12.4.4 Results

Cost-effectiveness analysis provides a quantitative basis to determine the value added by a new product, on the basis of costs invested in its development.

12.5 Application to practice

Two application areas, safety and productivity, are explored here to demonstrate how the method to determine cost-effectiveness described above can be used.

12.5.1 Safety effectiveness

Rugby players traditionally compete without using any protective gear. A time may come when public sentiment and insurance companies mandate the use of protective gear. Head protection and similar devices may pose a significant cost to teams. How can a sports equipment manufacturer demonstrate the cost-effectiveness of a new line of protective gear for rugby players?

The cost of injuries to competitive rugby team members can be identified through submitted health care insurance claims. Depending on severity, costs can be adjusted to reflect additional costs related to accidents such as lost time from work.

To estimate the new solution's performance improvement, identify the incidence of severe injury in similar sports that have already adopted protective gear (e.g. American football, hockey). To test the concept's performance improvement, develop a prototype of equipment that is designed to prevent the most severe injuries and collect data through a test program.

Subtract the number of severe injuries incurred while using the gear from current losses due to severe injuries (using no protective gear), divided by the cost of developing the gear. 'Gain' in this case will be a reduction in losses due to injuries for an acceptable cost of development.

There are a variety of issues that are involved in the final choice to produce, buy and use the protective gear. While there are a number of ways to reduce injuries, field tests may show that a particular design of headgear is the most cost-effective. All parties (the players, team management, the league and the insurers) may agree to the technical results. However, agreement on the technical results is not the end of the discussion. In the end, the insurer may choose not to mandate use of the gear, instead internalizing the cost by charging higher rates. Players may weigh the uninsurable potential losses (such as the loss of employment) against the pressures of customary practice that discourage the use of protective gear. The player who values rugby's customs may view gear as not being cost-effective. The players' organization might internalize the disabilities and loss of employment and, therefore, view the protective gear as cost-effective.

12.5.2 Productivity effectiveness

The checkout process in a supermarket can become a bottleneck for customers during peak demand hours. Customers who have few items and want to go about their business have a particular interest in avoiding long and slow checkout lines. A workstation that would enable customers to conduct a checkout on their own without having to wait in line could increase throughput and customer satisfaction. But would it be cost-effective?

Current data is available for the time and activities that are needed to checkout customers. Identify the time for those customers who need to purchase a few items over the course of a day's business. To estimate the benefit of the new solution, account for the time for a customer to complete checkout. To test the solution, develop a prototype that can be placed in operation and collect data on waiting times and throughput.

Subtract the time to complete checkout transactions using the new solution from the time that is currently required, less the cost of development. 'Gain' in this case would be a reduction of time needed to checkout.

Additional issues will involve customer satisfaction, protection from theft and error remediation, among others.

12.5.3 Further issues

Two further issues deserve consideration: negotiation and organizational influence.

12.5.3.1 NEGOTIATION

Stakeholders bring their best set of requirements to the discussion to negotiate a satisfactory outcome on behalf of each interest including finance, design, engineering, advanced manufacturing and customer service. It is through negotiation that trade-offs are made in order to satisfy those requirements. The strength of cost-effectiveness lies in its use as a tool to assist trade-off negotiations that are conducted among stakeholders. See Chapter 13—Effecting change for information on forging relationships with other stakeholders.

12.5.3.2 ORGANIZATIONAL INFLUENCE

This chapter has discussed cost-effectiveness from the project point of view. However, there are other points of view that also affect the final decision on what is cost-effective for an organization. For example, what may be cost-effective at the project level may not be at a management level. Van Daalen *et al.* (1999:1063–76) list additional considerations that can have an influence on whether a project is cost-effective:

- Impacts of a decision on a firm's strategic position
- Relations with other firms, organizations or individuals
- Issues of political gain or loss
- Long term environmental stability
- Concerns related to equity and legitimacy.

12.6 Summary

The value of something is what a person is willing to give up for it. It is in acting, making the choice to give up something, that value is realized. Traditional cost-benefit analysis compares an investment that is made (e.g. money) with the results that are obtained by the purchase (e.g. a component). Research and development creates benefits that are difficult to express in purely economic terms. Benefits are realized over many elements, may be realized over a long time and may cut costs that are not recognized by traditional economic models.

When benefits are difficult to express in purely economic terms, the analysis is referred to as 'cost-effectiveness' instead of 'cost-benefit.' In cost-effectiveness analysis, attention is focused on how well objectives are achieved. Cost-effectiveness can be determined by comparing data related to the costs and the performance of a current set of conditions and those that are related to the development of new solution. To do that, the investigator determines the purpose of performing study, determines costs and identifies variables of interest at a level that is specific enough to be quantified. The next step is to establish baseline standards for variables of interest, then create a test plan for comparison. Creating and operating a functional prototype of a new solution concept makes it possible to collect performance data. Compiling and reviewing performance data makes it possible to identify performance improvements and relate them to the development costs that were incurred to produce them. The results of the cost-effectiveness analysis can be used during reviews among stakeholders. The results of stakeholder negotiations are what will determine what is, ultimately, cost-effective for the organization.

13 Effecting change

How do others in my organization view human factors? Research and Development?
Research and development is change (13.1)

How do I 'sell' my idea to my team? My colleagues? Management?
Working relationships (13.2)

What team member characteristics improve how others perceive change?
Traits for change (13.3)

What activity is typical of effective change management?
Change management (13.4)

How does change occur in an organization?
Change evolution (13.5)

What steps are needed to effect change?
The change process (13.6)

How is change managed?
Example (13.7)

Development necessarily occurs within the context of an organization. Part of what amounts to success in development work, then, relies on success in working as a member of an organization. However excellent they may be, good ideas do not sell themselves. One of the most—some say *the* most—challenging aspects of research and development work is not the creation of products. It is working with all of the people who are necessary to the process that translates concepts into reality.

There is extensive business writing about teamwork and organizational change (e.g. 're-engineering') which will not be duplicated here. Instead, this chapter will focus on the nature of change, how people view and respond to it, what is necessary to introduce new thinking into an organization and how to shepherd a project to successful completion.

> *Most large, complex organizations are cauldrons of competing interests.*
> (C. Cappy and R. Anthony)

13.1 *Research and development* is change

The role of research and development is to bring new knowledge into an organization. That makes it, by its nature, a radical venture ('radical,' in the sense that it can affect the root of an organization's reason for being). Such new thinking can create change that results in a new product, department, division, business, or entire industry. Performed well, the research and development role always presents this potential for such change.

On the face of it, anyone might assume that change would be welcome. After all, it poses the potential for competitive advantage and future growth. Without the change brought on by new ventures, organizations will slowly lose their value to clients and customers. Why? The product that once matched what its clientele wanted will fail to keep up with what its clientele now needs or new market segments that have evolved.

There are other considerations to take into account. To effect change, it is necessary to understand how others view it. For instance, any new venture has an element of risk to it. The change it brings about can overturn previously accepted work roles, procedures, products, even entire businesses.

Organizations strive toward self-preservation. Those who work in organizations favor stability as a means to achieve it. Individuals who are faced with significant and frequent change in business, social and personal life seek to gain some control over daily events, rather than be controlled by them. In 1984, N.L. Hyer identified human resistance to change as the number one obstacle to implementation.

As Conner (1992:70) sees it,

> We view change as negative when we are unable to foresee it, when we dislike its implications and feel unprepared for its effects. Thus, a critical factor affecting our perception of change as positive or negative is the degree of control we exercise over our environment.

That sense of control comes from the ability to anticipate what is to come and to be prepared for it. What is known can be understood and planned for. Changes to that which is already known make the status quo less stable or predictable.

Even in the face of prospective failure, Kotter (1996:35–6) points out that 'needed change can still stall because of inwardly-focused cultures, paralyzing bureaucracy, parochial politics, a low level of trust, lack of teamwork, arrogant attitudes, a lack of leadership in middle management, and the general human fear of the unknown.' Instead, a cadre needs to be enlisted to create a pathway for the new concept(s) to evolve. 'In an organization of 100 employees, at least two dozen must go far beyond the normal call of duty to produce significant change. In a firm with 100,000 employees, the same might be required of 15,000 or more.' What will it take in order to enlist that scale of support? '... People in an organization will not make sacrifices, even if they are unhappy with the status quo, unless they think the potential benefits of change are attractive and unless they really believe that a transformation is possible.'

> *Only in retrospect ... does a successful innovation look inevitable. Up to that point, just about all innovating has a 'political' dimension.*
>
> (Rosabeth Moss Kantor)

13.2 Working relationships

The research and development staff is typically invested in the world of ideas and what the future may be. Others in the firm (including managers) are more invested in the world of people and what exists now. Hamel & Prahalad (1994a: 122–8) found 'on average, senior managers devote less than 3% of their time building a corporate perspective on the future. In some companies, the figure is less than 1%.' That requires those who work in research and development to create and implement change as a way of doing business, relying on an aptitude for building a consensus for the future. Indeed, management author Rosabeth Moss Kantor (1983) identifies three waves of activity that are needed to make innovation possible: problem definition, coalition building and mobilization. It is in coalition building and mobilization that working relationships play a vital role.

Many who succeed in research and development have done so through strong intellectual skills. Human factors, in particular, calls for substantial rational ability. While that ability is critical to a good research product, it may run counter to the creation of effective work relationships in concurrent engineering teams. Why? Kotter (1996:25) has found

> At the beginning, those who attempt to create major change with simple, linear, analytical processes almost always fail. The point is not that analysis is unhelpful. Careful thinking is always essential, but there is a lot more involved here than (a) gathering data, (b) identifying options, (c) analyzing and (d) choosing.

Instead, the leadership that is necessary to create the future 'defines what the future should look like, aligns people with that vision, and inspires them to make it happen despite the obstacles.'

In previous decades, functional organizations allowed departments to work in relative isolation, as each labored to satisfy the requirements of a detailed specification for a new product concept. Separated into functional groups, each department would perform its work and pass it to the next, in serial fashion. While departmental resources were used efficiently, the actual cost to the organization was high. Each department would need to fix oversights by previous departments that could have been anticipated if they had collaborated. Such reworking slowed progress, delayed time to market and produced suboptimal results.

More recently, matrix-style and product-specific teams have evolved under the umbrella of concurrent engineering. Dynamic, multi-disciplinary teams that 'get it right the first time' have by and large replaced the more parochial, inefficient serial development process that functional groups followed. Augmented by computing network and communications technology, knowledge and expertise can be assembled as a nearly custom-tailored organization at common or remote sites.

The concurrent engineering environment is clearly needed to meet the demands of a highly competitive environment. Because concurrent engineering consists of direct, person-to-person collaboration, the ability to develop effective working relationships is essential.

> There is no better way to reduce risk than to build trust.
> (R. Harmon and M. Toomey)

13.3 Traits for change

Human relationships are complex and varied. There are certain characteristics of working relationships that can be cultivated in order to improve how others perceive change. They include: trust, vision, awareness, empathy and altruism.

13.3.1 Trust

Others' willingness to have faith in new ideas determines whether they will rely on whomever works to bring them about. 'Trust helps enormously in creating a shared objective,' Kotter (1996:65) points out. 'One of the main reasons people are not committed to overall excellence is that they don't really trust other departments, divisions, or even fellow executives.'

Compressed development cycles may not allow for the gradual evolution of trust. Harmon and Toomey (1999:251–9) split trust into two types: earned and granted. 'Earned trust is built over time and based on performance' but may not develop to sufficient depth in time to realize necessary change. Instead, the change leader needs to learn how to generate granted trust, 'the trust individuals need to create with others and to obtain early action and results' and allow earned trust to evolve through time.

Informal contact in advance of more formal submission and reviews will make it possible to work out concerns ahead of time. This is informally referred to in the U.S. as 'greasing the skids,' referring to lessening the amount of resistance when one needs to push a major item forward. Such advance contact also makes it possible for departments to anticipate and adjust to change well before the need arrives. Advance contact provides others the chance to prepare for what is to come. Those who do enough advanced preparation demonstrate concern for others and their work and earn the trust that is necessary for the creation of a shared objective.

13.3.2 Vision

The ability to envision what the future can look like is a pivotal role in development. Each audience that considers new thinking will need cues regarding what is to come. Those cues are provided by the mock-ups and models the development team creates. How rough or complete does the mock-up or model need to be? Which specific parts need to be fully modeled or just blocked-in? That depends on:

- The audience (i.e. management, technical staff, marketing, sales, operations)
- Portions that pose a particular challenge to the organization's abilities or the users' interest (e.g. conversion to a 'glass cockpit' concept in an aircraft represents a radical departure in displays and controls compared to traditional dials and gauges).

13.3.3 Awareness

The distractions and demands of work can make it difficult to devote enough attention to understand issues thoroughly. Ensure that each aspect of the change is in the front of your audience's mind. Review with them what is new and unfamiliar in the project. Consider each department and representative whose support will be needed. Anticipate what may be a change for each. Understand that each participant will be interested in what the concept offers to benefit them. Listen for cues in conversations and correspondence which indicate doubt or lack of support (e.g. 'We are concerned about …').

13.3.4 Empathy

Consider the implications for others and other departments from their point of view. What will each be required to do? How similar, or different, is the current project from what has been done in the past? What unique aspects of your project will each pay particular attention to?

13.3.5 Altruism

The perception that an individual is pressing a project forward for personal gain erodes trust. Commitment to the idea because it is in the best interest of the organization and will benefit those who work in it is a source of trust. Conviction for the idea and its benefits will show an unselfish interest in the welfare of colleagues.

Make the relationship between the goals of the organization, the business plan and the project undeniably clear in the audience's mind. Show how the idea makes sense in terms of the firm's future.

Commitment to change relies on the recognition that the cost of the status quo is significantly higher than the cost of change.

(D. Conner)

13.4 Change management

Senge (1999:33) describes the basis of a program of continuous change through organizational learning. As 'organizations are products of the ways that people in them think and interact,' programs for change need to offer new ways to think and act. '... the practice of organizational learning involves developing tangible activities: new governing ideas, innovations in infrastructure, and new management methods and tools for changing the way people conduct their work.' To that end, Senge (1999:247–50, 287–90, 328–34, 371–80, 425–34, 496–500) presents six challenges to implementing change and recommends approaches to deal with them. Research and development professionals can take practical steps to counter the challenges of fear and anxiety, assessment and measurement, true believers and non-believers, diffusion, governance, strategy and purpose and create a program for change.

13.4.1 Fear and anxiety

Both Conner and Kotter referred to fear and anxiety earlier in this chapter as a response people have to the possibility of change.

Research and development team members can lessen the prospect of fear and anxiety by creating an atmosphere of mutual trust and understanding. Open discussion of concerns without confrontation can elicit participation by others in the change process. Disagreements can be used as an opportunity to learn about the views of all concerned. Starting with easily managed elements of the change and showing the link between current conditions and the vision of the future can make it easier to address more substantial issues later.

13.4.2 Assessment and measurement

One role of management is to gauge the nature of activity to see how activities are progressing and what efforts might be needed to influence them positively. That scrutiny can stifle the risk-taking that change requires. In addition, previous experience with change as a facade for other agendas (such as programs to reduce the size of the workforce) can cause further concern.

Effective change managers will encourage patience and value progress as it evolves. Enlisting the assistance of senior management can make it easier for many departments to understand change in the context of the whole organization instead of how individual department performance may be affected.

13.4.3 True believers and non-believers

Change can have rabid adherents and rabid opponents and both can present difficulties for change management.

Develop good relationships between research and development teams and other departments to make it easier to introduce the change to the larger organization. Ensuring development team members are flexible enough to challenge their own thinking can also open them to accepting the concerns of others. Demonstrating how the change can benefit the interests and agendas of others in the organization and how the change is related to commonly held organizational values can provide a common ground on which all parties can agree.

13.4.4 Diffusion

Departmental holdouts will impede the adoption of change and, ultimately, its success. Identification with departmental practice can lead to parochial attitudes.

The use of collaboration through such practices as concurrent engineering will encourage the building of consensus among all departments. Those teams can build a collective empathy by the use of appreciative inquiry to learn what thoughts others have to offer. Managers can help people gain genuine insights and network leaders can carry new ideas that serve as coaches.

13.4.5 Governance

Even though boundaries are ultimately arbitrary, it is a fact that others in an organization are responsible for different functions. Ignoring the differentiation in an organization can impede the progress of change.

Using business results as the case for change demonstrates how change affects the whole organization. Team member awareness of senior leader priorities will better enable them to relate the change to the vision for the organization. The use of collaborative inter-departmental teams makes it possible to represent each group's interests without violating boundaries.

13.4.6 Strategy and purpose

An organization that is habituated to focus on routine operational issues can have difficulty with the kind of strategic thinking which significant change requires.

Through training programs, management can encourage broader thinking around the way that external forces may affect the firm's future and what peoples' collective aspirations are. It can also encourage the use of scenario thinking to investigate blind spots and awareness of what signals might herald unexpected events. Such training can lead to an attitude of stewardship, which can embrace change as a means to promote the organization's well being and future growth.

13.5 Change evolution

Conner (1992:148) describes commitment to change as a process that evolves through three phases. 'Each of the three phases—preparation, acceptance and commitment—represents a critical juncture in the commitment process.' Conner's phases parallel Kantor's earlier-cited waves of problem definition, coalition building and mobilization activity that are needed to make innovation possible. Whether it is a research plan, research findings, or initial solution recommendations, it is up to the proponent to build support for the change through each phase, as Figure 13.1 shows. The amount of support for change can grow through each step, or decline if the change does not meet what is needed to proceed. If support does grow, increasing resources will need to be expanded in order to institutionalize the new approach. Cappy and Anthony (1999:199–215) put a keener edge on these issues in light of the need for decisive and quick 'productivity, innovation and performance improvements.' 'The best simultaneously get results, drive cultural change and build internal capability' by 'boundary busting,' which is the process of 'resolving conflicts between work groups or functions and between management levels.' Research and development staff members particularly need to assume such a role, rather than waiting for others to take the lead.

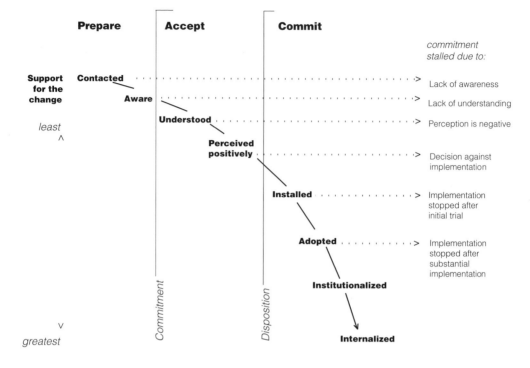

Figure 13.1 Stages of change commitment—Conner (1992:148) finds that support for change can grow while a person is exposed to it, as long as its proponents for the change respect what steps are necessary to allow for others to prepare for it, accept it and commit to it. Change starts with contact and progresses toward internalization. Success leads upward to the next stage. Failure at any step can retard or end the process by losing some or all support for the change.

Source: Adapted from *Managing at the Speed of Change.* Conner, D.

13.5.1 Preparation

At the outset, it is up to those who are in favor of a change to enlist support for it. This starts with making contact with audiences of interest and building awareness. The audience has to agree that there is a need for change. At this early stage in the process, the degree of support for the change is low. Failure to accomplish these steps will leave audiences of interest unaware of the change. Without this needed orientation, they will be reluctant to accept it.

13.5.2 Acceptance

After gaining commitment to their change, proponents need to build acceptance for their concept. This relies on the cultivation of understanding and a perception that the concept will result in a positive outcome if it is adopted. By fostering understanding, proponents can build a coalition of supporters that will make the change possible. Failure to build adequate understanding will leave room for confusion over the nature of what is proposed. Negative perceptions may result from skepticism or information that is contrary to the proponents' case.

13.5.3 Commitment

With the organization positively disposed to the change, it is up to the proponents to get resources allocated. This final phase of resource investment is the commitment necessary to make the change a reality. The amount of resources that are mobilized and the breadth of implementation depend on increasing support for the change through more and more members in the organization. With enough support over a long enough period of time, the change becomes part of the institution and its culture.

Failure to build commitment may result in a positive disposition regarding the change but a choice not to implement it. Without sufficient commitment, a change that is implemented may be cut short.

13.6 The change process

Conner (1992:78) has found that 'change is perceived to be positive when you feel in control by being able to accurately anticipate events and influence your immediate environment (or at least prepare for the consequences).' Individuals seek to control the future, because predicting the future helps to reduce the discomfort of uncertainty. Setting expectations, demonstrating the advantages of a new approach and showing how a new approach provides realizable benefits builds a path from the current to the new situation. Using Conner's concept, such a strategy shows that the future situation (like the present) will also be controllable. Failure to provide persuasive evidence in that regard may lead others to choose not to increase their level of commitment to the idea.

Each person in and outside of the organization with a stake in the outcome (stakeholder) must agree on how to proceed on a new project. That agreement will rest in good part on how each decision maker views his or her personal and departmental agendas within the organization. It is necessary to identify those agendas, demonstrate how the concept meets those requirements to the satisfaction of each participant in the decision making process. Agendas are usually hidden and need to be deduced. The following process provides some insight into how to elicit agendas and use such information to the benefit of the organization: prepare, craft a strategy, build a constituency and follow through.

13.6.1 Prepare

Determine the interests, motivation and satisfaction of each group that is expected to participate in the approval process.

Understand yourself. What do you and your department have to offer in terms of resources?

Understand others and what matters to them. Find out what they need. If what each person needs is not obvious to each stakeholder, you have the option to:

- Adjust your proposal to include it
- Reframe your presentation to demonstrate that it is actually there
- Disregard it, knowing that one runs the risk of losing the support of that constituent.

Look for behaviors that indicate others are comfortable with the change. Continue working with them until those signs appear. Determine biases and presumptions through informal means. Demonstrate aptitude in order to build credibility. Introduce the concept simply and clearly. Invite questions and welcome the responses, even though they may be challenges (e.g. 'Why is this even necessary?'). Each response is an individual's way to engage and learn about the new information. Listen intently to the responses, restate them in your own words and ask if the summation is accurate.

13.6.2 Devise a strategy

Identify the elements are that are essential to the concept. What aspects would be good to have, but can be eliminated in a crunch? To form a consensus, nearly every participant needs to give up some aspect of what was originally intended. Knowing what, where and how to change the concept through its development will enlist others' support for it.

Build a base of knowledge in advance in order to counter potential objections. Cost justification is one of the more difficult challenges in research and development. Tie the benefit of what is proposed to the firm's mission and its profitability.

For example, usability evaluation is one of the human factors activities that can suffer from the perception that it is simply 'research for the sake of research.' When proposing a usability program, find information that can be connected to the objectives for human factors work described in Chapter 1. Such information may include the cost of user dissatisfaction or difficulty (e.g. cost of production, cost of customer returns, cost of accidents). Use the proposed change to demonstrate how the research will target these costs to the organization. Rather than cutting product costs, new development initiatives seek to add value to the product. Demonstrate how the research effort will target areas that will build perceived value among customers.

13.6.3 Build a constituency

Cappy and Anthony (1999:199–215) recommend that change leaders encourage tolerance among others for solutions that are less than comprehensive (the Pareto solution described in Section 6.1). This tolerance is justifiable in the case of solutions that have 'the support and conviction of their most trusted people,' including the middle managers and supervisors who shape and implement most change agendas.

Expose the concept to those who are important to its future. Invite constructive criticism (e.g. ask 'Are there aspects that you would want changed?' 'What would have to be different for you to feel this is the best way to go?'). Take the other's point of view to see your concept through his or her eyes. Clarify expectations to spell out the benefits and the true costs of implementing the new change.

Use time before meetings to prepare participants. Invite comments and challenges. Account for and address any objections. Use meetings as a forum to demonstrate consensus.

13.6.4 Follow through

Review all of what you learn. Incorporate each aspect that will contribute to project goals, even if not the way you or your team originally conceived it. Reinforce others' contribution to your work. Informal thanks and noting where the original concept reflects new thinking contributed by others reinforces trust and improves your result. It also builds commitment among those whose support matters to the implementation of your work.

13.7 Example

The following scenario demonstrates how a number of elements in the foregoing chapter can be applied to the practice of human factors.

Apogee is the successor to A–L Electrical, a long-time manufacturer of electrical components. A–L has recently expanded into the production of personal communication and data products by purchasing the entrepreneurial start-up, E-World. The culture at Apogee reflects A–L's decades of business-to-business production and sales of engineered products. Because of this legacy, user orientation and other aspects of finished good are taken more on faith rather than understood or embraced.

'Tanager' is the code name for Apogee's newest product initiative, a 'make or break' effort to garner significant market share in the personal data communications market. Uday Patel, Kay Bennett, Glen Pinkus and Harume Sato comprise Apogee's human factors group. Department leader Sato has called a meeting to map out the group's strategy for Tanager, which she sees as an opportunity to advance the role of human factors at Apogee.

'Okay, thanks to each of you for drafting your sections of the project proposal,' she starts. 'Uday, I like your section on standards and requirements. Kay, your section on data collection and analysis should be an easy sell. Glen, you've got a good section on evaluation. I'm a bit concerned about the budget cap that we have to fit under, though. Evaluation might get cut unless we can make a strong enough case for it with senior management. What do you think?'

'I agree, because evaluation has been cut from Eagle, Warbler and Finch. Three projects this past year,' Glen said. 'Tanager is too big a risk if we release the product without knowing how the final version actually works when users operate it.'

Kay added, 'The issue is the Simple System concept. If it works with users, it'll be a huge leap forward. If it doesn't work, the cost and time to redesign that interface will do us in.'

'Uh-huh. You know, *that* is the message that we'll need to impress on senior management,' Harume added.

'Bill Phillips in particular,' Uday added. 'When I was in the A–L Research Group, I saw that he was an opinion leader, especially on budget issues. I think if we've got Phillips' support, then we'll be able to get the budget that Tanager deserves.'

'Good point, Uday,' Harume commented. 'Thanks for that. I'll schedule a meeting with Bill as soon as I can.'

A week later, Phillips and Sato are sitting in the passenger lounge of an airport terminal as each waits for their flight.

'Bill, Apogee's future depends on Tanager's success,' Harume said.

'No doubt about that,' he said, as his fingers fluttered over the keys on his laptop computer. 'What's your point?'

'I want to know if human factors has your support for the budget proposal that I sent last week for your review,' she said.

'I'm sure we'll fund human factors,' he said, looking up. 'With all of the software development that Tanager will take, I'm not sure we'll have all of the dollars that you say you need.'

'Bill, this is the one project that will fly or die based on the way users understand the Simple System concept,' she urged.

'Okay,' he said, raising his eyebrows. 'And so … ?'

'So, Tanager's success lies in how well we can do front end research and prototype evaluation,' she said.

Phillips frowned. 'Maybe, maybe not. I'm more concerned with the interface programmers and this FailSafe feature we're developing. Simple System should be easy enough that users will just figure it out … like people did with that graphical user interface Apple did in the '80s.'

'Bill,' Harume countered, 'Tanager is a complex product that we need to be simple to use. We won't know for sure that it really *is* easy to use unless we do adequate advance work *and* prototype evaluation. What will it take to get your support for the project budget for the scope and funding that we proposed?'

He looked down the concourse for a moment as the speaker overhead announced his flight was starting to board. 'Look,' he said, 'I'm not so sure that we need to test something that's supposed to be inherently simple to use. But I know that if Tanager stumbles we're going to fumble our chance at this market. So, I'll give you the benefit of the doubt. If you can trim your numbers … by scaling back on your sample size, for example … I'd back you up.'

Harume breathed a sigh of relief to herself as Phillips stood up to board his flight. 'Thanks, Bill,' she said. 'I'm sure we can work something out that you'll be satisfied with.'

Four months afterward, the human factors team is meeting again, this time to present the results of their research.

'Well,' Harume started, 'I want you to know how much I appreciate the hard work and long hours that each of you have put into this project so far.'

'Loooooong hours,' Kay smirked.

'Yes, and for good reason,' Harume replied. 'The design group really listened to our insights from the observation studies. Simple System's design is a lot better than it would have been otherwise. I got Maria Battista to make that point to Bill Phillips directly. So, he already knows that her team got more than they expected.'

'What about the design approval meeting, though?' Glen asked. 'You know that the design group is really big on those dumb symbols and the data show that users can't figure out what they mean.'

'Come on, Glen,' Kay interrupted. 'You know that the mean error rate for three of them was higher than what we expected. But none of them came close to causing a failure.'

'The data don't lie,' Glen scowled.

Harume cut in. 'This may be an issue in our presentation. Let's pick our battles. Remember Bill Phillips? We got our first budget and an add-on later when we showed him that we could be flexible without compromising our program. Glen, you're right about the data. But we can live with the mean error rate for those symbols. You know it's still within the limits that we set early in the project, right?'

Glen sputtered, 'Yes, but …'

'Bear with me on this one,' Harume said. 'We need to focus on bigger issues. We've proven our benefit and we'll be able to show that at the design review. I've already sounded out Marty Slatkin in Engineering and Maria in Design. So I know they're already fans of ours. Now that we've shown what we can do and have the key player support that we need, I want to raise the stakes.'

'Raise what?' Uday asked.

'I want to go for an additional phase of human factors work that we didn't ask for before,' Harume said. 'User validation.'

'We already did prototype evaluation,' Glen said.

Harume winked and said 'Yes, and we still need to show that the actual product does what we expected *and* how customers use the product outside of what we learned through Kay's focus groups, your observational studies, and Uday's data analysis.'

'Isn't that kind of pushing our luck?' Kay asked.

'Sure it is,' Harume replied. 'It's a gamble. But it's worth it if we're going to build our role here. Validation will draw us much closer to Sales and to Customer Service, which are both politically strong departments. It'll show what we need to know for Tanager's follow-on products. And, it will build a knowledge base for future personal data products that we don't have now.'

'What's your plan?' Uday asked.

'First, I'm going to talk with Phillips again in order to see what he thinks,' she said. 'And, I'll talk to Gail in Finance to see how Tanager's budget is doing. Uday and Glen, I'd like you to draft a rough proposal in case somebody wants to see what we have in mind. Run it past me when you're done.'

Kay jumped in, saying 'I've got a plan that I wrote for E-World two years ago that should start us in the right direction.'

'Great,' Harume said. 'If you get that to Uday and Glen, that would give them a good start. Then, please give them a hand with editing their rough. I'll let you know what I find out from Phillips and Gail by the end of this … no, the end of next week. Thanks for your time. We're on the way, folks,' she said, smiling, as she turned for the conference room door.

13.8 Summary

The role of research and development is to bring new knowledge into an organization. That makes it, by its nature, a radical venture; 'radical,' in the sense that it can affect the root of an organization's reason for being. Rosabeth Moss Kantor cautions that 'Innovations are not safe, bounded, or easy.' Innovative accomplishments are perceived as riskier, more controversial, requiring greater dollar investment, working across boundaries, getting peer support, benefit from higher management sponsorship, are more complex, taking more time and having larger dollar payoffs. Setting expectations, demonstrating the advantages of a new approach and showing how a new approach provides realizable benefits will build a path from the current situation to the new situation.

Trust, vision, awareness and empathy in working relationships can greatly improve how others perceive change. Others' willingness to have faith in new ideas determines whether they will rely on whomever works to bring them about. Each audience that considers new thinking will need cues regarding what is to come in order to envision what the future may look like. Ensure that each aspect of the change is in the front of the audience's mind. Consider the implications for others and other departments from their point of view. Stakeholders (each person in and outside of the organization with a stake in the outcome) must agree on how to proceed on a new project. To elicit information on agendas that stakeholders have and use it to the benefit of the organization: prepare, devise a strategy, build a constituency and follow through.

14 Communication

How do I make a case on behalf of our idea?
Purpose (14.1)

How do I explain our project to others?
Nature of research and development communications (14.2)

What are the ways that an organization communicates about research and development issues, needs, activity and results?
Types of human factors communications (14.3)

However powerful research findings are, the data alone will not result in support for a project. Insights, information, findings and more need to be forged into messages that generate understanding and support among key audiences. Organizations rely on written, visual and spoken communication to accomplish that. Communications convey the spirit and content of a message from an individual to a group, from a team to a firm, from one firm to other firms.

Chapter 13 described the need to establish effective relationships with those who have a stake in the outcome of the project. Effective communications are the *result* of those relationships as well as a contributing factor in building them. For example, a proposal reflects an understanding of what an organization needs based on personal contact with members of an organization. A plan reflects an agreement among those who are involved in accomplishing the work that is described. A report reflects the results of the efforts of those who have worked together.

The growth of remotely located work teams and telecommuting makes good communication even more necessary. How a project is perceived can have much to do with how it is presented. Good communication can convey insights that might otherwise go unnoticed. It can amount to a collective body of knowledge that provides an objective account of what is known about the problem and its possible or intended solutions. It can be used to clarify expectations, accomplishments, needs and intentions.

Even if the information itself is of high quality, poor communication can change the way it is perceived. Poorly expressed ideas can result in confusion, misperception, friction among colleagues and departments, inability to understand a problem and insufficient appreciation of an idea's value. Time and funding may be under-utilized or misused. Opportunities can be lost.

This chapter outlines the types of communication a research and development team can use, particularly when dealing with issues of human factors and research and development.

> *Without credible communication, and a lot of it, employees' hearts and minds are never captured.*
>
> (J. Kotter)

14.1 Purpose

Chapanis (1991:1–4) cautions 'The only kind of research that qualifies as human factors research is research that is undertaken with the aim of contributing to the solution of some design problem.' Compose and edit human factors writing with that purpose in mind. If the material does not show how the research contributes to the solution of a design problem, it does not belong.

Communications has become one of the hallmarks that separates average from superior technical professionals.

> Average performers think Show-and-Tell means getting noticed by upper management through slick presentations, long-winded memos, and public displays of affection for their own work. They focus primarily on their image and their message, not on the audience. Star producers use a series of skills involving selecting which information to pass on to which others and developing the most effective, user-friendly format for reaching and persuading a specific audience. At its highest level, Show-and-Tell involves selecting either the right message for a particular audience or the right audience for the particular message.
>
> (Kelley 1999:51–58)

The goal in communication is to develop understanding. Communication on technical topics, including human factors, can overemphasize detail and process description at the expense of understanding. Instead, effective technical communication should emphasize key intentions, insights or findings, then use detail to substantiate the main message.

Human factors communications are most often written to inform or persuade. Informative communication provides information that the audience would not otherwise know. It is created to answer a question that has to do with a design problem. Reports and presentations are the most common types of informative writing. Persuasive communication seeks support for an idea or project. A proposal sets forth a concept and solicits resources such as funding in order to accomplish it. Whether informative or persuasive, the communications must address the key question in the reader's or audience's mind: What will the change do for *me*?'

14.2 Nature of research and development communications

Six aspects of communications need to be taken into account when writing for research and development: tentativeness, audience and results orientation, method, style, organization and media.

14.2.1 Tentativeness

In reality research and development work is never done. There is always new information that can be evaluated and applied to understanding the problem. There are always additional development solutions that can be explored. Sooner or later, though, projects need to be brought to closure in order to deliver a product to the marketplace. This closure may need to occur before the final and conclusive data can be collected and accounted for. As a result, the research and development team may need to produce a set of tentative conclusions. To the traditional scientist, this can appear to be incomplete and even unduly hasty. This trade-off between timeliness and completeness is inherent in the research and development process. It can be managed by qualifying results so that the audience understands the basis that the team has used for its findings or recommendations. Another approach is to partition current work from that which is planned for further research.

Windsor (1996:32–9, 98–100) found that engineers tend to think of good work as being free from the engineer's preconceptions or wishful thinking. The engineers consider the 'primacy and purity of data are an ethical as well as a functional concern' and that the task of persuasion should be left to managers. This presumption can cause a lack of understanding among those who are the audience for technical communications but are not engineers. It can also give the impression that those technical professionals who present the information do not believe in it strongly. Windsor cautions that data that are presented but not explained do not persuade. Instead, the writer or presenter must make a persuasive argument that the data substantiate. Without such an argument, others in the audience will be free to make their own decisions on the meaning and use of the work. Their conclusions and the decisions that result may be very different from those that the writer or presenter sought.

14.2.2 Audience and results orientation

Develop a clear idea of the audience members who will read or hear the message. As Chapter 13 recommends, use the needs and interests of your audience to formulate how to build the message. For example, peers in research and development may focus strongly on methods and data. Senior management, though, may have more of an interest in long term customer trends. Decide on the end result the communication is intended to achieve (e.g. 'approval to fund the staff and equipment for data collection'). Include only the information that helps to achieve that result.

Anticipate any objections that might be raised. Think through a reasoned response and incorporate it in your communication. If a certain aspect will be 'a challenge' or 'may be difficult to accomplish,' be willing to acknowledge it. Follow it with a clear statement of how the team intends to achieve the best result under the conditions described.

Try the results out on a friendly critic such as someone who is a member of the intended audience and who knows the team. If possible, have someone who knows the audience well (e.g. a vice-president who has worked with the Chief Executive Officer) to review the communication before it is submitted or presented. Such guidance will likely help to adjust the material so that it will be as effective as possible for your intended audience.

14.2.3 Method

Rough drafts are a useful method to develop the understanding and agreement that is necessary for the consensus that Chapter 13 describes. The circulation of a rough draft invites reflection and discussion. It stimulates new thinking and encourages insights. It can also be used to discover questions and issues that can be resolved before the final draft is produced.

14.2.4 Style

Much information is available on how to write and that will not be reproduced here. There are a few considerations, though, that should be kept in mind. Table 14.1 presents key elements that Kotter (1996:9) recommends in order to effectively communicate a vision of change. To be effective, the message needs to be concise. Present the idea in a compelling and direct way, using as few words as needed to make a point. Kotter offers 'A useful rule of thumb: Whenever you cannot describe the vision driving a change initiative in five minutes or less and get a reaction that signifies both understanding and interest, you are in for trouble.'

Table 14.1 Key elements in effective communication of vision—Kotter (1996:9) provides seven considerations to take into account in order to effectively communicate a vision of change.

Simplicity	All jargon and technobabble must be eliminated
Metaphor, analogy, and example	A verbal picture is worth a thousand words
Multiple forums	Spread the word through meetings (both large and small), memos and newspapers, interaction (both formal and informal)
Repetition	Ideas sink in only after they have been heard many times
Leadership by example	Behavior from important people that is inconsistent with the vision overwhelms other forms of communication
Explanation of seeming inconsistencies	Unaddressed inconsistencies undermine the credibility of all communication
Give-and-take	Two-way communication is always more powerful than one-way communication

Visualization, the use of images to convey information, is an effective way to communicate efficiently. Images can include diagrams (e.g. charts, graphs, maps) and sketches, drawings, models and photos that are created by hand or by using computer supported media. Images used in coordination with words are a powerful means to show the big picture by making patterns and trends evident. Be careful with complexity, though. The goal is to convey an understandable message, not to impress with the amount of data that can be portrayed in one frame (Chapanis 1991:1–4).

14.2.5 Organization

Follow a pattern to lead the audience through the message, from background orientation, through need or purpose, to objectives, discussion and the conclusion. Address the facts, interpret the facts and relate both to the problem. In so doing, you will lead the audience to reach the same inevitable conclusion that you have reached—and they will think they have figured it out on their own.

The six types of communications that are discussed in Section 14.3 follow this general pattern of organization.

14.2.6 Media

The expansion of electronic communication has saved time, effort and expense. Their convenience has come at some expense to understanding by encouraging habits that impede rather than enhance messages. For example, text editing software and presentation software programs can help authors to organize a report or presentation. Visual features (e.g. animation) are intended to add entertainment value to presentations. Over-reliance on such tools, though, can result in stilted reports and distracting presentations. These tools should serve merely as aids to original thinking. Ensure that the audience understands what the author has to say. Choose media that improve and refine the message.[1]

14.3 Types of human factors communications

Six types of communication, summarized in Table 14.2, are most frequently used to convey information having to do with human factors research and development information: specifications, proposals, briefs, plans, reports and presentations.

14.3.1 Specifications

Lannon (1997:411–426) defines specifications as 'a particularly exacting type of description' that prescribes 'standards for performance safety and quality.' Governmental agencies such as the U.S. Consumer Product Safety Commission and the U.S. Department of Defense issue specifications in order to ensure safety and reliability. Military Standards are specifications that set requirements for the development of all equipment for use by those who are in the military. In particular, the 'Human Engineering Design Criteria for Military Systems, Equipment and Facilities' (MIL-STD-1472D) (1989) sets human factors requirements for both general (e.g. safety, ruggedness) and detailed (e.g. control display movement ratio) aspects of equipment. In industry, specifications are developed as a means to ensure essential needs for safety, reliability and quality are met. Lannon notes that specifications must be easily understood by the widest possible range of readers to gain agreement on what is to be done and how it is to be done. Readers can include customers, the designer, contractor or manufacturer, supplier, workforce, inspectors and (potentially) a judge and jurors. Use impartial language that is precise and informative to describe physical traits, the steps necessary to use the product and how the product is to be made.

An architect or contractor will avow that a good specification is no substitute for good management. On-site inspection of activities makes sure that what is being done will produce a result as it was specified. Catching an aspect of a project that is about to be completed incorrectly is far more effective than trying to argue who was right in court afterward.

Table 14.2 Types of human factors communications

Specification	Description that prescribes exacting standards for performance, safety and quality. Developed as a means to ensure essential needs for safety, reliability and quality are met
Proposal	Recommendation on how to do something. Seeks to persuade the reader to follow it. Can nominate an idea for consideration. Takes three forms: Planning—Suggests ways to solve a problem or bring about an improvement Research—Requests approval, funding, for research Sales—Offers a service or product
Brief	Statement of essential information. Description of project terms. Sets the nature of a client-team or client-consultant relationship
Plan	Sets objectives and provides resources to meet them
Report	Detailed written account Periodic—'Progress report'. Made to convey accomplishments Completion—Made at the end of activity. Takes two forms: • Informational—Informs readers, reports data, or documents a particular problem • Analytic—Recommends decisions or action, evaluates alternative feasibility, and reports results
Presentations	In-person reports made live, via video or audio conference, or video or audio recording

14.3.2 Proposals

A proposal is a recommendation on how to do something and seeks to persuade the reader to take the advice. Firms or government agencies will invite proposals by issuing a request for proposals (RFP) that is made public by mail and publication in periodicals such as *Commerce Business Daily*. RFPs mimic the same structure as a specification by inviting respondents to address how requirements for the product will be met. Proposals can also nominate an idea for consideration without a prior request. Proposals can originate from outside of an organization (e.g. when written by a consultant) or within an organization (e.g. when written by a colleague from another department).

14.3.2.1 TYPES OF PROPOSALS

Lannon (1997:522–41, 553–70) classifies proposals according to the intention in writing them: planning, research and sales. Planning and research proposals are most often used in research and development.

14.3.2.1.1 Planning—Planning proposals suggest ways to solve a problem or bring about an improvement. The planning proposal states a purpose, identifies a problem, proposes a solution, details what will be done and how, sets expectations and encourages a response from the reader.

14.3.2.1.2 Research—Research proposals are developed to request approval and often to request funding, for a research project. The research proposal provides background to set the stage, describes the problem or need, defines the scope of the study and how it will be conducted, describes what will be produced ('deliverables'), provides team membership and qualifications and recommends a schedule and structure for fees and expenses. Sims (1998:326–44, 374–89) recommends that the summary be written as a condensed version of the proposal because readers will look at the summary first to decide whether or not to consider it.

14.3.2.1.3 Sales—Sales proposals are written to offer a service or product.

14.3.2.2 PROPOSAL TRAITS

Recall that Chapter 1 described the role that problem definition plays in problem solution. The description of need or problem in the proposal is the problem definition step applied to a particular instance. The reader's confidence in the proposal can hinge on how insightful the need or problem description is.

The description of project activity is usually divided into phases with clearly defined deliverables. Sims recommends including a detailed description of what will be done, as well as what will not be done. Dividing the scope of work into phases has practical merit. When a phase has been completed, results can be evaluated and fees and expenses billed.

Proposals often serve as the basis for a contract between contractor and a firm or agency. A firm may use project proposals to think through how a project will be handled. This provides an opportunity for the proposal author to paint a picture of what should be done and how it should be done. The clearer the description, the easier it will be to understand what to expect. Ambiguous or vague descriptions invite others to read into what is said and can result in dissatisfaction later on.

14.3.3 Briefs

A brief is a statement of essential information that provides a starting point for a project by describing its terms. It sets the nature of a relationship between those who have a need (typically, a manager) and those who will satisfy it (typically, members of a development team). A brief often includes company background, project background (including user and technology information), a problem statement, objectives and criteria, resources, references, deliverables and a schedule.

Briefs for research and development projects require a deft touch. Give enough information and guidance to be substantial. At the same time, allow enough room for interpretation so that initiative and creative latitude are allowed to flourish.

14.3.4 Plans

Plans are used to set objectives and to provide the resources to meet them. While planning proposals (Section 14.3.2.1.1) attempt to persuade others how to solve a problem, plans are a statement of intention by those who have already made the decision to proceed.

Chapter 1 described the use of plans at higher levels in an organization. Management uses business plans to guide a company. Product plans are developed to guide a research and development organization to enact the business plan.

Plans can also be used to guide projects or parts of projects. A test plan, for example, describes the reason for a test, standards by which the test will be conducted, who will conduct the test, how it will be funded and how the test results should be reported.

14.3.5 Reports

Any report is a detailed written account. Two types of reports are of particular concern to research and development: periodic and completion.

14.3.5.1 PERIODIC

Periodic ('progress') reports are made to convey accomplishments. Increased use of telecommuting and working in distributed locations makes periodic progress reports a necessary tool for managers, employees and contractors.

Periodic progress or activity reports can be provided on a routine basis for a number of purposes, according to Windsor (1996:32–9, 98–100). Managers use progress reports to keep track of projects. Writers use them as a form of record keeping, to report on current activities and to plan future activity.

Periodic reports are not known as an exciting form of writing. They do offer an opportunity, though. When drafting a periodic report, Windsor recommends that the writer reflect on the significance of the work that has been accomplished. That can enrich the information in a report by adding a persuasive character to what might otherwise be dull material.

14.3.5.2 COMPLETION

Completion reports are made at the end of activity. Sims (1998:326–44, 374–89) describes two types of reports that are produced at the conclusion of activity: informational and analytic.

14.3.5.2.1 Informational—Informational reports 'inform readers, report data or document a particular problem of situation.'

Often, analysis is added in order to evaluate the provided information. Three types of reports combine both information and analysis. Progress reports contain information on completed work and may evaluate progress to date as well as recommend budget or schedule changes that may be necessary. Lab reports account for observations, such as human factors research activity and may also interpret what was observed. Personnel reports provide an account of employee performance and will evaluate prospects for advancement or promotion.

14.3.5.2.2 Analytic—Analytic reports go beyond the simple transmittal of information by analyzing information and data and interpreting them.

Analytic reports contain an executive summary (a one page summary), title page, table of contents, the basis for the report (e.g. citation of a contract or proposal authorization), introduction to provide problem background, methodology the investigator followed, terms definitions and the terms and scope of the study. Findings and interpretation of the findings follow. A conclusion is used to summarize and interpret what has been presented and to make recommendations. A list of references is included to show publications that were consulted to develop the report. Raw data and supporting materials are included in appendices. There are three types of analytic reports: recommendation, feasibility and research reports.

Recommendation—Recommendation reports pose questions, present, evaluate and interpret findings that were collected in research activity and recommend decisions or actions that will solve a problem or create an opportunity. In such reports it is important to interpret the information so that the reader knows what it means.

Feasibility—Feasibility reports evaluate alternatives on the basis of decision criteria (using the same approach as decision analysis in Chapter 7).

Research—Research reports are used to 'record and interpret the work the writer did to design a product or test a hypothesis.'

14.3.6 Presentations

Presentations are in-person reports that can be made live, via video or audio conference or video or audio recording. Formality can range from informal reviews among team members to formal management reviews and professional conferences. Presentations are the only formal communication described here that rely on the dynamics of human behavior. This offers the opportunity for using physical presence to dramatize one's message.

Structure is not as evident in spoken reports as it is in written reports. Riordan and Pauley (1996:439–49) recommend that presenters choose one of four principle formats in order to effectively organize what is presented: problem-solution, chronological, order of importance and spatial-order.

- *Problem-solution*—Relate problem context and nature to need and solution direction
- *Chronological*—Describe the evolution of what was done through time
- *Order of importance*—Identify key features or issues that the project addressed
- *Spatial order*—Use two- or three-dimensional space to lead the audience in a direction through a setting, such as inside to outside an object, facility or information structure.

Most audiences are willing to pay close attention to a presenter for about twenty to thirty minutes. Edit material to fit within that length of time. Allow time afterward to answer questions and address specific concerns. Watch audience members for signs of disinterest. Fidgeting and restlessness are non-verbal cues from members of the audience that the presenter should move on to new material.

Demonstrate personal conviction by showing a relaxed, forthright demeanor in order to build an audience's confidence in what is being presented. Two techniques will help to manage the anxiety that comes with such performances: review and rehearsal. A presenter who knows the material thoroughly will be less anxious about it. Giving the presentation to team members will help to sort through how to discuss the material. Rehearsal will also develop a sense of how an audience will react to what is said.

Use audio-visual materials (e.g. a videotape of the observation process or of the proposed solution being used in the field) in order to expand what is communicated beyond simple verbal description. Be careful to keep visual aids simple. If it is necessary to present complex data, it is better to split the information into a series of easily understood images than packing all of the data into one dense image.

After the presentation is over, provide a handout that audience members can take with them. It should *not* be hard copies of the presentation slides. Rather, it should be a two to four page summary of the presentation's main points and recommendations along with information on the presenter's qualifications and how to get in contact.

14.4 Summary

How a project is perceived can have much to do with how it is presented. Material that is included in a report or presentation needs to show how the research contributes to the solution of a design problem. Each communication should in one way or another provide an answer to the question: 'What will the proposed project do *for* each reader, or each member of the audience?'

Six aspects of communications need to be taken into account when writing for research and development: tentativeness, audience and results orientation, method, style, organization and media. Six types of communication are most frequently used to convey information having to do with human factors research and development information: specifications, proposals, briefs, plans, reports and presentations.

15 Examples

How do I use the human factors research methods in a project?
What results can I expect to get using human factors research methods?

Ultimately, the variety of methods in Part II needs to be put to practical use. This chapter provides examples for projects that cover each application area mentioned in the Preface.

Human Factors Methods for Design is intended for working professionals, scholars, undergraduate and graduate students in fields that involve people, including: architecture (including space planning), product ('industrial') design, communications ('graphic') design including multimedia, environmental design, human factors (ergonomics), roles involved in the design and development of machines and of computers (including interface development), systems engineering and engineering management.

The first example, bus workstation, shows the flow of activity in a straightforward human factors project. The example shows how to develop a problem statement and criteria, audit regulations, use basic anthropometric data, then formulate and evaluate solution recommendations. It also shows the importance of creating relationships with various project stakeholders.

The rest of the examples in the chapter are written in the general style of proposals. Each identifies a project, provides background and need, describes a scope of work that includes human factors research methods that would be employed and describes deliverables, which would be provided.

15.1 Bus workstation

The Chicago Transit Authority's plan to implement value-added fare cards across its fleet of 2,500 busses promised to save currency handling costs, provide fare flexibility, simplify fare transactions and improve customer service and operator safety. Installation of a farecard scanner in each bus required human factors planning and evaluation to ensure the new technology would benefit driver and rider alike. A 1995 study developed objectives, reviewed regulations, performed field audits, developed driver and passenger reach envelopes and assessed implications for unit installation. Following a successful pilot program, further research is focusing on prospective hazards, information needs and future bus workstation requirements.

15.1.1 Need

The Chicago Transit Authority (CTA) needs to become more efficient and flexible in its fare collection. The CTA intends to install a fare card reader throughout its system. Where should the reader be installed aboard its fleet of 2,500 busses to comply with federal regulations and meet the needs of drivers, maintainers and riders? There is a need to establish standards for installation that are based on human factors knowledge.

15.1.2 Background

The Chicago Transit Authority maintains a current fleet of 2,500 busses and hundreds of transit trains to serve much of the daily mass transportation needs of the Chicago, IL metropolitan area. In each bus, the driver must operate the vehicle and manage transactions with passengers from a single position, creating what amounts to a workstation, which extends from inside the windshield past the threshold leading to passenger seating.

The CTA planned to introduce value-added scannable fare cards in 1995 to minimize the cost of handling currency, improve fare flexibility, simplify fare transactions and improve customer service and operator safety. Figure 11.1 illustrated the bus ticket processing unit (BTPU) that was to be mounted in the front of each bus to collect fares. Passengers would be required to insert their card in a scanner port, which would debit a fare based on route, time of day and special (e.g. senior) rates. The driver would need to press any of three button controls. Both driver and rider would need to see either of two LED displays.

BTPU installation would affect thousands of Chicago area riders each day. In addition to customer satisfaction, the unit's installation would also influence operator safety, performance and comfort. Where and how the unit should be installed amounted to over a million-dollar question. Improper installation might incur penalties well above that amount if it:

- Ran afoul of federal regulations on accessibility for passengers with disabilities;
- Induced bus operator repetitive motion injuries, risking both time lost from work and claims which might be made against the CTA;
- Created a confusing and inefficient flow of transactions while passengers board the bus, costing customer goodwill and revenue.

There was a need to develop a plan for BTPU installation which was suited to the needs of the CTA, its employees and passengers, as well as regulation requirements. Two conditions complicated the problem. As a service provider, the CTA could only set the specification for manufactured components. The BTPU (already in the operating prototype stage) needed to be installed in a manner which provided least interference to bus operation and passenger activity. In addition, the project was a retrofit. The fleet had nine bus models and none had been designed with the addition of such equipment in mind.

In February 1995, the CTA retained a human factors consultant to create a plan for BTPU installation which was research-based. The intention was to ensure BTPU installation achieved intended benefits without compromising safety and efficiency. It would also quell the rising tide of conflicting subjective opinions, which was slowing acceptance within the CTA. The plan and project activities were organized into three phases: research, solution development and implementation. Table 15.1 summarizes the project human factors research activity.

15.1.3 Research

Project activity began with a review of current ADA regulations, observation of typical bus fare box use in the field, a review of recent CTA issues regarding unit installation requirements and an analysis of research findings in order to develop objectives, which could provide a basis to make trade-off decisions. Six objectives resulted:

Passenger safety—Minimize the opportunity for loss or damage to the person or property of CTA riders.

Table 15.1 Bus workstation—The Chicago Transit Authority intended to install a fare card reader throughout its system. Where should the reader be installed aboard its fleet of 2,500 busses to comply with federal regulations and meet driver, maintainer and rider needs?

Practices with an interest

The following practices would have an interest in this research:

Product design

Communications design

Environmental design

Human factors/ergonomics

Roles involved in the design and development of machines

Systems engineering, engineering management

Activities

Meet regularly with CTA departmental management and project staff

Determine objective and criteria

Observational study

Operational analysis—Determine system requirements

Audit each bus model configuration, reach and sight lines. Make video record to complement notes.

Evaluation study

Simulation—Develop scale illustrations for each bus model. Determine reach radii for 5th percentile female. Install plan view driver and user templates in each scale bus model illustration. Determine if bus operator visibility will be compromised, if bus operator and passengers will be able to see and reach it and if wheelchair access to the bus will be impeded

Workload assessment—Informal review with drivers to determine effect of BTPU location on bus operation, passenger loading and unloading and fare transactions

Meet with union representatives to solicit comments and demonstrate concern for driver safety and work preferences

Present report to representatives of participating CTA departments to solicit support for the solution recommendation

Deliverables

Video recordings demonstrating sight lines from driver and rider points of view

Study report including scale illustrations of bus workstations including driver and passenger reach envelopes

---

- Must not impede the entry/exit of passengers, including those with disabilities.
- Must present no opportunity for impact injury or snagging.

Operator/driver safety—Minimize the opportunity for injury through cumulative or impact trauma. Minimize the obstruction of visibility through the bus windshield.

- Must not require bus operator reach beyond normal seated arm extension.
- Should not increase obstruction to the field of view through the bus windshield.

Equipment ease of use—Enable drivers and riders to understand and use the BTPU with a minimum of difficulty.

- Should accommodate the reach and vision envelopes of passengers from smallest (5th percentile) female to largest (95th percentile) male, including those with disabilities.

- Should accommodate the reach and vision envelopes from smallest (5th percentile, female to largest (95th percentile) male bus operators.

Ease of installation and maintenance—Present no significant requirements beyond those already needed to install and maintain current CTA fare boxes.

- Should require no procedures beyond those already necessary for fare box installation and maintenance.

Cost of installation and maintenance—Minimize the costs incurred in installing the BTPU and the time needed to dismount/remove and reinstall the BTPU.

- Should incur the minimum cost to dismount/extract and reinstall/mount the BTPU.

Efficiency—Assist bus operator and rider to quickly board bus while collecting fares.

- Should incur no additional delay in fare collection, processing.

The need both for direct bus driver control over the unit and for passengers to use the unit while boarding led to mounting the BTPU near the fare box, which was already installed in each bus. The need for the BTPU and fare box to share power and control cabling led to mounting the BTPU on the fare box stand. Questions still remained:

- Would bus operator visibility be compromised?
- Would the bus operator and passengers be able to see and reach it?

15.1.4 Solution development

The second phase began with an audit of the implementation sites: the nine bus models in the fleet. Each bus manufacturer had originally provided small-scale plan view illustrations of each model to the CTA. Yet, the generic illustrations were inadequate for equipment specification. Plan orthographic line drawings were created for workstations of each of five current and two incoming, CTA bus models. Field visits were required to correct errors in the manufacturer-provided illustrations, to verify equipment location in each bus operator workstation and reflect configurations unique to the CTA fleet. During the visits, a videotape record was made of equipment variations and driver visibility limits for each model. Each of the models was then reviewed with regard to driver and passenger needs for fare box use.

Driver—Bus operator reach envelopes were developed, based on human factors anthropometric data (Diffrient *et al.* 1981b) using 97.5 percentile male and 2.5 percentile female for maximum and minimum stature. All CTA bus operator seats are adjustable front to back. At this point in the project, it was assumed that drivers would position the seat for a comfortable reach to the steering wheel in the same manner as the literature indicates.

Which reach distance to use was also an issue. Diffrient indicates the comfort zone (for hand grip) is best suited to frequently used controls (e.g. driver controls and working tool controls). The reach zone (for finger grip) is appropriate for infrequently used controls (e.g. starter, heater, vents, windshield wiper, windshield washer, lights). BTPU operations, which include ticket retrieval by fingertip and light button depression, were considered infrequent controls.

Passenger—As passengers using a wheelchair are more limited in reach, reach envelopes for 95th percentile male and 5th percentile female wheelchair-bound passenger were used to evaluate passenger access (Diffrient *et al.*1974). For ADA-compliant busses, wheelchair-bound passenger reach was specified in inches of radii, centered on shoulder pivots. While 'maximum forward and side reach' were possible, 'easy reach' was more appropriate to use for a consumer population and was adopted.

Compliance—The Americans with Disabilities Act exempts busses, vans and systems from the requirement to be retrofitted with boarding ramps (49 U.S. CFR Part 38, Subpart B, Article 38.21b). As a result, only four of the nine bus models in the CTA fleet were required to comply with the ADA regulations.

The ADA does not specify how to design for those who use wheelchairs. Instead, the Act sets a number of requirements which are to be met in order to enable wheelchair users to use mass transportation. The BTPU project identified significant requirements concerning:

- Platform surface: From 2 to 30 inches above the platform, provide a minimum clear width of 30 inches and length of 48 inches (49 U.S. CFR Part 38, Subpart B, Article 38.23.6) with the wheelchair either facing the device or parallel to it (49 U.S. CFR Part 38, 4.1.2.1).
- Fare box configuration: Do not obstruct wheelchairs or mobility aids (49 U.S. CFR Part 38, Subpart B, Article 38.33).
- Fare selling and collection facilities location: Ensure that individuals with disabilities remain with the flow of traffic of all passengers (49 U.S. CFR Part 38, 4.1.1).
- Equipment design: Must have a minimum clear opening of 32 inches and must permit passage of a wheelchair (49 U.S. CFR Part 38, 4.1.2).
- Controls location: Must be between 15 inches and 48 inches above the floor (49 U.S. CFR Part 38, 4.1.2.2).

Reach envelopes were imposed on workstation orthographic plan views for all nine models. The overlay showed that mounting the BTPU on the windshield side of the fare box:

- Fit within the driver reach zone;
- Did not impede wheelchair access;
- Presented some potential for minor vision obscurement toward the driver's lower right side.

Figures 15.1 and 15.2 shows how these reach, vision and accessibility requirements were demonstrated for busses that were and were not required to comply with ADA requirements.

Notes on compliance and recommendations for modification accompanied each illustration.

15.1.5 Implementation

Phase One and Two findings led to BTPU installation recommendations for each of the fleet's nine models. Five models that were not required to comply with ADA could be outfitted with a BTPU mounted on the windshield side of the GFI fare box. Models that did comply with ADA could also be outfitted with windshield side BTPUs, yet some required changes in order to maintain an unobstructed passage.

A pilot program installed the BTPU in a number of busses in regular service on Chicago's west side. Initial results showed that the installation plan is a success. The program did uncover one unforeseen practice: 'stashing' passengers next to the fare box. While waiting for a transfer as the bus pulls away from a stop, the 'stashed' rider cannot reach the BTPU. As transfers are not a part of the long-term program, the issue is not considered significant.

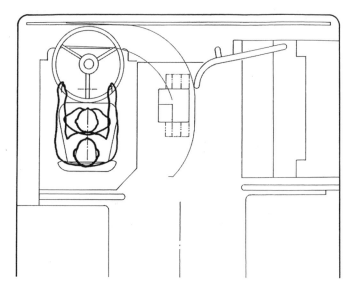

Figure 15.1 Bus workstation plan view—this plan view of the Flyer bus workstation, one of the nine CTA models to be outfitted with a bus ticket processing unit, shows the reach envelope for 5th percentile female drivers. Ninety-fifth percentile male driver position is also represented to account for vision and clearance issues. Safety railing clearance, potential visibility issues, compliance and recommended modifications were addressed in notes that accompanied each view.

Source: Reproduced with the permission of the Chicago Transit Authority.

15.1.6 Conclusion

Further work is now underway to explore further aspects of BTPU use in particular and the bus workstation in general. Field observation will be used to:

• Verify whether bus drivers actually adjust their seats and equipment as the literature indicates
• Determine how well drivers and passengers understand BTPU operation
• Assess how information displays affect riders and what changes will improve their comfort with embarkation and fare payment.

Workload analysis and operational sequence analysis will be used to evaluate the effect on bus operator performance of installing the BTPU in combination with the GFI fare box and related equipment. Particular attention will be paid to the discovery of potential hazards which might degrade driver performance. The accumulated results will be used to develop requirements and simulations for future bus workstations.

15.2 International currency exchange

International travel involves many activities, including the ability to perform transactions to obtain services and goods. Even though credit and smart cards are being used more often, there is still a need to obtain cash. Currently, banks and mobile kiosks perform currency exchange transactions. However, banks often have limited hours and are not necessarily

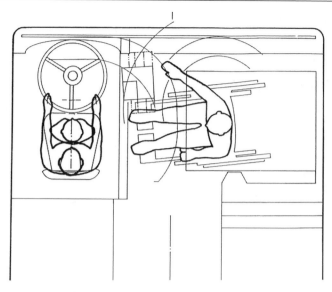

Figure 15.2 ADA-compliant bus workstation plan view—this plan view of the new flexible (low floor) bus workstation, one of the nine CTA models to be outfitted with a bus ticket processing unit, shows that because the low floor was designed to be ADA-compliant, BTPU placement could not impede wheelchair access. Reach envelopes are shown using the 5th percentile driver and rider using a wheelchair. The proposed location of the pedestal-mounted BTPU is shown to demonstrate how it could accommodate vision and reach needs for both.

Source: Reproduced with the permission of the Chicago Transit Authority.

located where a person may need to obtain currency. Kiosks are placed in certain airport locations, yet still have limited hours. Both banks and kiosks rely on paid staff to conduct transactions, which makes them costly.

15.2.1 Opportunity/need

Automated teller machines have been developed to handle single currency transactions. An automated currency exchange that can provide frequently used currencies in an area would be a boon to economic development as well as encourage travel. Such a unit would be continuously available and have a lower transaction cost per customer compared with manned exchanges.

15.2.2 Objective

Human factors research in this project will develop knowledge regarding two groups: users and maintainers. Research will examine user perceptions regarding comparative currency value, machine operation and error remediation. Research will also evaluate implications for user physical stature and configuration as it relates to unit operation (display legibility, control manipulation). Maintainer attitudes and practice with regard to currency stocking, machine diagnosis and repair will be addressed.

15.2.3 Scope of activity

Table 15.2 summarizes the project human factors research activity.

Table 15.2 International currency exchange—an automated currency exchange that can provide frequently used currencies in an area would be a boon to economic development and could also encourage travel.

Practices with an interest

The following practices would have an interest in this research:

Architecture, space planning

Product design

Communications design

Environmental design

Human factors/ergonomics

Roles involved in the design and development of computers

Systems engineering, engineering management

Activities

Determine objectives and criteria

Observational study

Error analysis—Detect error-inducing situations

Verbal protocol analysis—Understand various mental models that users bring to conduct of the transaction

Questionnaire—Examine user preferences. Develop insights into error causation

Analysis of similar systems—Assess vending machines including ATMs

Perform functional analysis—Model user/machine activity patterns

Perform task description and task analysis—Determine the steps users will follow and decision aids needed to perform them

Evaluation study

Failure modes and effects analysis—Determine the situations that might result in dysfunction or failure to perform

Simulation—Model user interface on a personal computer. Refine to two versions.

Develop control and display configuration based user stature information.

Operational sequence analysis—Map the flow of activity that needs to occur during prototype operation

Descriptive study

Assessment level usability tests—Assess final interface alternatives. Ensure transaction times are within tolerable limits for users. Collect user preferences regarding alternative interface designs

Deliverables

A summary report that describes methodology, project background, subject sample and research activity. Video recordings of the field observations

---

The human factors research group will provide the following support for the product development team through three phases.

15.2.3.1 PHASE 1—ANALYSIS

- Conduct field observations of ATM operation, including sites frequented by those who are international travelers. Perform error analysis, verbal protocol analysis. Conduct survey by brief questionnaire to develop inventory of preferences and insights into error causation. Additional attention will be paid to adding value for the customer, such as comparison in value between user's customary currency and currency to be vended.

- Perform analysis of similar systems on automated vending machines including automatic teller units.
- Review broadband wireless standards for systems data networking.

15.2.3.2 PHASE 2—SOLUTION DEVELOPMENT

- Perform functional analysis to model user/machine activity patterns.
- Perform task description and task analysis to determine the steps that users will follow and decision aids needed to perform them.

15.2.3.3 PHASE 3—EVALUATION

- Perform failure mode and effect analysis to determine the situations that might result in dysfunction or failure to perform.
- Simulate the user interface on a personal computer.
- Guide the development group in control and display configuration based user stature information.
- Perform operational sequence analysis to map the flow of activity that needs to occur during prototype operation.
- Perform usability tests at the assessment level between two final interface versions. Ensure transaction times are within tolerable limits for users. Collect user preferences regarding alternative interface designs.

15.2.4 Deliverables

Project activity will include the production of the following items.

- A summary report that describes methodology, project background, subject sample and research activity. A summary of findings and appendices with session notes will also be included.
- A presentation of findings based on the summary report that will be made to the venture capital group within four weeks of completing field research.
- Video recordings of the field observations, along with analyses of user patterns of behavior.

15.3 Web site design

Communications and transactions via the Worldwide Web continue to grow at a dramatic rate. Most designs tend to copy a few of the most popular sites, even if the site is not similar in its audience or goals. Many sites favor gimmicks and other attention-getting features at the expense of usability.

15.3.1 Opportunity/need

The performance of financial transactions on the Web requires a clear understanding by the user. Errors that are induced by web site operation erode confidence in the sponsor institution and have the potential to impose loss on the user. There is a need to create a transaction web site that provides users with accurate, reliable service free from error and confusion.

15.3.3 Objective

Human factors research activity will determine what structural and grammatical models minimize errors and how users perceive financial transactions, sponsoring organizations and their web sites. The team will also develop standards for the development and evaluation of a human-centered web site.

15.3.4 Scope of activity

Table 15.3 summarizes the project human factors research activity.

The team proposes the following activity, which will focus on implications for an effective design.

- Develop objective and criteria
- Observational study
    Analysis of similar systems—Review transaction web sites
    Error analysis—Determine frequency and type of transaction errors
    Verbal protocol analysis—Determine user perceptions, presumptions regarding transactions in general and existing site concepts in particular
    Functional flow analysis—Develop map of user, system activity patterns
    Link analysis—Configure element arrangement according to various transaction types
    Task description—Identify activities user is expected to perform
    Task analysis—Determine instructions, decision aids user needs to perform tasks
- Draft initial web site evaluation standard
- Evaluation study
    Operational sequence analysis—Map flow of information, transaction communication activity and user actions
    Simulation—Model alternative site design concepts
- Descriptive study
- Recruit and assign participants to sample using each web site alternative.
    Assessment usability test—Collect data on transaction type, time, error rate. Obtain pre-session and post-session user attitudes and preferences with regard to Web-based transactions and concept alternatives.
- Complete web evaluation standards (including time, error benchmarks) based on evaluation and descriptive study results.
    Validation level test—Determine user performance compared with time, error benchmarks for each concept under consideration.
- Decision analysis, cost-effective analysis—Review usability results. Choose preferred transaction concepts and implement final selection.

15.3.5 Deliverables

Documentation of error types and frequencies, user preferences and performance while operating current products. Video recordings and notes on verbal protocol sessions. Flow and link diagrams describing system and user activity patterns. Instructions, decision aids. Web site prototypes and usability test performance documentation. Draft and final standards for web evaluation.

15.4 Remote site collaborative communications tool

Increasingly, project team members are located at various sites across the country and in overseas locations. To date, team interaction has been supported by what amounts to found

Table 15.3 Web site design—communications and transactions via the Worldwide Web continue to grow at a dramatic rate. Most designs tend to copy a few of the most popular sites, even if the site is not similar in its audience or goals.

Practices with an interest

The following practices would have an interest in this research:

> Communications design
>
> Human factors/ergonomics
>
> Roles involved in the design and development of computers
>
> Systems engineering, engineering management

Activities

Develop objective and criteria

Observational study

> Analysis of similar systems—Review transaction web sites
>
> Error analysis—Determine frequency and type of transaction errors
>
> Verbal protocol analysis—Determine user perceptions, presumptions regarding transactions in general and existing site concepts in particular
>
> Functional flow analysis—Develop map of user, system activity patterns
>
> Link analysis—Configure element arrangement by transaction types
>
> Task description—Identify activities user is expected to perform
>
> Task analysis—Determine instructions, decision aids that user needs

Draft initial web site evaluation standard

Evaluation study

> Operational sequence analysis—Map flow of information, transaction communication activity and user actions
>
> Simulation—Model alternative site design concepts

Descriptive study

> Usability test—Assessment level test: Collect data on transaction type, time, error rate. Obtain pre- and post-session user attitudes and preferences

Complete web evaluation standards (including time, error benchmarks) based on evaluation and descriptive study results

> Usability test—Validation level test: Determine user performance compared with time, error benchmarks for each concept under consideration

Decision analysis, cost-effective analysis—Review usability results. Choose preferred transaction concepts and implement final selection

Deliverables

Video recordings, notes of verbal protocol sessions, map of user activity patterns, concept prototypes, operational sequence diagrams of transaction patterns, user attitude and preference documentation, draft and final standards for web evaluation

---

products that have been adapted to team use such as television, audio, video projection stands. Many of the systems are either expensive, require extensive installation, or impede team effectiveness.

15.4.1 Opportunity/need

There is a need for a product that links team members at remote sites and makes it possible for them to perform their work effectively and efficiently.

15.4.2 Objective

Human factors research will seek to understand team collaboration in person and at remote sites. An analysis will describe in-person and remote team dynamics. Human factors research efforts will strive to use technology to add value for users and enable seamless collaboration.

15.4.3 Scope of activity

Table 15.4 summarizes the project human factors research activity.

The team proposes the following activity, which will focus on implications for an effective design.

- Focus group—Determine attitudes and preferences regarding remote team products.
- Establish project objective, as well as criteria for user team success. Review published research on team knowledge such as Cooke *et al.* (2000). Select a model and measures of team knowledge to create a baseline and basis for alternative evaluation.
- Observational study
  - Activity analysis—Observe team interaction during in-person, remote sessions.
  - Analysis of similar systems—Assess existing remote team collaboration products.
- Evaluation and descriptive study
  - Failure modes and effects analysis—Determine potential failures, effects, remedies.
  - Simulation—Use rapid prototype collaboration to develop product concepts from commercial off-the-shelf (COTS) components based on focus group and observational studies. Use 'Wizard of Oz' method to model intelligent system attributes.
  - Exploratory usability test—Elicit user comments and insights using prototypes. Employ team knowledge measures as evaluation criteria.
  - Simulation—Develop prototypes for two to three alternative concepts.
  - Assessment usability test—Have teams use product alternative concepts. Compare performance against team success criteria.
  - Link analysis—Use assessment test observations to determine optimal control and display configuration.
  - Simulation—Develop final prototypes for two to three alternative concepts.
  - Alpha, beta tests—Have client teams use prototypes in actual project environment.
  - Focus group, questionnaires—Determine teams' attitudes and preferences regarding alternative concepts. Evaluate alternatives in terms of knowledge measures.
  - Decision analysis, cost-effective analysis—Compare alternatives and select final concept for recommendation.

15.4.4 Deliverables

Focus group session documentation, activity analysis of team interaction, comparative product assessment, tabular listing of failure modes, effects and remedial actions. Usability test documentation, including user preferences as well as performance information. Link analysis diagrams and prototypes of controls and displays for design concepts. Summary report providing documentation as a proof of concept for final design concept.

15.5 Sharps disposal container

The transmission of blood-borne pathogens is a particular concern among health care providers. An accidental needle stick can infect an individual with severe, even life-threatening, disease. For that reason, it is crucial to protect those who work with needles and other similar objects.

Table 15.4 Remote site collaborative communications tool—project team members are more often located at various sites across the country and overseas. There is a need for a product that links team members at remote sites and makes it possible for them to perform their work effectively and efficiently.

Practices with an interest

The following practices would have an interest in this research:

  Architecture, space planning

  Product design

  Communications design

  Environmental design

  Human factors/ergonomics

  Roles involved in the design and development of computers

  Systems engineering and engineering management

Activities

Focus group—Determine user attitudes about remote team collaboration products

Establish project objective, as well as criteria for user team success. Identify model and measures of team knowledge to use in evaluation

Observational study

  Activity analysis—Observe team interaction during in-person, remote sessions

  Analysis of similar systems—Assess existing team collaboration product

Evaluation study

  Failure modes and effects analysis—Assess potential failures, effects, remedies

  Simulation—Rapid prototype collaboration product concepts

  Exploratory usability test—Elicit user comments, insights while using prototypes

  Simulation—Develop prototypes for two to three alternative concepts

  Assessment usability test—Have teams use product alternative concepts. Compare performance against team success criteria

  Link analysis—Use assessment test observations to determine optimal control and display configuration

  Simulation—Develop final prototypes for two to three alternative concepts

  Alpha, beta tests—Have client teams use prototypes in actual project

  Focus group, questionnaires—Determine teams' attitudes and preferences on alternatives. Evaluate alternatives in terms of knowledge measures

  Decision analysis, cost-effective analysis—Compare and select final version

Deliverables

Focus group session documentation, activity analysis of team interaction, comparative product assessment, tabular listing of failure modes, effects and remedial actions. Usability test documentation, including user preferences as well as performance information. Link analysis diagrams and prototypes of controls and displays for design concepts. Summary report providing documentation as well as proof of concept for final design concept.

15.5.1 Opportunity/need

The development team will propose a plan for the development of two types of containers that are intended for the disposal of biohazards in the health care environment. A hard-sided sharps container will serve as a single use product. A liner-style suction canister presents an opportunity for cost savings by replacing a removable interior pouch within a durable housing.

15.5.2 Objective

There is a need to perform human factors research in two areas. Qualitative investigation will explore health care professional attitudes, preferences and experience regarding container requirements and opportunities. There are needs for additional knowledge which lie outside of the scope of qualitative sessions, including task mapping, regulations compliance, hazard-risk assessment, training, instructions for use and warnings and behavior and physiology related to safety, error minimization, ease of use and training/instructions.

15.5.3 Scope of activity

Table 15.5 summarizes the project human factors research activity.

Members of the human factors research team will perform the following research in order to focus on implications for an effective design.

- Participate in development team meetings. Address issues related to human factors aspects of projects.
- Perform operational analysis to identify and review pertinent regulations.
- Participate in competitive product review, focusing on human factors considerations.
- Perform analysis of similar systems.
- Identify, review pertinent regulations.
- Account for anticipated product operation and physical manipulation.
- Perform hazard analysis and fault tree analysis to determine the opportunity for exposure to hazards, particularly blood-borne pathogens.
- Conduct five focus group sessions of eight qualified participants over seven days in five separate metropolitan areas.
- Observe current products in use.
- Draft initial recommendations for training, instructions and warnings.
- Draft human factors contribution to team report.
- Model and observe use of solution alternatives by health care providers. Observe users following instructions.
- Interview simulation participants to determine preferences for alternatives, instructions, labels and warnings.
- Participate in team meetings.

There are human factors considerations which cannot be addressed until finished product concepts are available. They include a definitive hazard-risk assessment, training in actual product operation, instructions for actual product use and actual hazard warnings.

15.5.4 Deliverables

A final report will summarize team findings in order to guide design development. The team will contribute the human factors portion of the report which is provided to the client at the end of the phase. It will contain material which pertains to both the sharps container and liner container projects, including: description of activity, methodology, summary of findings and recommendations for hard and liner container design.

Table 15.5   Sharps disposal container—the transmission of blood-borne pathogens is a particular concern among health care providers. An accidental needle stick can infect an individual with severe, even life-threatening, disease. For that reason, it is crucial to protect those who work with needles and other similar objects.

Practices with an interest

The following practices would have an interest in this research:

Architecture

Product design

Communications design

Environmental design

Human factors/ergonomics

Roles involved in the design and development of machines

Systems engineering, engineering management

Activities

Develop objectives and criteria

Observational study

Operational analysis—Identify, review pertinent regulations

Analysis of similar systems—Competitive product review, focusing on human factors considerations

Activity analysis—Observe current products in use. Account for anticipated product operation and physical manipulation

Management oversight and risk tree analysis and fault tree analysis—Determine the probability for exposure to hazards, particularly blood-borne pathogens

Focus group—Five sessions of eight qualified participants over seven days in five separate metropolitan areas

Draft initial recommendations for training, instructions and warnings

Evaluation study

Simulation—Model and observe use of solution alternatives by health care providers. Observe users following instructions

Interview—Determine preferences for alternatives, instructions, labels and warnings

Deliverables

A final report will summarize team findings to guide design development

Initial and final drafts will be provided for user training as well as instructions and warnings that are to be provided with the product

---

Initial and final drafts will be provided for user training as well as instructions and warnings that are to be provided with the product.

15.6 Infusion pump

The development team has been assigned to propose a plan for the development of a new generation infusion pump to provide anesthesia in the surgical operating room.

15.6.1 Opportunity/need

The new generation unit will replace a well-considered, reliable product that faces obsolescence. In addition, the new unit holds the promise for application in new areas outside of its traditional use in anesthesia which also require reliable, measured dispensing of medications. The development of a new generation infusion pump presents an opportunity to improve product versatility and expand its potential uses and markets. Human behavioral and physical limits and abilities play a significant role in the prospects for the new pump within the operator-task-equipment-environment (OTEE) system.

There is a need to ensure the new pump is suited to use requirements, through the collection, analysis, interpretation and use of human factors knowledge.

15.6.2 Objective

Initiate a design concept that accommodates a wider range of drug protocols, expands use of the pump to applications outside of the operatory, allows for multiple pump use and can be configured for use in countries other than the United States. Perform research and guidance on both behavior and physiology related to safety, error minimization, ease of use, reliability and training/instruction.

15.6.3 Scope of activity

Table 15.6 summarizes the project human factors research activity.

The team proposes the following activity, which will focus on implications for an effective design.

15.6.3.1 PHASE ONE—ANALYSIS

- Participate in project team meetings; initiate discussion of human factors-related issues and respond to team needs for further information and concept guidance.
- Review current information (e.g. U.S. FDA, ANSI) and European (CE) regulations/standards) for compliance requirements. Identify key performance criteria.
- Perform critical incident study, analysis of similar systems, function analysis to identify opportunities for improvement, as well as current/potential threats to product reliability. Review existing reports such as USFDA MAUDE database.
- Develop representative sample of observation sites in concert with client. Perform extensive field observations of procedures on site. Perform focused interviews with practitioners.
- Summarize key performance issues and requirements for project team.

15.6.3.2 PHASE TWO—CONCEPT DEVELOPMENT

- Participate in project team meetings.
- Recommend interface configuration, to include structure, symbolic and alphanumeric information, controls and displays, their arrangements for U.S. use. Provide additional interface configuration for European market and Asian market.
- Arrange and conduct qualitative individual and group evaluations of alternative concept rough prototypes, to obtain preferences and response to proposed features.
- Summarize key performance issues and requirements for project team.

15.6.3.3 PHASE THREE—EVALUATION

- Participate in project team meetings.
- Evaluate concept controls and displays and their arrangements, in terms of performance.
- Conduct failure modes and effects analysis, fault tree analysis, to determine potential human-machine sources of hazard, risk and failure. Provide guidance for remediation.
- Develop warnings for product and product documentation.

Table 15.6 Infusion pump—the development team has been assigned to propose a plan for the design of a new generation infusion pump to provide anesthesia in the clinical setting.

Practices with an interest

The following practices would have an interest in this research:

    Product design

    Communications design

    Environmental design

    Human factors/ergonomics

    Roles involved in the design and development of computers

    Roles involved in the design and development of machines

    Systems engineering, engineering management

Activities

Participate in project team meetings

Observational study

    Operational analysis—Review current regulations, standards for compliance Identify key performance criteria. Review pertinent published literature

    Critical incident study, analysis of similar systems, function analysis—Identify opportunities for improvement, current/potential threats to reliability

Develop representative sample of observation sites in concert with client. Perform field observations of anesthesia procedures

    Interviews—Determine and summarize key performance issues, requirements

Contact clinicians to schedule operating prototype evaluations

Recommend interface configuration, to include structure, symbolic and alphanumeric information, controls and displays, their arrangements for U.S. use. Provide additional interface for markets other than U.S.

Evaluation study

    Interviews—Arrange and conduct qualitative individual and group evaluations of alternative concept rough prototypes. Obtain preferences and response to proposed features

    Failure modes and effects analysis, fault tree analysis—Determine potential human-machine sources of hazard, risk, and failure

Develop training and documentation for product

    Verbal protocol analysis—Confirm actual use by clinicians

Develop training materials and instructions for use. Collaborate with manual authors on content concerning human-machine hazards. Review manuscript with selected clinicians

Deliverables

Provide a brief summary report at the end of each phase including a description of activity, summary of findings and conclusions with implications for pump design

---

- Contact clinicians to schedule operating prototype evaluations. Conduct verbal protocol analysis to confirm actual use by clinicians. Collect background information for training, instructions.
- Summarize key performance issues and requirements for project team.

15.6.3.4 PHASE FOUR—TRAINING/DOCUMENTATION

- Develop manuscript for training materials and instructions for use, based on results of Phases Two and Three.

- Collaborate with maintenance and operation manual authors on content concerning human-machine hazards.
- Review manuscript with selected clinicians, revise and submit to client.

15.6.4 Deliverables

The team will provide a brief summary report at the end of each phase, which will include a description of activity, summary of findings and conclusions with implications for pump design. Focus group recordings will be included.

15.7 Reference laboratory assay system

The provision of timely, quality health care relies on the services of clinical laboratories to analyze tissue and fluid specimens. Assay machines rapidly perform tests on a large quantity of small specimens using small amounts of chemical reagents to prepare samples for spectroscopic analysis.

15.7.1 Opportunity/need

Human behavioral and physical limits and abilities play a significant role in the prospects for a new immunoassay unit within the operator-task-equipment-environment (OTEE) system. There is a need to ensure such a product is suited to human use through observation, analysis, interpretation and the use of human factors knowledge.

15.7.2 Objective

Ensure that the immunoassay system will be able to efficiently process routine batch specimen samples as well as immediate samples in the clinical lab. Ensure that the unit operator is not exposed to contamination hazard.

15.7.3 Scope of activity

Table 15.7 summarizes the project human factors research activity.
   The team proposes the following activity, which will focus on implications for an effective design.

15.7.3.1 PHASE 1—ANALYSIS

In addition to the steps below, the members of the human factors research group will participate in design team meetings and contribute the human factors section of the final report and presentation.

- Review request for proposal.
- Perform operational analysis. Review and assess information collected to date on procedures, equipment and work environment. Review and assess safety and work environment regulations, standards as specified in RFP Section 3.1.
- Perform activity analysis and operations sequence analysis. Observe and analyze current operator lab work patterns identified in RFP.
- Perform function allocation and task description. Create flow chart that represents user sample manipulation and control/display interface interaction, based on activity analysis and failure modes and effects analysis findings. Create task list to describe procedures operator will follow during preparation, operation, retrograde and maintenance. Develop initial structure of training content.

Table 15.7 Reference laboratory assay system—the provision of timely, quality health care relies on the services of clinical laboratories to analyze tissue and fluid specimens. Assay machines rapidly perform tests on a large quantity of small specimens using small amounts of chemical reagents to prepare samples for spectroscopic analysis.

Practices with an interest

The following practices would have an interest in this research:

Product design

Communications design

Environmental design

Human factors/ergonomics

Roles involved in the design and development of computers

Roles involved in the design and development of machines

Systems engineering and engineering management

Activities

Review request for proposal (RFP)

Observational study

Operational analysis—Review and assess information on procedures, equipment and work environment. Review/assess regulations, standards

Activity analysis and operations sequence analysis—Observe and analyze current operator lab work patterns identified in RFP

Functional flow and task description—Create flow chart that represents user sample manipulation and control/display interface interaction. Create task list to describe procedures operator will follow

Failure modes and effects analysis—Account for all known and potential biosafety threats to operator throughout preparation, operation, retrograde and maintenance

Evaluation study

Simulation—Develop structure for control/display interface and depict it in diagram form to guide interactive screen software development

Interview—Obtain health care professional preferences regarding interface and equipment. Review mock-up and model of solutions, and provide recommendations for improvement

Exploratory usability tests—Design, conduct and evaulate the results to determine the effectiveness of the design concept

Descriptive study

Validation usability tests—Verify actual user performance, cycle time

Deliverables

Diagrams representing flow of user/specimen and machine activity, tables accounting for failure modes, effects and remedies, interface simulations and guidance diagrams for software development, and summary report including usability results

- Perform failure modes and effects analysis. Account for all known and potential biosafety threats to operator throughout preparation, operation, retrograde and maintenance in accordance with RFP. Assess potential opportunities for system failure, implications for system performance and design remedies.

15.7.3.2 PHASE 2—CONFIGURATION

Collaborate with design team members in the development of equipment, control/display interface, packaging and labeling based on function allocation, task description and requirements analysis findings, as well as RFP. Develop structure for control/display interface and depict it in diagram form to guide interactive Windows NT screen software development. Review mock-up and model of solutions and provide recommendations for improvement.

15.7.3.3 PHASE 3—EVALUATION

Design, conduct and evaluate the results of exploratory usability tests to determine the effectiveness of the design concept. Tasks will include sample and waste manipulation, product operation (physical item and user control/display) and use of training and documentation.

15.7.4 Deliverables

Diagrams representing flow of user/specimen and machine activity, tables accounting for failure modes, effects and remedies, interface simulations and guidance diagrams for software development and a summary report including usability results.

Notes

1 The human-made environment

1 The adaptation of life cycle to include cost implications was developed by Patrick Whitney and the Institute of Design faculty, Illinois Institute of Technology in 1995.

6 Analysis methods

1 Random number tables can be found in texts that deal with research and statistical methods, such as Bordens and Abbott (1998).

7 Design guidance methods

1 The reader is invited to review Luczak (1997:341–416) for a detailed account of current task analysis methods according to analyses that consider physiological costs, time consumption, human machine interaction and components and combinations of a one-person job, work organizations, company organization and inter-company relations and societal demands.
2 For tasks that pertain to nuclear power plant operations, consult Swain and Guttmann (1983).

9 Surveys: interviews and questionnaires

1 For further information on questionnaire design, the reader is encouraged to consult:

- Aaker, Kumar, and Day (1998) *Marketing Research*, 6th edn. John Wiley & Sons: New York.
- Blankenship, A., Breen, G. and Dutka, A. (1998) *State of the Art Marketing Research*. McGraw Hill: New York.
- Hair, J., Bush, R.P. and Ortinau, D.J. (2000) *Marketing Research: A Practical Approach for the New Millennium* (with Data Disk Package). McGraw Hill: New York.
- Kinnear, T.C., Taylor, J.R. (1996) *Marketing Research: An Applied Approach*. McGraw Hill: New York.

14 Communication

1 Edward R. Tufte's series of books on information graphics are a rich resource on the visual display of data and information:

- Tufte, E. (1983) *The Visual Display of Quantitative Information*. Graphics Press: Cheshire, CT.
- Tufte, E. (1990) *Envisioning Information*. Graphics Press: Cheshire, CT.
- Tufte, E. (1997) *Visual Explanations*. Graphics Press: Cheshire, CT.

Appendix

Organizations, tools and publications are available to assist human factors research in research and development. The reader is encouraged to use these resources as a starting point to first understand what resources exist and then explore other resources such as organizations in countries other than the U.S.

As web site addresses change with some frequency, it is difficult to guarantee that each URL listed in this appendix will remain correct. Web addresses are accurate at the time of publication and if they are changed the sponsor will likely provide a guide to the new location.

The following resources are drawn in part from Gawron, Dennison and Biferno (1996) and Andre (1998).

A.1 Organizations

Professional organizations promote the advancement of knowledge in the field of human factors and ergonomics. Each organization can provide information on conferences, publications and services.

These organizations are specific to human factors:

- Board of Certification in Professional Ergonomics (BCPE)—P.O. Box 2811, Bellingham, WA 98227-2811, <http://www.bcpe.org>. (BCPE certifies professionals who have the educational background, knowledge and skills that are commensurate with the practice of ergonomics/human factors.)
- International Ergonomics Association (IEA)—<http://ergonomics-iea.org/> (IEA is the association of ergonomics and human factors societies around the world.)
- Human Factors and Ergonomics Society (HFES)—Box 1369, Santa Monica, CA 90406 (310) 394-1811 <http://hfes.org>. (HFES is the professional organization for human factors in the United States.)
- The Ergonomics Society—4 John Street, London WC1N 2ET,<http://www.ergonomics. org.uk/>. (The Ergonomics Society is the U.K. organization for professionals using information about people to design for comfort, efficiency and safety.)

Other organizations deal with professional development issues including human factors:

- American Institute of Industrial Engineers (AIIE)—25 Technology Park, Norcross, GA 30092 <http://www.iienet.org>
- American National Standards Institute (ANSI)—1819 L Street, NW, 6th Fl., Washington, DC, 20036 <http://web.ansi.org/default_js.htm>
- Association for Computing Machinery (ACM)—11 West 42nd Street, New York, NY 10036 http://www.acm.org/sigchi
- Association of Aviation Psychologists (AAP)— <http://userwww.sfsu.edu/~kmosier/aap.html>

- American Society of Heating, Refrigerating and Air Conditioning Engineers (ASHRAE)—<http://www.ashrae.org/>
- Canadian Center for Occupational Health and Safety (CCOHS)—250 Main Street East, Hamilton, Ontario, Canada L8N 1H6. <http://www.ccohs.ca/oshanswers/ergonomics/ergonomics.htm>
- Human Systems Information Analysis Center (HSIAC)—AFRL/HEC/HSIAC, Building 196, 2261 Monahan Highway, Wright-Patterson Air Force Base, OH 45433-7022 <http://www.iac.dtic.mil/hsiac>
- U.S. Consumer Product Safety Commission (CPSC)—4330 East-West Highway Bethesda, Maryland 20814-4408 <http://www.cpsc.gov/>
- Design Research Society (DRS) http://www.designresearchsociety.org—This United Kingdom-based international organization promotes the advancement of knowledge in design and research. DRS sponsors an active discussion of design and research issues among all interested participants on the DRS Listserv. Subscriptions to the DRS News monthly digital newsletter can be requested through the Society's web site.
- Institute of Electrical and Electronic Engineers (IEEE) (Systems, Man and Cybernetics)—345 East 47th Street, New York, NY 10017 <http://www.ieee.org>
- National Institute for Occupational Safety and Health (NIOSH)—Hubert H. Humphrey Bldg., 200 Independence Ave., SW , Room 715H, Washington, DC 20201, <http://www.cdc.gov/niosh/ergoweb.html>. (NIOSH is the U.S. Federal agency responsible for conducting research and making recommendations for the prevention of work-related disease and injury.)
- National Institute of Standards and Technology (NIST)—100 Bureau Drive, Stop 3460, Gaithersburg, MD 20899-3460 <http://www.nist.gov/>
- National Society for Performance and Instruction (NSPI)—1126 16th Street, NW, Washington, DC 20036 <http://www.ispi.org>
- The Society for Human Performance in Extreme Environments (SHPEE)—2951 Marina Bay Drive #130-411, League City, TX 77573 <http://www.hpee.org>
- Society for Information Display (SID)—654 Sepulveda Boulevard, Los Angeles, CA 90049 <http://www.sid.org>
- Illumination Engineering Society of North America (IES)—120 Wall Street, Floor 17 New York, NY 10005 <http://www.iesna.org/>
- Society for Risk Analysis (SRA)—1313 Dolley Madison Blvd., Suite 402, McLean, VA 22101 <http://www.sra.org/>
- Society for Technical Communication (STC)—815 15th Street, NW, Washington, DC 20005 <http://www.stc.org>
- Underwriter Laboratories, Inc. (UL)—333 Pfingsten Road, Northbrook, IL 60062-2096 <http://www.ul.com/mas/ergo.htm>. (Underwriters Laboratories is an independent, not-for-profit product safety testing and certification organization.)
- Usability Professionals' Association (UPA)—10875 Plano Road, #115, Dallas, TX 75238 <http://www.upassoc.org>

A.2 Periodicals

Journals and magazines provide a forum to present, examine and discuss human factors knowledge and issues.

> *Applied Ergonomics*
> *Behavior and Information Technology*
> *Communications of the ACM*
> *HSIAC Gateway*
> *Ergonomics*
> *Ergonomics in Design*
> *Ergonomic Abstracts*
> *Human Factors*
> *Human-Computer Interaction*
> *International Journal of Human-Computer Studies*
> *Journal of Aviation Psychology*
> *The Journal of Human Performance in Extreme Environments*

Newsletters that are published by technical groups within human factors organizations can also provide valuable information. Such publications discuss current issues in professional practice of concern to technical group membership.

A.3 Web sites

A wealth of institutional and commercial web sites exist to provide information and to promote products related to human factors. Application areas that are represented here include the office, industry, injury prevention and treatment, computing and transportation.

A.3.1 Office

- Office Ergonomics by Chris Grant <http://www.office-ergo.com/>
- Tips for Healthy Computer Use! <http://www.ME.berkeley.edu/ergo/tips/tips.html>
- Ergonomics Training and Resources < http://keats.admin.virginia.edu/ >

A.3.2 Industrial

- ErgoWeb <http://www.ergoweb.com>

A.3.3 Injury prevention and treatment

- The Typing Injury FAQ <http://www.tifaq.com>
- CTD News <http://www.ctdnews.com>

A.3.4 Products

- Safe Computing <http://www.safecomputing.com>

A.3.5 Human-computer interaction

- Association for Computing Machinery Special Interest Group on Computer and Human Interaction (SIGCHI) <www.acm.org>

- The HCI Bibliography <http://www.hcibib.org/>
- The Alertbox: Current Issues in Web Usability <http://www.useit.com/alertbox>
- Frequently Asked Questions: comp.human.factors <http://www.dgp.toronto.edu/people/ematias/faq/contents.html>
- University of Maryland Human-Computer Interaction Lab <http://www.cs.umd.edu/projects/hcil/index.html>

A.3.6 Qualitative research tools

- Scolari, the software division of social science publisher Sage Publications. < http://www.scolari.com>

A.3.7 Transportation

- Society of Automotive Engineers (SAE) <www.sae.org>
- U.S. National Transportation Safety Board <www.ntsb.gov>
- The Human Factors Research and Technology Division, Ames Research Center, U.S. National Aeronautics and Space Administration (NASA) <http://human-factors.arc.nasa.gov/>

A.3.8 Search topics

As a guide to web-based searches, here are recent frequently used topical search patterns where sites related to human factors and ergonomics can be found.

- Science/Social_Sciences/Psychology/Industrial_and_Organizational/Human_ Factors_and_Ergonomics/
- Human-Computer Interaction
- Arts > Design > Industrial Design
- Health > Conditions and Diseases > Neurological Disorders > Peripheral Nervous System > Nerve Compression Syndromes > Carpal Tunnel Syndrome
- Health > Conditions and Diseases > Musculoskeletal Disorders > Repetitive Strain Injuries
- Health > Occupational Health and Safety > Ergonomics
- Science > Technology > Safety Engineering

A.4 Tools

Human factors research can benefit from the use of tools, including instruments, two-dimensional figure models, three-dimensional figure models and electronic simulations.

A.4.1 Instruments

Instruments can be used to collect data for use in such methods as activity analysis (Section 6.4) and workload assessment (Section 8.5). Eye tracking instrumentation is available from:

- Applied Science Laboratories, 175 Middlesex Turnpike, Bedford, MA 01730 <http://www.a-s-l.com>
- ISCAN, Inc. 89 Cambridge Street, Burlington, MA 08103 <http://www.iscaninc.com>

A.4.2 *Human figure models*

Software-based or physical models of the human figure can be used as a template while configuring or evaluating concepts.

A.4.2.1 *Two-dimensional figures*

Templates that describe the human form in profile in two dimensions can be used to determine clearance, fit and reach, as well as compliance with specifications and regulations. Made from plexiglas, the figures are made to scale and hinged to allow for movement at the major joints. MIL-HDBK-759 provides instuctions and templates to produce the two-dimensional figures for 5th and 95th percentile male figures. Mannekin, the first version of the two-dimensional figure template, was developed at the U.S. Air Force Armstrong Aeromedical Research Laboratory.

A.4.2.2 *Three-dimensional figures*

Full-size mannequins that are similar to human size, weight and density can be outfitted with instruments and subjected to events (e.g. vehicle impact) to evaluate products and equipment (e.g. safety harness). Independent test labs are often used as the resource for the knowledge and equipment that are necessary to manage the complex instrumentation and statistical models used to accurately collect and analyze the data. Government agencies have also developed mannequins for test and evaluation studies. For example, the U.S. Department of Transportation regularly uses figures (crash test dummies) and the U.S. Air Force has developed figures to test ejection seats.

A.4.3 *Software human simulations*

A variety of three-dimensional software models have been developed by commercial firms and governmental agencies to represent the human figure in computer-aided design (CAD) systems. The models are based on human anthropometry, movement and selected aspects of performance (e.g. vision). Imposing the human model within a product (e.g. environment, vehicle) makes it possible to evaluate such aspects as fit, reach, sight lines, strength and energy expenditure. For example, two versions of COMBIMAN software models (one for a pilot, one for a maintenance worker) are used at the Aeromedical Research Laboratory at Wright-Patterson Air Force base. MannequinPro is a commercial 'ergonomic modeling' product that is available through HumanCAD Systems <www.mqpro.com>.

Bibliography

Anderson, B. (2000) Where the Rubber Hits the Road: Notes on the Deployment Problem in Workplace Studies. In Luff, P., Hindmarsh, J. and Heath, C. (eds), *Workplace Studies*. Cambridge University Press: New York. 223.

Andre, A. (1998) Effective Use of the Worldwide Web for HF/E Consultants. *Proceedings of the Human Factors and Ergonomics Society 42nd Annual Meeting*. Human Factors and Ergonomics Society: Santa Monica, CA. 714–8.

Bahr, N.J. (1997) *System Safety Engineering and Risk Assessment*. Taylor & Francis: London.

Bailey, R.W. (1971) *Human Performance Engineering*. PTR Prentice Hall: Englewood Cliffs, NJ.

Bailey, R.W. (1989) *Human Performance Engineering*, 2nd edn. PTR Prentice Hall: Englewood Cliffs, NJ.

Bailey, R.W. (1996) *Human Performance Engineering*, 3rd edn. PTR Prentice Hall: Englewood Cliffs, NJ.

Bannon, L.J. (2000) Simulating Workplace Studies Within the Human-Computer Interaction Field. In Luff, P., Hindmarsh, J. and Heath, C. (eds), *Workplace Studies*. Cambridge University Press: New York. 230–241.

Barnes, R. (1968) *Motion and Time Study*. John Wiley & Sons: New York.

Bateman, R.P. (2000a) ANOVA. E-mail. (30 June 2000).

Bateman, R.P. (2000b) Emotion. E-mail. (4 July 2000).

Bateman, R.P. (2000c) Kendall's coefficient. E-mail. (23 August 2000).

Beevis, D. (ed) (1999) *Analysis Techniques for Human-Machine Systems Design*. Crew Systems Ergonomics/Human Systems Technology Information Analysis Center. Wright Patterson AFB, Ohio.

Berg, M. (1997) *Rationalizing Medical Work*. The MIT Press: Cambridge, MA.

Blanchard, B. (1998) Cost Management. In Sage, A. and Rouse, W. (eds), *Handbook of Systems Engineering and Management*. John Wiley & Sons: New York. 235–68.

Blumer, H. (1969) *Symbolic Interactionism*. University of California Press: Berkeley.

Boff, K. and Lincoln, J. (eds) (1998) *Engineering Data Compendium*. H. G. Armstrong Aerospace Medical Research Laboratory. Wright-Patterson AFB, Ohio. Vol. I, II, III.

Bordens, K. and Abbott, B. (1998) *Research Design and Methods*. Mayfield Publishing Company: Mountain View, CA.

Bradsher, K. (2000) Ford Is Conceding S.U.V. Drawbacks. *New York Times*, 12 May. A1–C2.

Brody, J. (1998) Facing Up to the Realities of Sleep Deprivation. *New York Times*, 31 March.

Browne, M. (1995) Scientists Deplore Flight From Reason. *New York Times*, June 6.

Buede, D. M. (1999) Functional Analysis. In Sage, A.P. and Rouse, W.B. (eds) *Handbook of Systems Engineering and Management*. John Wiley & Sons: New York. 997–1036

Cappy, G. and Anthony, R. (1999) Leading Successful Change Initiatives. In Hesselbein, F., Goldsmith, M., and Somerville, I. (eds) *Leading Beyond the Walls*. Jossey-Bass Publishers: San Fransisco. 199–215

Carnahan, E. (2002) Software Development Process. Personal communication, 6 January 2002.

Charlton, S.G. (1996a) Mental Workload Test and Evaluation. In O'Brien, T.G. and Charlton, S.G. (eds) *Handbook of Human Factors Testing and Evaluation*. Lawrence Erlbaum and Associates: Mahwah, NJ. 181–97.

Charlton, S.G. (1996b) Questionnaire Techniques for Test and Evaluation. In O'Brien, T.G., and Charlton, S.G., eds. *Handbook of Human Factors Testing and Evaluation*. Lawrence Erlbaum and Associates: Mahwah, NJ. 81–99.

Charlton, S.G. and O'Brien, T.G. (1996) The Role of Human Factors Testing and Evaluation in Systems Development. In O'Brien, T.G. and Charlton, S.G. (eds), *Handbook of Human Factors Testing and Evaluation*. Lawrence Erlbaum and Associates: Mahwah, NJ.

Chapanis, A. (1959) *Research Techniques in Human Engineering*. Johns Hopkins Press: Baltimore, MD.

Chapanis, A. (1985) Methodological Considerations in Human Factors Engineering. *Human Factors Engineering: Engineering Summer Conferences*. University of Michigan: Ann Arbor, MI.

Chapanis, A. (1991) To Communicate the Human Factors Message, You Have to Know What the Message Is and How to Communicate It. *Bulletin*. Human Factors Society: Santa Monica, CA. Vol.34, No.11. 1–4.

Chapanis, A. (1996) *Human Factors in Systems Engineering*. Wiley Interscience: New York.

Chapanis, A. and Van Cott, H.P. (1972) Human Engineering Tests and Evaluations. In Van Cott and Kinkade, R.G. (eds), *Human Engineering Guide to Equipment Design*. U.S. Government Printing Office: Washington, D.C. 701–28.

Christensen, J. (1985a) The Nature of Systems Development. *Human Factors Engineering: Engineering Summer Conferences*. University of Michigan: Ann Arbor, MI.

Christensen, J. (1985b) Human Factors in Hazard Risk Evaluation. *Human Factors Engineering: Engineering Summer Conferences*. University of Michigan: Ann Arbor, MI.

Christensen, J. (1987) The Human Factors Profession. In Salvendy, G. *The Handbook of Human Factors*. John Wiley & Sons: New York. 3–16.

Clegg, C., Ravden, S., Corbett, M. and Johnson, S. (1989) Allocating Functions in Computer Integrated Manufacturing: A Review and a New Method. *Behaviour and Information Technology*, 8. Taylor & Francis: London.

Collins, J. and Porras, J. (1994) *Built to Last*. Harper Collins: New York.

Conner, D. (1992) *Managing at the Speed of Change*. Villard Books: New York.

Cooke, N., Salas, E., Cannon-Bowers, J. and Stout, R. (2000) *Measuring Team Knowledge*. Human Factors. Vol.42, No.1. Human Factors and Ergonomics Society. 151–73.

Coppleston, F. (1960) *A History of Philosophy*: Vol.4 Doubleday: Garden City, NY.

Coppleston, F. (1964) *A History of Philosophy*: Vol.5, Part II. Doubleday: Garden City, NY.

Coppleston, F. (1967) *A History of Philosophy*: Vol.8, Part II. Doubleday: Garden City, NY.

Cross, N. (1994) *Engineering Design Methods*. John Wiley & Sons: Chichester, U.K.

Cross, N., Christiaans, H. and Dorst, K. (eds) (1996) *Analysing Design Activity*. John Wiley & Sons: Chichester, U.K.

Cushman, W. and Rosenberg, D. (1991) *Human Factors in Product Design*. Elsevier: New York.

Czaja, S. (1997) Systems Design and Evaluation. In Salvendy, G. (ed), *Handbook of Human Factors and Ergonomics*. John Wiley & Sons: New York. 17–40.

DiBiasi, G. (2002) Software Development. E-mail. 22 January 2002.

Diffrient, N., Tilley, A. and Bardagjy, J. (1974) *Humanscale 1/2/3*. MIT Press: Cambridge, MA.

Diffrient, N., Tilley, A. and Harmon, D. (1981a) *Humanscale 4/5/6*. MIT Press: Cambridge, MA.

Diffrient, N., Tilley, A. and Harmon, D. (1981b) *Humanscale 7/8/9*. MIT Press: Cambridge, MA.

Dorst, K. (1996) The Design Problem and Its Structure. In Cross, N., Christiaans, H. and Dorst, K. (eds), *Analysing Design Activity*. John Wiley & Sons: Chichester, U.K.

Downie, N.M. and Heath, R.W. (1959) *Basic Statistical Methods*. Harper: New York.

Drury, C., Paramore, B., Vancott, H., Grey, S., Corlett, E. (1987) Task Analysis. In Salvendy, G. *Handbook of Human Factors*. New York: John Wiley & Sons. 370–401.

Dumas, J. and Redish, J. (1994) *A Practical Guide to Usability Testing*. Ablex Publishing Company: Norwood, NJ.

Dunlap, W. and Kennedy, R. (1995) Testing for Statistical Power. *Ergonomics in Design*. Human Factors and Ergonomics Society: July 1995. 6–7, 31.

Eberts, R. (1997) Cognitive Modeling. In Salvendy, G. (ed), *Handbook of Human Factors and Ergonomics*, 2nd edn. John Wiley & Sons: New York. 1328–74.

Endsley, M. (1996) Situational Awareness Measurement in Test and Evaluation. In O'Brien, T.G. and Charlton, S.G. (eds), *Handbook of Human Factors Testing and Evaluation*. Lawrence Erlbaum and Associates: Mahwah, NJ. 159–80.

Feather, N. (ed.) (1982) *Expectations and Actions: Expectancy-Value Models in Psychology*. Lawrence Earlbaum Associates: Hillsdale, NJ.

Ferry, T.S. (1988) *Modern Accident Investigation and Analysis*, 2nd edn. John Wiley & Sons: New York.

Fogler, H. and LeBlanc, S. (1995) *Strategies for Creative Problem Solving*. Prentice Hall PTR: Englewood Cliffs, NJ.

Folkman, S., Schaefer, C. and Lazarus, R. (1979) Cognitive Processes as Mediators of Stress and Coping. In Hamilton, V. and Warburton, D. *Human Stress and Cognition: An Information Processing Approach*. John Wiley & Sons: New York.

Freeman, S. (1991) *Injury and Litigation Prevention*. Van Nostrand Reinhold: New York.

Friedman, D. (1996) *Hidden Order: The Economics of Everyday Life*. Harper Collins: New York.

Friedman, K. (2000) Creating Design Knowledge: From Research Into Practice. IDATER 2000. Loughborough University: United Kingdom.

Fu, L., Salvendy, G. and Turley, L. (1998) Who Finds What in Usability Evaluation. *Proceedings of the Human Factors and Ergonomics Society 42nd Annual Meeting*. Human Factors and Ergonomics Society: Santa Monica, CA.1341–5.

Gardner, H. (1983) *Frames of Mind: The Theory of Multiple Intelligences*. Basic Books: New York.

Gawron, V., Dennison, T. and Biferno, M. (1996) Mockups, Physical and Electronic Human Models, and Simulations. In O'Brien, T.G. and Charlton, S.G. (eds), *Handbook of Human Factors Testing and Evaluation*. Lawrence Erlbaum and Associates: Mahwah, NJ. 43–80.

Gee, E. and Tyler, C. (1976) *Managing Innovation*. John Wiley & Sons: New York.

Gertman, D. and Blackman, H. (1994) *Human Reliability and Safety Analysis Data Handbook*. John Wiley & Sons: New York.

Gleitman, H. (1986) *Psychology*, 2nd edn. W.W. Norton & Company: New York.

Goodstein, L., Anderson, H. and Olsen, S. (eds.) (1988) *Tasks, Errors, and Mental Models*. Taylor & Francis: London.

Gould J.D. and Lewis, C. (1985) Designing for Usability. *Communications of the ACM*. Vol.28, No.3. 300–11.

Grandjean, E. (1980) *Fitting the Task to the Man*. 3rd edn. Taylor and Francis: London.

Green, P. (1995) Automotive Techniques. In Weimer, J. (ed.) *Research Techniques in Human Engineering*. Prentice-Hall: Englewood Cliffs, NJ. 165–208.

Green, R.G, Muir, H., James, M., Gradwell, D. and Green, R.L. (1991) *Human Factors for Pilots*. Avebury: Aldershot, U.K.

Harmon, R. and Toomey, M. (1999) Creating a Future we Wish to Inhabit. In Hessebein, F., Goldsmith, M., and Somerville, I. (eds.) *Leading Beyond the Walls*. Jossey-Bass Publishers: San Fransisco.

Hamel, G. and Prahalad, C.K. (1994a) Competing for the Future. *Harvard Business Review*. July–August. 122–8.

Hamel, G. and Prahalad, C.K. (1994b) Seeing the Future First. *Fortune*. September 5. 64–70.

Hamilton, V. and Warburton, D. (1979) *Human Stress and Cognition: An Information Processing Approach*. John Wiley & Sons: New York.

Hammer, W. (1980) *Product Safety Management and Engineering*. Prentice-Hall: Englewood Cliffs, NJ. 204–5.

Hawkins, F.H. (1993) *Human Factors in Flight*, 2nd edn. Ashgate: Aldershot, U.K.

Heath, C. and Luff, P. (2000) *Technology in Action*. Cambridge University Press: New York.

Hesselbein, F., Goldsmith, M. and Somerville, I. (eds.) (1999) *Leading Beyond the Walls*. Jossey-Bass Publishers: San Francisco.

Hoffman, R., Crandall, B. and Shadbolt, N. (1998) Use of the Critical Decision Method to Elicit Expert Knowledge: A Case Study in the Methodology of Cognitive Task Analysis. *Human Factors*. Vol.40, No.2. June. 254–76.

Hoffman, R.R., Ford, K., Feltovich, P.J., Woods, D.D., Klein, G. and Feltovich, A. (2002). A Rose by Any Other Name ... Would Probably be Given An Acronym. *IEEE Intelligent Systems*, July/August, 72–9.

Hyer, N.L. (1984) Management's Guide to Team Technology. In Hyer, N.L. (ed.) *Team Technology at Work*. Society of Manufacturing Engineers.

Jensen, R.S. (1989) *Aviation Psychology*. Gower: Aldershot, U.K.

Jones, J.C. (1992) *Design Methods*. Van Nostrand Reinhold: New York.

Jones, P.M. (1999) Human Error and Its Amelioration. In Sage, A.P. and Rouse, W.B. (eds), *Handbook of Systems Engineering and Management*. John Wiley & Sons: New York. 687–702.

Kalton, G. (1983) *Introduction to Survey Sampling*. Sage Publications: Beverly Hills.

Kantor, R. (1983) *The Change Masters*. Simon and Schuster: New York.

Kantowitz, B. and Sorkin, R. (1983) *Human Factors: Understanding People-System Relationships*. John Wiley & Sons: New York.

Kelley, R. (1999) How to be a Star Engineer. *IEEE Spectrum*. October. 51–8.

Kirkwood, C.W. (1999) Decision Analysis. In Sage, A.P. and Rouse, W.B. (eds) *Handbook of Systems Engineering and Management*. John Wiley & Sons: New York. 1119–46.

Klein, G. (1999) *Sources of Power: How People Make Decisions*, 2nd edn. The MIT Press: Cambridge, MA.

Knott, B. (2000) Virtual Reality. *Gateway*. Human Systems IAC. United States Department of Defense Information Analysis Center. Air Force Research Laboratory Human Effectiveness Directorate. Wright-Patterson AFB, Ohio. Vol.XI, No.2. 6–7.

Koberg, D. and Bagnall, J. (1991) *The Universal Traveler*. William Kaufmann: Los Altos, CA.

Konz, S. and Johnson, S. (2000) *Work Design: Industrial Ergonomics*, 5th edn. Holcomb Hathaway: Scottsdale, AZ.

Kotter, J. (1996) *Leading Change*. Harvard Business School Press: Boston.

Kreifeldt, J. (1992) Ergonomics of Product Design. In Salvendy, G. (ed.) *Handbook of Industrial Engineering*. John Wiley & Sons: New York. 1144–63.

Kroemer, K.H.E., Kroemer, H.J., and Kroemer-Elbert, K.E. (1990) *Engineering Physiology: Bases of Human Factors/Ergonomics*. Van Nostrand Reinhold: New York.

Kurzweil, R. (1999) *The Age of Spiritual Machines*. Penguin Books: New York.

Lannon, J. (1997) *Technical Writing*. Addison Wesley Longman: New York.

Latino, R. (2000) Automating Root Cause Analysis. In Spath, P. (ed.) *Error Reduction in Health Care*. Jossey-Bass: San Francisco. 155–64.

Laughery, K. (1993) Everybody Knows—Or Do They? *Ergonomics in Design*. January. 8–13.

Laughery, K.R. and Laughery K.R. (1997) Analytic Techniques for Function Analysis. In Salvendy, G. (ed.) *Human Factors Handbook*. New York: John Wiley & Sons. 329–54.

Leo, J. (1998) Nothing But the Truth? *U.S. News and World Report*, 6 July.

Levis, A. (1999) System Architectures. In Sage, A. and Rouse, W. (eds), *Handbook of Systems Engineering and Management*. John Wiley & Sons: New York. 427–8.

Lewis, C. (1982) Using the Thinking Aloud Method in Cognitive Interface Design. *Computer Science*. RC 9265 (40713). February 17.

Lowell, S. (1999) Standards in Systems Engineering. In Sage, A. and Rouse, W. (eds), *Handbook of Systems Engineering and Management*. John Wiley & Sons: New York.

Luczak, H. (1997) Task Analysis. In Salvendy, G. *The Handbook of Human Factors and Ergonomics*. John Wiley & Sons: New York. 340–416.

Luff, P., Hindmarsh, J. and Heath, C. (eds), (2000) *Workplace Studies*. Cambridge University Press: New York.

Lund, A. (1998) The Need for a Standardized Set of Usability Metrics. *Proceedings of the Human Factors and Ergonomics Society 42nd Annual Meeting*. Human Factors and Ergonomics Society: Santa Monica, CA. 688–91.

Malaskey, S. (1974) *System Safety*. Spartan Books: Rochelle Park, NJ.

Malone, T. (1996) Human Factors Test Support Documentation. In O'Brien, T.G. and Charlton, S.G. (eds). *Handbook of Human Factors Testing and Evaluation*. Lawrence Erlbaum and Associates: Mahwah, NJ. 101–16.

Matthews, R.A. (1997) The Science of Murphy's Law. *Scientific American*. 276(4) April. 88–91.

Maurino, D.E., Reason, J., Johnston, N. and Lee, R.B. (1995) *Beyond Aviation Human Factors*. Avebury: Aldershot, U.K.

McCracken, G. (1991) *The Long Interview*. Sage Publications: Newbury Park, CA.

Medsker, G. and Campion, M. (1997) Job and Team Design. In Salvendy, G. (ed.), *Handbook of Human Factors and Ergonomics*, 2nd edn. John Wiley & Sons: New York. 451–89.

Meister, D. (1971) *Human Factors Theory and Practice*. Wiley Interscience: New York.

Meister, D. (1985) *Behavioral Analysis and Measurement Methods*. John Wiley & Sons: New York.

Meister, D. (1989) *Conceptual Aspects of Human Factors*. The John Hopkins University Press: Baltimore. 141–216.

Meister, D. (1991) *Psychology of System Design*. Elsevier: New York.

Meister, D. (1998) The Editor's Comments. *Newsletter of the Test and Evaluation Technical Group*. Human Factors and Ergonomics Society: Santa Monica, CA. Spring.

Meister, D. and Enderwick, T. (eds) (1992) Special Issue: Measurement in Human Factors. *Human Factors*. 34 (4) August.

Melsa, J.L. (1999) Total Quality Management. In Sage, A. and Rouse, W. (eds),, *Handbook of Systems Engineering and Management*. John Wiley & Sons: New York. 269–302.

Miller, G. (1956) The Magical Number Seven, Plus or Minus Two. *The Psychological Review*. Vol. 63, No. 2. March.

Miller, D.P. and Swain, A.D. (1997) Human Error and Human Reliability. In Salvendy, G. (ed.) *Human Factors Handbook*. New York: John Wiley & Sons. 219–47.

Miner, M. and Rawson, H. (1996) *American Heritage Dictionary of American Quotations*. Penguin Books: New York.

Mitchell, P. (1995) Consumer Products Techniques. In Weimer, J. (ed.), *Research Techniques in Human Engineering*. Prentice-Hall: Englewood Cliffs, NJ. 246–67.

Mitchell, T. (1982) Expectancy-Value Models in Organizational Psychology. In Feather, N. (ed.), *Expectations and Actions: Expectancy-Value Models in Psychology*. Lawrence Earlbaum Associates: Hillsdale, NJ. 293–312.

Morgan, J.S. (1972) *Managing Change*. McGraw-Hill: New York.

Morgan, C., Cook, J., Chapanis, A. and Lund, M. (eds), (1963) *Human Engineering Guide to Equipment Design*. McGraw-Hill Book Company: New York.

Moroney, W. and Bittner, A. (1995) Military Systems Techniques. In Weimer, J. (ed.), *Research Techniques in Human Engineering*. Prentice-Hall: Englewood Cliffs, NJ. 361–438.

Nakayama, O., Futami, T. Nakamura, T. and Boer, E. (1998) Development of a Steering Entropy Method for Evaluating Driver Workload. *Proceedings of the 1999 Society of Automotive Engineers International Congress*. Society of Automotive Engineers (draft).

NASA (undated) Task Load Index (NASA-TLX). V 1.0. Human Performance Research Group. Ames Research Center. National Air and Space Administration: Moffett Field, CA.

Neisser, U. (1976) *Cognition and Reality*. W.H. Freeman and Company: San Francisco.

Nemeth. C. (1995) Shared Solutions: Rapid Prototyping's Practice and Promise. *Proceedings of the International Ergonomics Association World Congress*. Rio de Janiero, Brazil. October.

Nemeth, C. (1996a) Getting to the Heart of the Matter: A Cardiology Lab Human Factors Investigation. *Newsletter of the Test & Evaluation Technical Group*. Human Factors and Ergonomics Society: Santa Monica, CA. Summer.

Nemeth, C. (1996b) Design for Use: Increasing the User Role in Product and Service Development. *Proceedings of the Industrial Designers Society of America National Conference*. Orlando, FL. September.

Nemeth, C. (1998) Enacting Design for Use. *Proceedings of the Sixth International Conference on Human Aspects of Advanced Manufacturing: Agility and Hybrid Automation*. International Ergonomics Association: Hong Kong, P.R. China. July 5–8.

Nielsen, J. (1997a) Usability Testing. In Salvendy, G. (ed.), *The Handbook of Human Factors and Ergonomics*. John Wiley & Sons: New York. 1543–68.

Nielsen, J. (1997b) *The Use and Misuse of Focus Groups*. <www.useit.com> (Obtained 23 July 2002).

Norman, D. (1988) *The Design of Everyday Things*. Doubleday: New York.

Norman, D. (1995) On the Difference Between Research and Practice. *Ergonomics in Design*. Human Factors and Ergonomics Society: Santa Monica, CA. April.

Norman, D. (2002) Emotion and Design: Attractive Things Work Better. *Interactions*. Association for Computing Machinery. July–August. 36–42.

Olson, J. (1985) Human Memory. *Human Factors Engineering: Engineering Summer Conferences*. University of Michigan: Ann Arbor, MI.

Orasanu, J. and Shafto, M. (1999) Designing for Cognitive Task Performance. In Sage, A. and Rouse, W. (eds), *Handbook of Systems Engineering and Management*. John Wiley & Sons: New York.

Osborne, L. (2002) Consuming Rituals of the Suburban Tribe. *New York Times Magazine*. January 13. 28–31.

Park, K. (1987) *Human Reliability: Analysis, Prediction and Prevention of Human Errors*. Elsevier: New York.

Park, K. (1997) Human Error. In Salvendy, G. (ed.), *Handbook of Human Factors and Ergonomics*. 2nd edn. John Wiley & Sons: New York. 150–71.

Passell, P. (1996) Why the Best Doesn't Always Win. *New York Times Magazine*. May 5. 60–1.

Pew, R. (1985a) Introduction to Human Factors Engineering. *Human Factors Engineering: Engineering Summer Conferences*. University of Michigan: Ann Arbor, MI.

Pew, R. (1985b) Human Skills and Their Utilization. *Human Factors Engineering: Engineering Summer Conferences*. University of Michigan: Ann Arbor, MI.

Pew, R. (1985c) Human Information Processing and Cognitive Psychology. *Human Factors Engineering: Engineering Summer Conferences*. University of Michigan: Ann Arbor, MI.

Phillips, J.L. (1982) *Statistical Thinking*. W.H. Freeman and Co: San Francisco.

Potter, S.S., Roth, E.M., Woods, D.D., and Elm, W. (1998) *Toward the Development of a Computer-Aided Cognitive Engineering Tool to Facilitate the Development of Advanced Decision Support Systems for Information Warfare*. Report prepared for Air Force Research Laboratory (AFRL/HEC).

Praetorius, N. and Duncan, K. (1988) Verbal Reports: A Problem in Research Design. In Goodstein, L., Anderson, H. and Olsen, S. (eds), *Tasks, Errors, and Mental Models*. Taylor & Francis: London. 293–314.

Prahalad, C.K. and Ramaswamy, V. (2000) Co-Opting Customer Competence. *Harvard Business Review*. January–February. 79–87.

Rasmussen, J. (1983) Skills, Rules and Knowledge: Signals, Signs and Symbols, and Other Distinctions in Human Performance Models. *IEEE Transactions on Systems, Man and Cybernetics*. Vol. SMC-13. No.3. May. 257–66.

Rasmussen, J. (1986) *Information Processing and Human-Machine Interaction*. Elsevier Science Publishing: New York.

Reason, J. (1988) Framework Models of Human Performance and Error: A Consumer Guide. In Goodstein, L., Anderson, H. and Olsen, S. (eds), *Tasks, Errors, and Mental Models*. Taylor & Francis: London. 35–49.

Reason, J. (1990) *Human Error*. Cambridge University Press: New York.

Reason, J. (1997) *Managing the Risks of Organizational Accidents*. Ashgate: Brookfield, VT.

Reid, G.B., Shingledecker, C.A., Nygren, T.E. and Eggemeier, F.T. (1981). Development of Multidimensional Subjective Measures of Workload. *Proceedings of the International Conference on Cybernetics and Society*. Institute of Electrical and Electronic Engineers: New York.

Rifkin, J. (1995) *The End of Work*. G.P. Putnam's Sons: New York.

Riordan, D. and Pauley, S. (1996) *Technical Report Writing Today*. Houghton Mifflin Company: New York.

Robertson, S. (ed.) (1996) *Contemporary Ergonomics 1996: Proceedings of the Annual Conference of the Ergonomics Society*, University of Leicester 10–12 April 1996. Taylor & Francis: London.

Rodgers, S. (1997) Work Physiology—Fatigue and Recovery. In Salvendy, G. (ed.), *Handbook of Human Factors and Ergonomics*, 2nd edn. John Wiley & Sons: New York. 269–88.

Rogers, W., Meyer, B., Walker, N. and Fisk, A. (1998) Functional Limitations to Daily Living Tasks in the Aged: A Focus Group Analysis. *Human Factors*. Vol.40, No.1. March. 111–25.

Roosenburg, N.F.M and Eeckels, J. (1995) *Product Design: Fundamentals and Methods*. John Wiley & Sons: New York.

Roscoe, S.N. (1980) *Aviation Psychology*. Iowa State University Press: Ames, IA.

Rothstein, E. (1996) Connections: Science and Scientists are Facing Some Rough Going as the Millenium Approaches. *New York Times*, 19 August. C3.

Rothstein, E. (2001) The Unforeseen Disruptions of Moving Ahead. *New York Times*, December 22. A 17, 19.

Rouse, W.B. and Boff, K.R. (1997) Assessing Cost/Benefits of Human Factors. In Salvendy, G. (ed.), *Handbook of Human Factors and Ergonomics*, 2nd edn. John Wiley & Sons: New York. 1617–32.

Rubin, J. (1994) *Handbook of Usability Testing*. John Wiley & Sons: New York.

Sage, A. and Rouse, W. (eds) (1999) *Handbook of Systems Engineering and Management*. John Wiley & Sons: New York.

Salvendy, G. (ed.) (1987) *Handbook of Human Factors*. John Wiley & Sons: New York.

Salvendy, G. (ed.) (1992) *Handbook of Industrial Engineering*. John Wiley & Sons: New York.

Salvendy, G. (ed.) (1997) *Handbook of Human Factors and Ergonomics*. 2nd edn. John Wiley & Sons. New York.

Salvendy, G. and Carayon, P. (1997) Data Collection and Evaluation of Outcome Measures. In Salvendy, G. (ed.), *Handbook of Human Factors*, 2nd edn. John Wiley & Sons: New York. 1458–65.

Sanders, M.S. and McCormick, E.J. (1993) *Human Factors in Engineering and Design*, 7th edn. McGraw-Hill: New York.

Scerbo, M. (1995) Usability Testing. In Weimer, J. (ed.), *Research Techniques in Human Engineering*. Prentice-Hall: Englewood Cliffs, NJ. 72–111.

Schön, D. (1983) *The Reflective Practitioner*. Basic Books: New York.

Schwartz, J. (1991) Facing the Human Factor. *Newsweek*, September 30. 48.

Senge, P. (1999) *The Dance of Change: The Challenges of Sustaining Momentum in Learning Organizations*. Doubleday: New York.

Shackel, B. *Applied Ergonomics Handbook*. IPC Business Press: Surrey, U.K. 1974.

Shappell, S.A. and Wiegmann, D.A. (2000) *The Human Factors Analysis and Classification System (HFACS)*. DOT/FAA/AM-00/7. Federal Aviation Administration. U.S. Department of Transportation. February.

Sharit, J. (1997) Allocation of Functions. In Salvendy, G. (ed.), *Handbook of Human Factors*. 2nd edn. John Wiley & Sons: New York. 301–39.

Shattuck, L. and Woods, D. (1995) The Critical Incident Technique: 40 Years Later. *Test & Evaluation Technical Group Newsletter*. Human Factors and Ergonomics Society. No.1.

Sheridan, T. (1998) Allocating Functions Rationally Between Humans and Machines. *Ergonomics in Design*. July. Human Factors and Ergonomics Society: Santa Monica, CA. 20–5.

Sheridan, T. (1999) Human Supervisory Control. In Sage, A. and Rouse, W. (eds) *Handbook of Systems Engineering and Management*. John Wiley & Sons: New York. 591–628.

Shostack, G. (1984) Designing Services That Deliver. *Harvard Business Review*. January–February. 133–9.

Siegel, S. (1956) *Nonparametric Statistics for the Behavioral Sciences*. McGraw-Hill: New York.

Simon, H. (1998) *The Sciences of the Artificial*. The MIT Press: Cambridge, MA.

Sims, B. (1998) *Technical Writing for Readers and Writers*. Houghton Mifflin Company: New York.

Smith, M. and Beringer, D. (1997) Human Factors in Occupational Injury Evaluation and Control. In Salvendy, G. (ed.), *Handbook of Human Factors*, 2nd edn. John Wiley & Sons: New York. 768–89.

Spath, P. (ed.) (2000) *Error Reduction in Health Care*. Jossey-Bass: San Francisco.

Spiegel, M.R. (1988) *Statistics*, 2nd edn. Schaum's Outline Series in Mathematics: McGraw-Hill. New York.

Stephenson, J. (1991) *System Safety 2000*. Van Nostrand Reinhold: New York. 274–5.

Stewart, D. and Shamdasani, P. (1990) *Focus Groups: Theory and Practice*. Sage Publications: London.

Stross, R. (2000) Silicon Valley Killjoy. *U.S. News and World Report*, April 3. 44.

Stuster, J. and Chovil, A. (1994) Tracking Disease Outbreak. *Ergonomics in Design*. July. Human Factors and Ergonomics Society: Santa Monica, CA. 8–15.

Suchman, L.A. (1987) *Plans and Situated Actions*. Cambridge University Press: New York.

Suchman, L.A. (2001a) Human-Machine Reconsidered. Department of Sociology, Lancaster University. <www.comp.lancs.uk/sociology/soc040ls.html> (accessed 29 December 2001).

Suchman, L.A. (2001b) I Have, More Than Ever, a Sense of the Immovability of These Institutions. Department of Sociology, Lancaster University. <www.comp.lancs.uk/sociology/soc040ls.html> (accessed 29 December 2001).

Swain, A.D. and Guttmann, H.E. (1983) *Handbook of Human Reliability Analysis with Emphasis on Nuclear Power Plant Applications*. NUREG/CR-1278. U.S. Nuclear Regulatory Commission.

Thomson, R.M. (1972) Design of Multi-Man–Machine Work Areas. In Van Cott, H. and Kinkaide, R.G. (eds) *Human Engineering Guide to Equipment Design*. U.S. Government Printing Office: Washington. 419–66.

Urban, G., Hauser, J. and Dholakia, N. (1986) *Essentials of New Product Management*. Prentice-Hall: Englewood Cliffs, NJ.

Van Cott, H.P. and Kinkade, R.G. (eds.) (1972) *Human Engineering Guide to Equipment Design*. U.S. Government Printing Office: Washington.

Van Daalen, C., Thissen, A. and Verbraeck, A. (1999) Methods for Modeling and Analysis of Alternatives. In Sage, A. and Rouse, W. (eds.), *Handbook of Systems Engineering and Management*. John Wiley & Sons. New York. 1063–76.

Vincoli, J.W. (1994) *Accident Investigation and Loss Control*. Van Nostrand Reinhold: New York. 184–5.

Virzi, R. (1992) Refining the Test Phase of Usability Evaluation: How Many Subjects is Enough? *Human Factors*. 34(4). 457–68.

Voss, C.A. (1992) Measurement of Innovation and Design Performance in Services. *Design Management Journal*. Winter. 40–6.

Weimer, J. (ed.) (1995) *Research Techniques in Human Engineering*. Prentice-Hall: Englewood Cliffs, NJ.

Weiner, E.L. and Nagel, D.C.(1998) *Human Factors in Aviation*. Academic Press: San Diego, CA.

White, A. (2001) Chapter 9 Observations. Personal correspondence. December 28.

Wickens, C. (1992) *Engineering Psychology and Human Performance*. Harper-Collins: New York.

Wickens, C. (1998) Commonsense Statistics. *Ergonomics in Design*. Human Factors and Ergonomics Society: Santa Monica, CA. October. 18–22.

Williges, R.C. (1985) Introduction to Experimental Design. *Human Factors Engineering: Engineering Summer Conferences*. University of Michigan: Ann Arbor, MI.

Windsor, D. (1996) *Writing Like an Engineer—A Rhetorical Education*. Lawrence Earlbaum Associates: Mahwah, NJ.

Winograd, T. (2001) Human Factors Manuscript. E-mail. 9 February.

Winograd, T. and Woods, D.D (1997) Challenges for Human-Centered Design. In J. Flanagan, T. Huang, P. Jones and S. Kasif (eds) *Human-Centered Systems: Information, Interactivity, and Intelligence*. July National Science foundation: Washington, DC.

Wittenborg, L. (1981) *Good Show: A Practical Guide for Temporary Exhibitions*. Smithsonian Institution: Washington, DC.

Wolcott, H. (1995) *The Art of Fieldwork*. Altamira Press: Walnut Creek, CA.

Wood, C.T., McCarthy, R.L. and Arndt, S.R. (1998) Using Risk Analysis to Reduce the Hazards on Playgrounds. *National Manufacturing Week Conference Proceedings '98*. Vol. 3. March 18–19. 103–9.

Woods, D. (1988) Coping With Complexity: The Psychology of Human Behavior in Complex Systems. In Goodstein, L.P., Andersen, H.B. and Olsen, S.E. (eds), *Tasks, Errors and Mental Models*. Taylor & Francis: New York. 128–48.

Woods, D.D. and Tinapple, D. (1999) W³: Watching Human Factors Watch People at Work. Presidential Address. *Proceedings of the 43rd Annual Meeting of the Human Factors and Ergonomics Society*. September 28. Multimedia Production at URL http.//csel.eng.ohio-state.edu.hf99/

Woodson, W. (1981) *Human Factors Design Handbook*. McGraw-Hill: New York.

Woodson, W. and Conover, D. (1960) *Human Engineering Guide for Equipment Designers*. University of California Press: Los Angeles.

Yang, D. (2000) Leaving Moore's Law in the Dust. *U.S. News and World Report*, July 10. 37–8.

Harvard Business Review. (1994) Product Development: Empathic Design Helps Understand Users Better. March/April. 10.

Institute of Design (1994) *The Strategic Value of Human-Centered Design*. Illinois Institute of Technology: Chicago.

Joint Commission on Accreditation of Healthcare Organizations. (1999) Sentinel Events: Evaluating Cause and Planning Improvement. 2nd Edn.

National Institutes of Health. (2002) <www.nih.gov> Accessed 23 February.

National Safety Council (2001) *Injury Facts 2001*. Itasca, IL.

National Safety Council (2001) Report on Injuries in America, 2001. <www.nsc.org> Accessed 2 January 2002.

U.S. Department of Defense (1994) Human Engineering Design Criteria for Military Systems, Equipment and Facilities (MIL-STD-1472D). March 1989, incl. Notice 3.

Americans with Disabilities Act (ADA). 49 U.S. Code of Federal Regulations, Part 38. Accessibility Specifications for Transportation Vehicles. Subpart B—Busses, Vans, and Systems. USGPO. Washington, DC.

Index